T0235237

Practical Numerical Mathematics with MATLAB

A Workbook

Other World Scientific Titles by the Author

Numerical Linear Algebra
ISBN: 978-981-122-389-1
ISBN: 978-981-122-484-3 (pbk)

Practical Numerical Mathematics with MATLAB
A Workbook

Myron M Sussman
University of Pittsburgh, USA

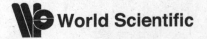

World Scientific

NEW JERSEY · LONDON · SINGAPORE · BEIJING · SHANGHAI · HONG KONG · TAIPEI · CHENNAI · TOKYO

Published by

World Scientific Publishing Co. Pte. Ltd.
5 Toh Tuck Link, Singapore 596224
USA office: 27 Warren Street, Suite 401-402, Hackensack, NJ 07601
UK office: 57 Shelton Street, Covent Garden, London WC2H 9HE

British Library Cataloguing-in-Publication Data
A catalogue record for this book is available from the British Library.

PRACTICAL NUMERICAL MATHEMATICS WITH MATLAB
Vol. 1: A Workbook/Vol. 2: Solutions

ISBN 978-981-124-600-5 (set_hardcover)
ISBN 978-981-124-518-3 (set_paperback)

ISBN 978-981-124-035-5 (vol. 1_hardcover)
ISBN 978-981-124-519-0 (vol. 1_paperback)
ISBN 978-981-123-918-2 (vol. 1_ebook for institutions)
ISBN 978-981-124-070-6 (vol. 1_ebook for individuals)

ISBN 978-981-124-069-0 (vol. 2_hardcover)
ISBN 978-981-124-520-6 (vol. 2_paperback)
ISBN 978-981-124-433-9 (vol. 2_ebook for institutions)
ISBN 978-981-124-434-6 (vol. 2_ebook for individuals)

For any available supplementary material, please visit
https://www.worldscientific.com/worldscibooks/10.1142/12340#t=suppl

Desk Editor: Yumeng Liu

Printed in Singapore

Dedication

This workbook is dedicated to my granddaughters, Rachel and Ruby Arnt. They are part of our bright future.

Preface

This workbook is intended for advanced undergraduate or beginning graduate students as a supplement to a traditional two-semester course in numerical mathematics and as preparation for independent research involving numerical mathematics. Upon completion of this workbook, students will have a working knowledge of MAT-LAB [MathWorks (2019)] programming, they will have themselves programmed illustrative examples of some algorithms presented in numerical mathematics courses, and they will know how to check and verify their own programs even when there is no existing program with which to compare results. Students are not assumed to have programming experience and all necessary features of MATLAB are presented as needed.

This workbook arose from a MATLAB laboratory class accompanying a lecture class on numerical mathematics at the University of Pittsburgh. These two classes, aimed at advanced undergraduate or beginning graduate students were taken together and worth three credits each. They were presented independently, but the order of presentation of topics was coordinated. The lecture course included its own homework assignments, generally involving proofs and hand calculations. The laboratory class was presented in a computer laboratory room with MATLAB installed. Personal versions of MATLAB were also available for students, and many did their work on their own personal computers.

This workbook, then, is intended to accompany classroom lectures and contains little theoretical development. Students are expected to have a working knowledge of undergraduate mathematics through calculus and differential equations. No programming experience is expected and necessary features of MATLAB are presented as needed. This workbook is not, however, intended as a programming tutorial because only those features of MATLAB needed for numerical mathematics at this level are presented. Some students preferred to either use Gnu Octave [Eaton *et al.* (2019)] on their own computers or on the octave online [Octave (2019)] web site. Essentially all of the exercises can be done using Octave.

The objectives of the exercises in this workbook are threefold:

(i) To provide illustrative examples of common algorithms in numerical mathematics;

(ii) To raise students to a level of competence using MATLAB sufficient for them to develop their own code for their own research; and,

(iii) To teach students how to check that their code is correct even when there are no reference codes with which to compare.

To accomplish the third objective, references are made to theoretical results, and special polynomial solutions and other specialized results are employed. Whenever possible, students are required to show not only that convergence is achieved, but also that the *rate of convergence* agrees with theory. When polynomials are compared, enough sample points must be chosen to guarantee that the polynomials agree. Iterative stopping criteria based on theory are employed when convenient. Furthermore, frequent testing is emphasized and every effort has been made to break exercises into steps that are small enough to comprehend yet large enough to be tested.

The majority of exercises in this workbook include enough testing and consistency checking that the student can tell, while working the exercises, that his or her work is correct. In consequence, this workbook is suitable for self-study.

This workbook is divided into two parts, corresponding roughly to two semesters of material. The exercises in each chapter are designed to be completed in approximately 4 to 7 hours in front of a computer using the MATLAB program, although programming and debugging inexperience can greatly increase the required time. Generally, the difficulty increases within each chapter and as the chapters progress. The second part is largely independent of the first and the first few chapters in the second part again require less programming sophistication than later chapters.

Much MATLAB code is included in these exercises. Most of the code included in the workbook is also provided for download in order to save the trouble of copying from print and to avoid bugs arising from copying errors. Such bugs are extremely difficult to discover and correct. The author has made every effort to avoid errors in these code samples, but the chance of errors remains. My sincere apologies to those encountering errors.

A set of solution templates accompany this workbook. These templates make it clear what responses are expected from the exercises and the format of the templates is intended to simplify grading. Students are expected to complete the templates as they are doing their exercises on a computer. The templates are intended to be filled in using cut-and-paste to minimize copying efforts. The instructor can then grade the templates without printing them. If the instructor wishes to grade printed versions of the students' work, care should be taken that the printed versions are complete, especially for copies of code. If not, additional sheets of paper should be submitted.

Unfortunately, the author knows of no way to paste plots into a pdf template. As a result, students using pdf templates must include additional files or hardcopy for plots.

Acknowledgments

Many people contributed to this book in many different ways. Dr. John Burkardt originated the general content and organization of several of the chapters and some of the MATLAB function files when he taught the numerical analysis laboratory before I did. This workbook would be very different if I had not had his model to follow.

Several of my colleagues at the University of Pittsburgh, particularly Profs. W. J. Layton, C. Trenchea and M. J. Neilan and the many students who took the numerical analysis computer laboratory provided suggestions and other valuable feedback that helped me construct this workbook.

My family, and especially my wife Jill, provided support and encouragement for my efforts in this long project.

Finally, I must acknowledge the effort of my vigilant but patient editor, Liu Yumeng, at World Scientific Publishing Co. for her efforts to hammer my drafts into professional form.

Contents

Differential Equations and Linear Algebra 227

PART 1

Rootfinding, Interpolation, Approximation and Quadrature

Chapter 1

Introduction to MATLAB

1.1 Introduction

The exercises throughout this workbook use the MATLAB [MathWorks (2019)] mathematical programming system. Alternatively, all of the exercises can be completed using either Gnu Octave [Eaton *et al.* (2019)], a MATLAB work-alike program that can be freely installed on personal computers, or the Octave online [Octave (2019)] web site. The MATLAB program comes with extensive documentation and employs a very convenient desktop work environment. Before starting the exercises in this workbook, you would be well served by becoming familiar with basic use of the MATLAB desktop environment. The Mathworks makes a number of videos and tutorials available for your use.

This chapter is focussed on the MATLAB features commonly used in numerical linear algebra and writing scripts (programs) to implement your particular linear algebra tasks. The selection of topics in the exercises below is intended to reinforce the particular commands and techniques that are used in later chapters. After you finish the exercises in this chapter, you will have a good foundation in writing your own scripts and programs.

1.2 MATLAB files

One good way to use MATLAB is to use its scripting (programming) facility. With sequences of MATLAB commands contained in files, it is easy to see what calculations were done to produce a certain result, and it is easy to show that the correct values were used in producing a graph. It is terribly embarrassing to produce a very nice plot that you show to your teacher or advisor only to discover later that you cannot reproduce it or anything like it for similar conditions or parameters. When the commands are in clear text files, with easily read, well-commented code, you have a very good idea of how a particular result was obtained. And you will be able to reproduce it and similar calculations as often as you please.

The MATLAB comment character is a percent sign (%). That is, lines starting with % are not read as MATLAB commands and can contain any text. Similarly, any text

3

on a line following a % can contain textual comments and not MATLAB commands. Similarly, a group of lines starting with a line containing only %{ and ending with a line containing only %} will be treated as a comment. It is important to include comments in script and function files to explain what is being accomplished.

MATLAB commands are sometimes terminated with a semicolon (;) and sometimes not. The difference is that the result of a calculation is printed to the screen when there is no semicolon but no printing is done when there is a semicolon. It is a good idea to put semicolons at the ends of all calculational lines in a function file.

There are three kinds of files that MATLAB can use:

(1) Script files (or "m-files"),
(2) Function files (or "m-files"), and
(3) Data files.

1.2.1   *Script m-files*

A MATLAB script m-file is a text file with the extension .m. MATLAB script files should always start off with comments that identify the author, the date, and a brief description of the intent of the calculation that the file performs. MATLAB script files are invoked by typing their names without the .m at the MATLAB command line or by using their names inside another MATLAB file. Invoking the script causes the commands in the script to be executed, in order.

1.2.2   *Function m-files*

MATLAB function files are also text files with the extension .m, but the first non-comment line *must* start with the word `function` and be of the form

   `function` *output variable(s)* = *function name* (*parameters*)

For example, the function that computes the sine would start out

```
function y=sin(x)
```

and the name of the file would be `sin.m`. The defining line, starting with the word `function` is called the "signature" of the function. If a function has no input parameters, they, and the parentheses, can be omitted. Similarly, a function need not have output parameters. There can be more than one output parameter, and the syntax for several output parameters is discussed later in this chapter. The name of the function must be the same as the file name. It is best to have the first line of the function m-file be the signature line, starting with the word "function." The lines following the signature should contain comments with the following information.

(1) Repetition of the signature of the function (useful as part of the help message),
(2) A brief description of the intent of the calculation the function performs,
(3) Brief descriptions of the input and output variables, and

(4) The author's name and date.

Part of the first of these lines is displayed in the "Current directory" windowpane, and the lines themselves comprise the response to the MATLAB command `help` *function name.*

The key differences between function and script files are that

- Functions are intended to be used repetitively,
- Functions can accept parameters, and
- Variables used inside a function are invisible outside the function.

This latter point is important: variables used inside a function (except for output variables) are invisible after the function completes its tasks while variables in script files remain in the workspace.

When starting a new task, it is often convenient to start out using script files. As the task evolves, it becomes clear just which tasks are repetitive or when the same calculation is repeated for different parameters. It might also be that so many intermediate variables are needed that new variable names are hard to devise. In these cases, it is best to switch to function files. In these chapters, function or script files will be specified when it is important, and you are free to use what you like otherwise

Because function files are intended to be used multiple times, it is a bad idea to have them print or plot things. Imagine what happens if you have a function that prints just one line of information that you think might be useful, and you put it into a loop that is executed a thousand times. Do you plan to read those lines?

1.2.3 *Data files*

MATLAB also supports data files. The MATLAB `save` command will cause every variable in the workspace to be saved in a file called "`matlab.mat`". You can also name the file with the command `save filename` that will put everything into a file named "`filename.mat`". This command has many other options, and you can find more about it using the help facility. The inverse of the `save` command is `load`.

1.3 Variables

MATLAB uses variable names to represent data. A variable name represents a matrix containing complex double-precision data. If you simply tell MATLAB `x=1`, MATLAB will understand that you mean a 1×1 matrix and it is smart enough to print `x` out without its decimal and imaginary parts, but make no mistake: they are there. And `x` can just as easily represent a matrix.

A variable can represent some important value in a program, or it can represent some sort of dummy or temporary value. Important quantities should be given names longer than a few letters, and the names should indicate the meaning of the quantity. For example, if you were using MATLAB to generate a matrix containing a table of squares of numbers, you might name the table `tableOfSquares`. (The convention used here is that the first part of the variable name should be a noun and it should be lower case. Modifying words follow with upper case letters separating the words. This rule is similar to the recommended naming of Java variables.)

Once you have used a variable name, it is bad practice to re-use it to mean something else. It is sometimes necessary to do so, however, and the statement

```
clear variableOne variableTwo
```

should be used to clear the two variables `variableOne` and `variableTwo` before they are re-used. This same command is critical if you re-use a variable name but intend it to have smaller dimensions.

MATLAB has a few reserved variable names. You should not use these variables in your m-files. If you do use such variables as `i` or `pi`, they will lose their special meaning until you clear them. Reserved names include

`ans`: The result of the previous calculation.

`computer`: The type of computer you are on.

`eps`: The smallest positive number ϵ that both satisfies the expression $1 + \epsilon > 1$ and can be represented on this computer.

`i, j`: The imaginary unit ($\sqrt{-1}$). In this course you should avoid using `i` as a subscript or loop index.

`inf`: Infinity (∞). This will be the result of dividing 1 by 0.

`NaN`: "Not a Number." This will be the result of dividing 0 by 0, or `inf` by `inf`, multiplying 0 by `inf`, *etc.*

`pi`: π.

`realmax, realmin`: The largest and smallest real numbers that can be represented on this computer.

`version`: The version of MATLAB you are running.

Exercise 1.1. Start up MATLAB and use it to answer the following questions.

(1) What are the values of the reserved variables `pi`, `eps`, `realmax`, and `realmin`?

(2) Use the "`format long`" command to display `pi` in full precision and "`format short`" (or just "`format`") to return MATLAB to its default, short, display.

(3) No matter how it is printed, the internal precision of any variable is always about 15 decimal digits. The value for `pi` printed in short format is 3.1416. What is `pi-3.1416`? You should see that this value is not zero. You found the value of `pi` printed using `format long` in the previous part of this exercise. What is the difference between the printed value and `pi`? This value might not be zero, but it is still much smaller than the value of `pi-3.1416`.

(4) Set the variable `a=1` and the variable `b=1+eps`. What is the difference in the way that MATLAB displays these values? Can you tell from the form of the printed value that `a` and `b` are different?

(5) Will the command **format long** cause all the decimal places in b to be printed, or is there still some missing precision?

(6) Set the variable **c=2** and the variable **d=2+eps**. Are the values of **c** and **d** different?

(7) Choose a value and set the variable **x** to that value.

(8) What is the square of **x**? Its cube?

(9) Choose an angle θ and set the variable **theta** to its value (a number).

(10) What is $\sin\theta$? $\cos\theta$? Angles can be measured in degrees or radians. Which of these has MATLAB used?

(11) MATLAB variables can also be given "character" or "string" values. A string is a sequence of letters, numbers, spaces, *etc.,* surrounded by single quotes ('). In your own words, what is the difference between the following two expressions?

```
a1='sqrt(4)'
a2=sqrt(4)
```

(12) The MATLAB **eval** function is used to **eval**uate a string as if it were typed at the command line. If **a1** is the string given above, what is the result of the command **eval(a1)**? Of **a3=6*eval(a1)**?

(13) Use the command **save myfile.mat** to save all your variables. Check your "Current Directory" to see that you have created the file **myfile.mat**. Always use ".**mat**" for MATLAB data files.

(14) Use the **clear** command. Check that there are no variables left in the "current workspace" (windowpane is empty).

(15) Restore all the variables with **load myfile.mat** and check that the variables have been restored to the "Current workspace" windowpane.

1.4 Vectors and matrices

You know that MATLAB treats all its variables as though they were matrices. Important subclasses of matrices include row vectors (matrices with a single row and possibly several columns) and column vectors (matrices with a single column and possibly several rows). One important thing to remember is that you don't have to declare the size of your variable; MATLAB decides how big the variable is when you try to put a value in it. The easiest way to define a row vector is to list its values inside of square brackets, and separated by spaces or commas:

```
rowVector = [ 0, 1, 3, -6, pi ]
```

The easiest way to define a column vector is to list its values inside of square brackets, separated by semicolons or line breaks.

```
columnVector1 = [ 0; 1; 3; -6; pi ]
```

```
columnVector2 = [ 0
                  1
                  9
                  36
                  100 ]
```

(It is not necessary to line the entries up as done here, but it looks nicer.) Note that `rowVector` is *not* equal to `columnVector1` even though each of their components is the same.

Matrices can be written using both commas and semicolons. The matrix

$$\mathcal{A} = \begin{bmatrix} 1\,2\,3 \\ 4\,5\,6 \\ 7\,8\,9 \end{bmatrix} \tag{1.1}$$

can be generated with the expression

```
A=[1,2,3;4,5,6;7,8,9]
```

or, more clearly, as

```
A=[ 1 2 3
    4 5 6
    7 8 9];
```

MATLAB has a special notation for generating a set of equally spaced values, which can be useful for plotting and other tasks. The format is:

```
start : increment : finish
```

or

```
start : finish
```

in which case the increment is understood to be 1. Both of these expressions result in row vectors. So you could define the even values from 10 to 20 by:

```
evens = 10 : 2 : 20
```

Sometimes, you'd prefer to specify the *number* of items in the list, rather than their spacing. In that case, you can use the `linspace` function, which has the form

```
linspace( firstValue, lastValue, numberOfValues )
```

in which case you could generate six even numbers with the command:

```
evens = linspace ( 10, 20, 6 )
```

or fifty evenly-spaced points in the interval [10,20] with

```
points = linspace ( 10, 20, 50 )
```

As a general rule, use the colon notation when `firstValue`, `lastValue` and `increment` are integers, or when you would have to do mental arithmetic to get `increment`, and `linspace` otherwise.

Another nice thing about MATLAB vector variables is that they are *flexible*. If you decide you want to add another entry to a vector, it's very easy to do so. To add the value 22 to the end of our **evens** vector:

```
evens = [ evens, 22 ]
```

and you could just as easily have inserted a value 8 before the other entries, as well.

Even though the number of elements in a vector can change, MATLAB always knows how many there are. You can request this value at any time by using the **numel** function. For instance,

```
numel ( evens )
```

should yield the value 7 (the 6 original values of 10, 12, ... 20, plus the value 22 tacked on later). In the case of matrices with more than one nontrivial dimension, the **numel** function returns the *total* number of entries. The **size** function returns a vector containing two values: the number of rows and the number of columns. To get the number of rows of a variable v, use `size(v,1)` and to get the number of columns use `size(v,2)`. For example, since **evens** is a row vector, `size( evens, 1)=1` and `size( evens, 2)=7`, one row and seven columns.

To specify an individual entry of a vector, you need to use index notation, which uses *round* parentheses enclosing the index of an entry. *The first element of an array has index 1* (as in Fortran, but not C or Java). Thus, if you want to alter the third element of **evens**, you could say

```
evens(3) = 2
```

There is special syntax for the final index value. Since **evens** is a vector with length 7, you can write `evens(end)` to mean the same thing as `evens(7)`.

Exercise 1.2.

(1) Use the `linspace` function to create a row vector called **meshPoints** containing exactly 1000 values with values evenly spaced between -1 and 1. Do not print all 1000 values!

(2) What expression will yield the value of the 95th element of **meshPoints**? What is this value?

(3) Double-click on the variable **meshPoints** in the "Current workspace" window-pane to view it as a vector and confirm its length is 1000.

(4) Use the **numel** function to again confirm the vector has length 1000.

(5) Produce a plot of a sinusoid on the interval $[-1, 1]$ using the command

```
plot(meshPoints,sin(2*pi*meshPoints))
```

(6) Create a file named `exer2.m`. You can use the MATLAB command `edit`, type the commands for this exercise into the window and use "Save as" to give it a name, or you can highlight some commands in the history windowpane and use a right mouse click to bring up a menu to save the commands as an m-file. The following first lines of the file should added:

```
% Chapter 1, exercise 2
% A sample script file.
% Your name and the date
```

Follow the header comments with the commands containing exactly the commands you used in the earlier parts of this exercise. Test your script by using `clear` to clear your results and then execute the script from the command line by typing `exer2`.

1.5   Vector and matrix operations

MATLAB provides a large assembly of tools for matrix and vector manipulation. You will investigate a few of these by trying them out.

Exercise 1.3. Define the following vectors and matrices (you can use cut-and-paste from the file `exer3.txt`):

```
rowVec1 = [ -1 -4 -9]
colVec1 = [ 2
            9
            8 ]
mat1 = [ 1  3  5
         7  9  0
         2  4  6 ]
```

(1) You can multiply vectors by constants. Compute

```
colVec2 = (pi/4) * colVec1
```

(2) The cosine function can be applied to a vector to yield a vector of cosines. Compute

```
colVec2 = cos( colVec2 )
```

Note that the values of `colVec2` have been overwritten. Are these the values you expect?

(3) You can add vectors. Compute

```
colVec3 = colVec1 + colVec2
```

(4) The Euclidean norm of a matrix or a vector is available using **norm**. Compute

```
norm(colVec3)
```

(5) You can do row-column matrix multiplication. Compute

```
colvec4 = mat1 * colVec1
```

(6) A single quote following a matrix or vector indicates a transpose.

```
mat1Transpose = mat1'
rowVec2 = colVec3'
```

Warning: The single quote really means the complex-conjugate transpose (or Hermitian adjoint). If you want a true transpose applied to a complex matrix you must use ".'" (dot-single quote), or use the function **transpose**.

(7) Transposes allow the usual operations. Even if A is a non-symmetric matrix, AA^T is always symmetric, and you might find $\mathbf{u}^T\mathbf{v}$ a useful expression to compute the dot (inner) product of two column vectors $\mathbf{u} \cdot \mathbf{v}$ (although there is a MATLAB **dot** function).

```
mat2 = mat1 * mat1'      % symmetric matrix
rowVec3 = rowVec1 * mat1
dotProduct = colVec3' * colVec1
euclideanNorm = sqrt(colVec2' * colVec2)
```

(8) Matrix operations such as determinant and trace are available, too.

```
determinant = det( mat1 )
tr = trace( mat1 )
```

(9) You can pick certain elements out of a vector, too. Use the following command to find the smallest element in a vector **rowVec1**.

```
min(rowVec1)
```

(10) The **min** and **max** functions work along one dimension at a time. They produce vectors when applied to matrices.

```
max(mat1)
```

(11) You can compose vector and matrix functions. For example, use the following expression to compute the max norm of a vector.

```
max(abs(rowVec1))
```

(12) How would you find the single largest element of a matrix?

(13) As you know, a magic square is a matrix all of whose row sums, column sums and the sums of the two diagonals are the same. (One diagonal of a matrix goes

from the top left to the bottom right, the other diagonal goes from top right to bottom left.) Consider the matrix

```
A=magic(201);    % please do not print all 40,401 entries.
```

The matrix A has 201 row sums (one for each row), 201 column sums (one for each column) and two diagonal sums. These 404 sums should all be exactly the same, and you **could** verify that they are the same by printing them and "seeing" that they are the same. It is easy to miss small differences among so many numbers, though. *Instead*, verify that A is a magic square by constructing the 201 column sums (without printing them) and computing the maximum and minimum values of the column sums. Do the same for the 201 row sums, and compute the two diagonal sums. Check that these six values are the same. If the maximum and minimum values are the same, the flyswatter principle says that all values are the same.

Hints:

- Use the MATLAB min and max functions.
- Recall that sum applied to a matrix yields a row vector whose values are the sums of the columns.
- The MATLAB function diag extracts the diagonal of a matrix, and the composed function sum(diag(fliplr(A))) computes the sum of the other diagonal.

(14) Suppose you want a table of integers from 0 to 10, their squares and cubes. You could start with

```
integers = 0 : 10
```

but now you will get an error when you try to multiply the entries of integers by themselves.

```
squareIntegers = integers * integers
```

Realize that MATLAB deals with vectors, and the default multiplication operation with vectors is row-by-column multiplication. What you want here is *element-by-element* multiplication, so you need to place a *period* in front of the operator:

```
squareIntegers = integers .* integers
```

Now you can define cubeIntegers in a similar way.

```
cubeIntegers = squareIntegers .* integers
```

Finally, you might like to print them out as a table. squareIntegers, integers, *etc.* are row vectors, so make a matrix whose **columns** consist of these vectors and allow MATLAB to print out the whole matrix at once.

```
tableOfPowers=[integers', squareIntegers', cubeIntegers']
```

(15) Watch out when you use vectors. The multiplication, division and exponentiation operators all have two possible forms, depending on whether you want
to operate on the arrays, or on the elements in the arrays. In all these cases,
you need to use the **period** notation to force elementwise operations. Compute
the squares of the values in `integers` alternatively using the exponentiation
operator as:

```
sqIntegers = integers .^ 2
```

and check that the two calculations agree with the command

```
norm(sqIntegers-squareIntegers)
```

which should result in zero.

Remark: Addition, subtraction, and division or multiplication by a scalar do
not require the dot in front of the operator, although you will get the correct
result if you use one.

(16) The index notation can also be used to refer to a subset of elements of the
array. With the *start:increment:finish* notation, you can refer to a range of
indices. Two-dimensional vectors and matrices can be constructed by leaving
out some elements of our three-dimensional ones. For example, submatrices can
be constructed from `tableOfPowers`. (The **end** function in MATLAB means the
last value of that dimension.)

```
tableOfCubes = tableOfPowers(:,[1,3])
tableOfEvenCubes = tableOfPowers(1:2:end,[1:2:3])
```

Note: `[1:2:3]` is the same as `[1,3]`.

(17) You have already seen the MATLAB function `magic(n)`. Use it to construct a
10×10 matrix.

```
A = magic(10)
```

What commands would be needed to generate the four 5×5 matrices in the
upper left quarter `AUL`, the upper right quarter `AUR`, the lower left quarter `ALL`,
and the lower right quarter `ALR`.

(18) It is possible to construct vectors and matrices from smaller ones in the same
way they can be constructed from numbers. Reconstruct the matrix `A` in the
previous exercise as

```
B=[AUL AUR
   ALL ALR];
```

and show that `A` and `B` are the same by showing that `norm(A-B)` is zero.

(19) Sometimes MATLAB syntax generates results that are surprising from the standpoint of linear algebra. Compute

```
surprise = colVec1 + rowVec1
```

This result is consistent with the way that MATLAB will add a scalar to a vector by "broadcasting" the scalar into a vector, but in the linear algebra context it is almost always the wrong thing to do. Be careful to avoid this trap.

Remark 1.1. Exercise 1.3(19) is a warning that adding row vectors and column vectors results in a matrix. This surprising result is one of the most common sources of errors that students make in MATLAB m-files. If you are testing a newly written m-file and you find matrices where you expect only vectors, use the debugger to highlight places where you might have added a row vector to a column vector.

You will see that the function m-files presented in this workbook check that input vectors conform to their expected row or column shape. Most vectors in this workbook will be column vectors, and this check avoids the errors that will arise when the function is mistakenly applied to the wrong shape vector. When you write your own function m-files, you should likewise check.

1.6   Flow control

It is critical to be able to ask questions and to perform repetitive calculations in m-files. These topics are examples of "flow control" constructs in programming languages. MATLAB provides two basic looping (repetition) constructs: `for` and `while`, and the `if` construct for asking questions. These statements each surround several MATLAB statements with `for`, `while` or `if` at the top and `end` at the bottom.

Remark 1.2. It is an excellent idea to indent the statements between the `for`, `while`, or `if` lines and the `end` line. This indentation strategy makes code immensely more readable. Your m-files will be expected to follow this convention.

	Syntax	Example

for loop	for *control-variable*=*start* : *increment* : *end*
	  one or more statements
	end
		```nFactorial=1;```
		```for i=1:n```
		``` nFactorial=nFactorial*i;```
		```end```

while loop	*statement initializing a variable*
	while *logical condition involving the variable*
	  one or more statements
	  statement changing the variable
	end
		```nFactorial=1;```
		```i=1; % initialize i```
		```while i <= n```
		``` nFactorial=nFactorial*i;```
		``` i=i+1;```
		```end```

simple if	if *logical condition*
	  one or more statements
	end
		```if x ~= 0 % ~ means "not"```
		``` y = 1/x;```
		```end```

compound if	if *logical condition*
	  one or more statements
	elseif *logical condition*
	  one or more statements
	else
	  one or more statements
	end
		```if x ~= 0```
		``` y=1/x;```
		```elseif sign(x) > 0```
		``` y = +inf;```
		```else```
		``` y = -inf;```
		```end```

Note that `elseif` is one word! Using two words `else if` changes the statement into two nested `if` statements with possibly a *very* different meaning, and a different number of `end` statements.

Exercise 1.4. The trapezoid rule for the approximate value of the integral of e^x on the interval $[x_0, x_1]$ can be written as

$$\int_{x_0}^{x_1} e^x dx \approx \frac{h}{2}e^{x_0} + h\sum_{k=2}^{N-1} e^{x_k} + \frac{h}{2}e^{x_N}$$

where $h = 1/(N-1)$ and $x_k = 0, k, 2k, \ldots, 1$.

The following MATLAB code computes the trapezoid rule for the case that $N = 40$, $x_0 = 0$ and $x_1 = 1$. You can find it in the file `trapz.txt`.

```
% Use the trapezoid rule to approximate the integral from 0 to 1
% of exp(x), using N intervals
% Your name and the date
N=40;
h=1/(N-1);
x=-h;   % look at this trick
approxIntegral=0.;
for k=1:N
  % compute current x value
  x=x+h;

  % add the terms up
```

```
if k==1 | k==N
  approxIntegral=approxIntegral+(h/2)*exp(x); % ends of interval
else
  approxIntegral=approxIntegral+h*exp(x);    % middle of interval
end
end
```

(1) Use cut-and-paste to put the code directly into the MATLAB command window to execute it. Is the final value for `approxIntegral` nearly equal to $e^1 - e^0$? (You can get the value of a variable by simply typing its name at the command prompt.)

(2) Notice the indentation. Typically, statements inside `for` loops and `if` tests are indented for readability. (There isn't much to be gained by indentation when you are typing at the command line, but when you put commands in a file you should always use indentation.)

(3) What is the complete sequence of all values taken on by the variable `x`?

(4) How many times is the following statement executed?

```
approxIntegral=approxIntegral+(h/2)*exp(x); % ends of interval
```

(5) How many times is the following statement executed?

```
approxIntegral=approxIntegral+h*exp(x);    % middle of interval
```

1.7   M-files and graphics

If you have to type everything at the command line, you will not get very far. You need some sort of scripting capability to save the trouble of typing, to make editing easier, and to provide a record of what you have done. You also need the capability of making functions or your scripts will become too long to understand. In this section you will consider first a script file and later a function m-file. You will be using graphics in the script file, so you can pick up how graphics can be used in your work.

The Fourier series for the function $y = x^3$ on the interval $-1 \le x \le 1$ is

$$y = 2 \sum_{k=1}^{\infty} (-1)^{k+1} \left(\frac{\pi^2}{k} - \frac{6}{k^3} \right) \sin kx. \tag{1.2}$$

You are going to look at how this series converges.

Exercise 1.5. Copy and paste the following text (a copy of which you will find in the file `exer5.txt`) into a file named `exer5.m` and then answer the questions about the code.

```
% compute NTERMS terms of the Fourier Series for y=x^3
% plot the result using NPOINTS points from -1 to 1.
% Your name and the date

NTERMS=20;
NPOINTS=1000;
x=linspace(-1,1,NPOINTS);
y=zeros(size(x));
for k=1:NTERMS
  term=2*(-1)^(k+1)*(pi^2/k-6/k^3)*sin(k*x);
  y=y+term;
end
plot(x,y,'b');  % 'b' is for blue line
hold on
plot(x,x.^3,'g'); % 'g' is for a green line
axis([-1,1,-2,2]);
hold off
```

It is always good programming practice to define constants symbolically at the beginning of a program and then to use the symbols within the program. Sometimes these special constants are called "magic numbers." By convention, symbolic constants are named using all upper case.

(1) Add your name and the date to the comments at the beginning of the file.
(2) How is the MATLAB variable x related to the dummy variable x in Equation (1.2)? (Please use no more than one sentence for the answer.)
(3) How is the MATLAB statement that begins y=y... inside the loop related to the summation in Equation (1.2)? (Please use no more than one sentence for the answer.)
(4) In your own words, what does the line

    y=zeros(size(x));

    do? **Hint:** You can use MATLAB help for `zeros` and `size` for more information.
(5) Execute the script by typing its name exer5 at the command line. You should see a plot of two lines, one representing a partial sum of the series and the green line a plot of x^3, the limit of the partial sums.
(6) What would happen if the two lines `hold on` and `hold off` were omitted?
    Note: The command hold, without the "on" or "off", is a "toggle." Each time it is used, it switches from "on" to "off" or from "off" to "on." Using it that way is easier but you have to remember which state you are in.

In the following exercise you are going to modify the calculation so that it continues to add terms so long as the largest component of the next term remains larger in absolute value than, say, 0.05. Since the series is of alternating sign, this quantity is a legitimate estimate of the difference between the partial sum and the limit. The `while` statement is designed for this case. It would be used in the following way, *replacing* the `for` statement.

```
TOLERANCE=0.05;    % the chosen tolerance value
  <<some lines of code from above>>
k=0;
term=TOLERANCE+1;  % bigger than TOLERANCE
while max(abs(term)) > TOLERANCE
  k = k + 1;
    <<some lines of code from above>>
end
disp( strcat('Number of iterations =',num2str(k)) )
  <<some lines of code to plot results>>
```

Exercise 1.6.

(1) Copy the file `exer5.m` to a new file called `exer6.m` or use "Save as" from the File menu. Change the comments at the beginning of the file to reflect the objective of this exercise.

(2) Modify `exer6.m` by *replacing* the `for` loop with a `while` loop as outlined above.

(3) What is the purpose of the statement

```
term=TOLERANCE+1;
```

(4) What is the purpose of the statement

```
k = k + 1;
```

(5) Try the script to see how it works. How many iterations are required? Does it generate a plot similar to the one from Exercise 6?

Remark 1.3. If you try this script and it does not quit (it stays "busy") you may be able to interrupt the calculation by holding the Control key ("CTRL") and pressing "C". There is also an option to halt a running calculation in the Debug menu. Failure to stop indicates you have a bug. For some reason, the value of the variable `term` is not getting small. The debugger can help you see why the value of `term` is not getting small.

The next task is to make a function m-file out of `exer6.m` in such a way that it takes as argument the desired tolerance and returns the number of iterations required to meet that tolerance.

Exercise 1.7.

(1) Use the following steps to write a function m-file named `exer7.m`.

  (a) Copy the file `exer6.m` to a new file called `exer7.m`, or use "Save As" from the File menu. Turn it into a function m-file by placing the signature line first and adding a comment for its usage

```
function k = exer7( tolerance )
  % k = exer7( tolerance )
  % more comments
  % Your name and the date
```

  (b) Replace the upper-case `TOLERANCE` with a lower-case `tolerance` because it no longer is a constant, *and discard* the line fixing its value.

  (c) Note that the function name *must* agree with the file name. Add comments just after the signature and usage lines to indicate what the function does.

  (d) Delete the lines that create the plot and print (`disp(...)`) so that the function does its work silently.

  (e) Add an additional `end` statement at the bottom of the file and indent all lines between the signature line and the final line. Lines inside the loop will have double indentation.

(2) Invoke this function from the MATLAB command line by choosing a tolerance and calling the function. Using the command

```
exer7(0.05)
```

will cause MATLAB to print the result. Using the function on the right side of an equal sign with a variable on the left side of the equal sign causes the variable to take the value given by the function.

```
numItsRequired=exer7(0.05)
```

causes *both* a printed value and assignment of a value to `numItsRequired`.

(3) Use the command `help exer7` to display your help comments. Make sure they describe the purpose of the function.

(4) How many iterations are required for a tolerance of 0.05? This value should agree with the value you saw in Exercise 6.

(5) To observe convergence, how many iterations are required for tolerances of 0.1, 0.05, 0.025, and 0.0125?

Unlike ordinary mathematical notation, MATLAB allows a function to return two or more values instead of a single value. The syntax to accomplish this trick is similar to that of defining a vector, but the meaning is nothing like a vector. For a function named "`funct`" that returns two variables depending on a single variable as input, the signature line would look like:

```
[y,z] = funct( x )
```

and to use the function, for the value at x = 3,

```
[y,z] = funct( 3 )
```

If you have the same function but wish only the first output variable, y, you would write

```
y = funct( 3 )
```

and if you wish only the second output variable, z, you would write

```
[~,z] = funct( 3 )
```

Exercise 1.8.

(1) Copy the file `exer7.m` to a new file called `exer8.m`, or use "Save As" from the File menu. Modify the function so that it returns first the converged value of the sum (a vector) and, second, the number of iterations required. Be sure to change the comments to include a description of all input and output parameters.

(2) What form of the command line would return *only* the number of iterations for a tolerance of 0.05? Try it: does the number of iterations agree with the result from Exercises 1.6 and 1.7?

(3) What form of the command line would return *only* the (vector) value of the sum to a tolerance of 0.03. What is the norm of this vector, using `format long` to get at least 14 digits of accuracy?

(4) What form of the command line would return *both* the (vector) value of the sum and the number of iterations required to achieve a tolerance of 0.02? How many iterations were taken, and what is the norm of the (vector) value of the sum, using `format long` to get at least 14 digits of accuracy?

Exercise 1.9. You will often find it useful in this course to have function names (more precisely, "function handles") as variables. For example, the sine function in `exer8` could be replaced with an arbitrary function. While the series might not converge for some choices of functions, it would converge for others

(1) Copy the file `exer8.m` to a new file called `exer9.m`, or use "Save As" from the File menu. Modify the function so that it accepts a second parameter: `exer9(tolerance, func)` and replace the `sin` function inside the sum with `func`. Do not forget to modify the comments in the file to reflect this change.

(2) Test that `exer9` and `exer8` give equivalent results when `func` is really `sin` with the following code:

```
y8=exer8(0.02);
y9=exer9(0.02, @sin);
```

```
% following difference should be small
norm(y8-y9)
```

(3) The @ symbol is used to identify **sin** as a function instead of an ordinary variable. If you forget the @, you will receive an error message seems difficult to interpret. What is the result of the following command?

```
y=exer9(0.02, sin)
```

What the error message is telling you is that the **sin** that appears in the argument is being interpreted as a function that needs an argument (like **sin(pi)**) but doesn't have one. The @ symbol tells the interpreter that what follows is a function name.

(4) Use **exer9** to compute the (vector) sum of the series for **exer9(.02, @cos)** and plot the result. Please include this plot with your summary.

Chapter 2

Roots of equations

2.1 Introduction

The *root finding* problem is the task of finding one or more values for a variable x so that an equation

$$f(x) = 0$$

is satisfied. Denote the desired solution as x_0. As a computational problem, you are interested in effective and efficient ways of finding a value x that is "close enough" to x_0. There are two common ways to measuring "close enough."

(1) The "residual error," $|f(x)|$, is small, or
(2) The "approximation error" or "true error," $|x - x_0|$ is small.

Of course, the true error is not known, so it must be approximated in some way. In this chapter you will see some nonlinear equations and some simple methods of finding an approximate solution. You will use an estimate of the true error when it is readily available and will use the residual error the rest of the time.

You will see in this chapter that error estimates can be misleading and there are many ways to adjust and improve them. More sophisticated error estimation will not be considered in this chapter.

This chapter is at first devoted to discussion of the bisection method. This method is also used to introduce (to presumed novices) MATLAB programming and the first seven exercises form a sequence to this effect. The final three exercises address the secant method, Regula Falsi, and Muller's method and are largely independent of one another.

2.2 Remarks on programming style

You can find MATLAB code on the internet and in books ([Quarteroni *et al.* (2007)] is an example). This code appears in a variety of styles. The coding style used in examples in this workbook has an underlying consistency and is, in my humble opinion, one of the easiest to read, understand and debug. Some of the style rules followed throughout this workbook include:

23

- One statement per line, with minor exceptions.
- Significant variables are named with long names starting with nouns and followed by modifiers, and with beginnings of modifiers capitalized.
- Loop counters and other insignificant variables are short, often a single letter such as k, m or n. i and j are avoided as variable names.
- The interior blocks of loops and if-tests are indented.
- Function files always have comments following the `function` statement so that they are available using the `help` command. Function usage is indicated by repeating the signature line among the comments.

These rules of style improve clarity of the code so that people and not just computers can make sense of it.

Clarity is important for debugging because if code is harder to read then it is harder to debug. Debugging is one of the most difficult and least rewarding activities you will engage in, so *anything* that simplifies debugging pays off. Clarity is also important when others read your code. Since a grader needs to read your code in order to give you a grade, please format your code using the above guidelines.

2.3   A sample problem

Suppose you want to know if there is a solution to

$$\cos x = x \, ,$$

whether the solution is unique, and what its value is. This is not an algebraic equation, and there is little hope of forming an explicit expression for the solution.

Since you can't solve this equation exactly, it's worth knowing whether there is anything you can say about it. In fact, if there is a solution x_0 to the equation, you can *probably* find a computer representable number x which approximately satisfies the equation (has low residual error) and which is close to x_0 (has a low approximation error). Each of the following two exercises uses a plot to illustrate the existence of a solution.

Remark 2.1. MATLAB has the capability of plotting some functions by name. You won't be using this capability in this course because it is too specialized. Instead, you will construct plots the way you probably learned when you first learned about plotting: by constructing a pair of vectors of x-values (abscissæ) and corresponding y-values (ordinates) and then plotting the points with lines connecting them.

Exercise 2.1. The following steps show how to use MATLAB to plot the functions $y = \cos x$ and $y = x$ together in the same graph from $-\pi$ to π.

(1) Define a MATLAB (vector) variable `xplot` to take on 200 evenly-spaced values between `-pi` and `pi`. You can use `linspace` to do this.
(2) Define the (vector) variable `y1plot` by `y1plot=cos(xplot)`. This is for the curve $y = \cos x$.

(3) Define the (vector) variable `y2plot` by `y2plot=xplot`. This is for the line $y = x$.

(4) Plot both `y1plot` and `y2plot` by plotting the first,

```
plot(xplot,y1plot)
```

then `hold on`, plot the second, and `hold off`. **Note**: If you read the help files for the "plot" function, you will find that an alternative *single* command is `plot(xplot,y1plot,xplot,y2plot)`.

(5) The two lines intersect, clearly showing that a solution to the equation $\cos x = x$ exists. Read an approximate value of x at the intersection point from the plot.

Exercise 2.2.

(1) Write an m-file named `cosmx.m` ("COS Minus X") that defines the function
$$f(x) = \cos x - x.$$

Recall that a function m-file is one that starts with the word `function`. In this case it should look something like

```
function y = cosmx ( x )
  % y = cosmx(x) computes the difference y=cos(x)-x

  % your name and the date

  y = ???

end
```

(This is very simple — the object of this exercise is to get going.) You can find this code in the file `cosmx.txt` and copy it.

(2) Now use your `cosmx` function to compute `cosmx(0.5)`. Your result should be about 0.37758 and should not result in extraneous printed or plotted information.

(3) Now use your `cosmx` function in the same plot sequence

```
plot(???
hold on
plot(???
hold off
```

as above to plot `cosmx` and a line representing the horizontal axis on the interval $-\pi \le x \le \pi$.

Hint: The horizontal axis is the line $y = 0$. To plot this as a function, you need a vector of at least two x-values between $-\pi$ and π and a corresponding vector of y-values all of which are zero. Think of a way to generate an appropriate set of x-values and y-values when all the y-values are zero. You could use the MATLAB function `zeros`.

(4) Does the value of x at the point where the curve crosses the x-axis agree with the previous exercise?

2.4 The bisection method

The idea of the bisection method is very simple. Assume that you are given two values $x = a$ and $x = b$ $(a < b)$, and that the function $f(x)$ is positive at one of these values (say at a) and negative at the other. (When this occurs, $[a, b]$ is called a "change-of-sign interval" for the function.) Assuming f is continuous, there must be at least one root in the interval.

Intuitively speaking, if you divide a change-of-sign interval in half, one or the other half must end up being a change-of-sign interval, and it is only half as long. Keep dividing in half until the change-of-sign interval is so small that any point in it is within the specified tolerance.

More precisely, consider the point $x = (a + b)/2$. If $f(x) = 0$ you are done. (This is pretty unlikely.) Otherwise, depending on the sign of $f(x)$, you know that the root lies in $[a, x]$ or $[x, b]$. In any case, your change-of-sign interval is now half as large as before. Repeat this process with the new change of sign interval until the interval is sufficiently small and declare victory.

You are *guaranteed* to converge. You can even compute the maximum number of steps this could take, because if the original change-of-sign interval has length ℓ then after one bisection the current change-of-sign interval is of length $\ell/2$, *etc.* You know in advance how well you will approximate the root x_0. These are very powerful facts, which make bisection a *robust* algorithm — that is, it is very hard to defeat it.

Exercise 2.3.

(1) If you know the start points a and b and the interval size tolerance ϵ, you can predict beforehand the number of steps required to reach the specified accuracy. The bisection method will always find the root in that number or fewer steps. What is the formula for that number?

(2) Give an example of a continuous function that is equal to zero only once in the interior of the interval $[-2, 1]$, but for which bisection could not be used.

2.5 Variable function names in MATLAB

Before you look at a sample bisection program, let's discuss some programming issues. If you haven't done much programming before, this is a good time to try to understand the logic behind how to choose variables to set up, what names to give them, and how to control the logic.

First of all, you should write the bisection algorithm as a MATLAB `function`. There are two reasons for writing it as a function instead of a script m-file (without the function header) or by typing everything at the command line. The first reason is a practical one: you will be executing the algorithm several times, with differing end points, and functions allow us to use temporary variables without worrying about whether or not they have already been used before. The second reason is pedantic: you should learn how to do this now because it is an important tool in your toolbox.

A precise description of the bisection algorithm is presented, for example, in [Quarteroni *et al.* (2007)] in Section 6.2.1, on pages 250–251, and in [Atkinson (1978)] on page 56. The algorithm can be expressed in the following way. (Note: the product $f(a) \cdot f(b)$ is negative if and only if $f(a)$ and $f(b)$ are of opposite sign and neither is zero.)

Given a function $f : \mathbb{R} \to \mathbb{R}$ and $a, b \in \mathbb{R}$ so that $f(a) \cdot f(b) < 0$, then sequences $a^{(k)}$ and $b^{(k)}$ for $k = 0, \ldots$ with $f(a^{(k)}) \cdot f(b^{(k)}) < 0$ for each k can be constructed by starting out with $a^{(1)} = a$, $b^{(1)} = b$ then, for each $k > 0$,

(1) Set $x^{(k)} = (a^{(k)} + b^{(k)})/2$.
(2) If $|x^k - b^{(k)}|$ is small enough, or if $f(x^{(k)}) = 0$ exit the algorithm.
(3) If $f(x^{(k)}) \cdot f(a^{(k)}) < 0$, then set $a^{(k+1)} = a^{(k)}$ and $b^{(k+1)} = x^{(k)}$.
(4) If $f(x^{(k)}) \cdot f(b^{(k)}) \leq 0$, then set $a^{(k+1)} = x^{(k)}$ and $b^{(k+1)} = b^{(k)}$.
(5) Return to step 1.

The bisection algorithm can be described in a manner more appropriate for computer implementation in the following way. Any algorithm intended for a computer program *must* address the issue of when to stop the iteration. In the following algorithm,

Algorithm Bisect(f,a,b,x,ϵ)

(1) Set $x:=(a + b)/2$.
(2) If $|b - x| \leq \epsilon$, or if $f(x)$ happens to be exactly zero, then accept x as the approximate root, and exit.
(3) If `sign`($f(b)$) * `sign`($f(x)$) < 0, then $a:=x$; otherwise, $b:=x$.
(4) Return to step 1.

Remark 2.2.
• The latter form of the algorithm is expressed in a form that is easily translated into a loop, with an explicit exit condition.

- The latter form of the algorithm employs a signum function rather than the value of f in order to determine sign. This is a better strategy because of roundoff errors. If $f(a) > 0$ and $f(b) > 0$ are each less than the MATLAB quantity $\sqrt{\texttt{realmin}}$, then their product is zero, not positive.

The language of this description, an example of "pseudocode," is based on a computer language called Algol but is intended merely to be a clear means of specifying algorithms in books and papers. (The term "pseudocode" is used for any combination of a computer coding language with natural language.) One objective of this chapter is to show how to rewrite algorithms like it in the MATLAB language. As a starting point, fix $f(x)$ to be the function cosmx that you just wrote. Later you will learn how to add it to the calling sequence. Let's also fix $\epsilon = 10^{-10}$.

Here is a MATLAB function that carries out the bisection algorithm for your cosmx function. Returned values in MATLAB are separated from arguments, so the variables x and itCount are written on the left of an equal sign.

```
function [x,itCount] = bisect_cosmx( a, b)
  % [x,itCount] = bisect_cosmx( a, b) uses bisection to find a
  % root of cosmx between a and b to tolerance of 1.0e-10
  % a=left end point of interval
  % b=right end point of interval
  % cosmx(a) and cosmx(b) should be of opposite signs
  % x is the approximate root found
  % itCount is the number of iterations required.

  % your name and the date

  EPSILON = 1.0e-10;
  fa = cosmx(a);
  fb = cosmx(b);

  for itCount = 1:(???)  % fill in using the formula from Exercise 3

    x = (b+a)/2;
    fx = cosmx(x);

    % The following statement prints the progress of the algorithm
    disp(strcat(  'a=' , num2str(a), ', fa=' , num2str(fa), ...
                ', x=' , num2str(x), ', fx=' , num2str(fx), ...
                ', b=' , num2str(b), ', fb=' , num2str(fb)))

    if ( fx == 0 )
      return; % found the solution exactly!
```

```
    elseif ( abs ( b - x ) < EPSILON )
      return;  % satisfied the convergence criterion
    end

    if ( sign(fa) * sign(fx) <= 0 )
      b = x;
      fb = fx;
    else
      a = x;
      fa = fx;
    end

  end

  error('bisect_cosmx failed with too many iterations!')
end
```

Remark 2.3. This code violates the rule about having functions print things, but the `disp` statement merely prints progress of the function and will be discarded once you are confident the code is correct.

Compare the code and the algorithm. The code is similar to the algorithm, but there are some significant differences. For example, the algorithm generates sequences of endpoints $\{a_i\}$ and $\{b_i\}$ and midpoints $\{x_i\}$. The code keeps track of only the most recent endpoint and midpoint values. This is a typical strategy in writing code — use variables to represent only the most recent values of a sequence when the full sequence does not need to be retained.

A second difference is in the looping. There is an error exit in the case that the convergence criterion is never satisfied. If the function ever exits with this error, it is because either your estimate of the maximum number of iterations is too small or because there is a bug in your code.

Note the symbol `==` to represent equality in a test. Use of a single `=` sign causes MATLAB to display a syntax error. In languages such as C no syntax error is produced but the statement is wrong nonetheless and can cause serious errors that are very hard to find.

Exercise 2.4. Create the file `bisect_cosmx.m` by typing it yourself or downloading the file `bisect_cosmx.txt`

(1) Replace the "???" with your result from Exercise 2.3. This value should be an integer, so be sure to increase the value from Exercise 2.3 to the next larger integer. (If you are not confident of your result, use 10000. If the code is correctly written, it will work correctly for any value larger than your result from Exercise 2.3.) **Hint:** Recall that $\log_2(x) = \log(x)/\log(2)$.

(2) Why is the name EPSILON all capitalized?

(3) The variable EPSILON and the reserved MATLAB variable eps have similar-sounding names. As EPSILON is used in this function, is it related to eps? If so, how?

(4) In your own words, what does the sign(x) function do? What if x is 0? The sign function can avoid representation difficulties when fa and fm are near the largest or smallest numbers that can be represented on the computer.

(5) The disp command is used only to monitor progress. Note the use of ... as the continuation character.

(6) What is the result if the MATLAB error function is called?

(7) Try the command:

```
[z,iterations] = bisect_cosmx ( 0, 3 )
```

Is the value z close to a root of the equation cos(z)=z?

Warning: You should not see the error message! If you do, your value for the maximum number of iterations is too small.

(8) How many iterations were required? Is this value no larger than the value from your formula in Exercise 2.3?

(9) In the unusual circumstance that a and b are already closer together than the tolerance when the function starts, you will see the error message. The error message in this case is misleading, but the situation is sufficiently rare that it does not pay to fix it. Based on your understanding of the code, however, what would the final value of itCount be?

(10) Type the command help bisect_cosmx at the MATLAB command line. You should see your comments reproduced as a help message. This is one reason for putting comments at the top.

(11) For what input numbers will this program produce an incorrect answer (*i.e.*, return a value that is not close to a root)? Add a check before the loop so that the function will call the MATLAB error function instead of returning an incorrect answer.

(12) In your opinion, why is abs used in the convergence test?

2.6   Convergence criteria

Now you have written an m-file called bisect_cosmx.m that can find roots of the function cosmx on any interval, but you really want to use it on an arbitrary function. To do this, you *could* create an m-file for the other function, but you would have to call that m-file cosmx.m even if there were no cosines in it, because that is what the bisect_cosmx function expects. Suppose that you have three files, called f1.m through f3.m, each of which evaluates a function that you want to use bisection on. You *could* rename f1.m to cosmx.m, and change its name inside the file as well, and repeat this for the other files. But who wants to do that?

It is more convenient to introduce a dummy variable for the function name, just as you used the dummy variables a and b for numerical values. Function names in MATLAB can be passed as "function handles."

In the bisection function in the next exercise, the name of the function is specified in the function call using the @ character as:

```
[x, itCount] = bisect ( @cosmx , a, b )
```

Warning: If you forget the @ in front of the function name, MATLAB will produce errors that seem to make no sense at all (even less than usual). Keep in mind that function names in calling sequences *must* have the @ character in front of them. Check for this first when you get errors in a function.

Exercise 2.5.

(1) Write a new function m-file bisect.m.

    (a) Copy the file bisect_cosmx.m to a new file named bisect.m, or use the "Save as" menu choice to make a copy with the new name bisect.m.

    (b) In the declaration line for bisect, include a dummy function name. You might as well call it func, so the line becomes

```
function [x, itCount] = bisect ( func, a, b )
```

    (c) Change the comments describing the purpose of the function, and explain the variable func.

    (d) Where you evaluated the function by writing cosmx(x), you now write func(x). For instance, the line

```
fa = cosmx(a);
```

    must be rewritten as:

```
fa = func(a);
```

    Make this change for fa, and a similar change for fb and fx.

(2) When you use bisect, you must include the function name preceded by @

```
[z, iterations] = bisect ( @cosmx, 0, 3 )
```

Execute this command and make sure that you get the same result as before. (This is called "regression testing.")

Warning: If you forget the @ before cosmx and use the command

```
[z, iterations] = bisect ( cosmx, 0, 3 )   % error!
```

you will get the following mysterious error message:

```
>> [z, iterations] = bisect ( cosmx, 0, 3 )
Not enough input arguments.

Error in cosmx (line 6)
y = cos(x)-x;
```

(3) It is always a good idea to check that your answers are correct. Consider the simple problem $f_0(x) = 1 - x$. Write a simple function m-file named f0.m (based on cosmx.m but defining the function f_0) and show that you get the correct solution (x=1 to within EPSILON) starting from the change-of-sign interval [0,3].

Remark 2.4. MATLAB allows a simple function to be defined without a name or an m-file: You can then use it anywhere a function handle could be used. For example, for the function $f_0(x) = 1 - x$, you could write

```
[z, iterations] = bisect ( @(x) 1-x, 0, 3 );
```

or even as

```
f0=@(x) 1-x;
[z, iterations] = bisect ( f0, 0, 3 );
```

Be very careful when using this last form because there is no @ appearing in the bisect call but it is a function handle nonetheless.

Exercise 2.6. The following table contains formulas and search intervals for four functions. Write four simple function m-files, f1.m, f2.m, f3.m and f4.m, each one similar to cosmx.m but with formulas from the table. Then use bisect to find a root in the given interval, and fill in the table. Since you should be confident that your code is correct, turn the disp statement into a comment so that you do not get distracting printed lines.

Name	Formula	Interval	approxRoot	No. Steps
f1	x^2-9	[0,5]		
f2	x^5-x-1	[1,2]		
f3	x*exp(-x)	[-1,2]		
f4	2*cos(3*x)-exp(x)	[0,6]		
f5	(x-1)^5	[0,3]		

You found above that the bisection algorithm admits an excellent stopping criterion, and the number of iterations can be predicted before beginning the algorithm. This feature is remarkable and highly unusual. The much more common case is that it takes as much or more ingenuity to design a good stopping criterion as it does to design the algorithm itself. Furthermore, stopping criteria are likely to be problem-dependent.

One stopping criterion that is usually available is to stop when the residual of the function becomes small ($|f(x)| \leq \epsilon$). This criterion is better than nothing, and often serves as a good starting point, but it can have drawbacks.

Exercise 2.7.

(1) Copy your `bisect.m` file to one named `bisect0.m`, or use "Save as" from the "File" menu. Be sure to change the name of the function inside the file. Change the line

```
elseif ( abs ( b - x ) < EPSILON )
```

to read

```
elseif ( abs ( fx ) < EPSILON )
```

Since there is no longer a theoretical expression for the maximum number of iterations, you should also increase the maximum number of iterations to 10000.

(2) Consider the function `f0=x-1` on the interval `[0,3]`. Does your new function `bisect0` find the correct answer? (It should.)

(3) The amount the residual differs from zero is called the *residual error* and the difference between the solution you found and the true solution (found by hand or some other way) is called the *true error*. What are the residual and true errors when you used `bisect0` to find the root of the function `f0` above? How many iterations did it take?

(4) What are the residual and true errors when using `bisect0` to find the root of the function `f5=(x-1)^5`? How many iterations did it take?

(5) To summarize your comparison, fill in the following table

Name	Formula	Interval	Residual error	True error	Number of steps
f0	x-1	[0,3]			
f5	(x-1)^5	[0,3]			

(6) The `bisect0` function has the value `EPSILON = 1.0e-10`; in it. Does the table show that the residual error is always smaller than `EPSILON`? What about the true error?

2.7 The secant method

The secant method is described by [Atkinson (1978)] on page 66 or [Quarteroni *et al.* (2007)] in Section 6.2.2. Instead of dividing the interval in half, as is done in the bisection method, it regards the function as approximately linear, passing through the two points $x = a$ and $x = b$ and then finds the root of this linear function. The interval $[a, b]$ need no longer be a change-of-sign interval, and the next value of x need no longer lie inside the interval $[a, b]$. In consequence, residual convergence must be used in the algorithm instead of true convergence.

It is easy to see that if a straight line passes through the two points $(a, f(a))$ and $(b, f(b))$, then it crosses the x-axis $(y = 0)$ when $x = b - \frac{b-a}{f(b)-f(a)} f(b)$.

The secant method can be described in the following manner.

Algorithm Secant(f,a,b,x,ϵ)

(1) Set

$$x = b - \frac{b-a}{f(b) - f(a)} f(b)$$

(2) If $|f(x)| \leq \epsilon$, then accept x as the approximate root, and exit.
(3) Replace a with b and b with x.
(4) Return to step 1.

The secant method is sometimes much faster than bisection, but since it does not maintain an interval inside which the solution must lie, the secant method can fail to converge at all. Unlike bisection, the secant method can be generalized to two or more dimensions, and the generalization is usually called Broyden's method.

Exercise 2.8.

(1) Write an m-file named `secant.m`.

    (a) Base `secant.m` partly on `bisect0.m`, and beginning with the lines

```
function [x,itCount] = secant(func, a, b)
  % [x,itCount] = secant(func, a, b)
  % more comments explaining what the function does and
  % the use of each variable

  % your name and the date

  << your code >>

end
```

    (b) Because residual error is used for convergence, leave `EPSILON = 1.0e-10` and choose 1000 as the maximum number of iterations.
    (c) Comment any "`disp`" statements so they do not provide a distraction.
    (d) Complete `secant.m` according to the secant algorithm given above.

(2) Test `secant.m` with the linear function `f0(x)=x-1`. You should observe convergence to the exact root in a single iteration. If you do not, go back and correct your code. Explain why it should only take a single iteration.
(3) Repeat the experiments done with bisection above, using the secant method, and fill in the following table.

Name	Formula	Interval	Secant approxRoot	Secant Steps	Bisection approxRoot	Bisection Steps
f1	x^2-9	[0,5]				
f2	x^5-x-1	[1,2]				
f3	x*exp(-x)	[-1,2]				
f4	2*cos(3*x) -exp(x)	[0,6]				
f5	(x-1)^5	[0,3]				

(4) In the above table, you can see that the secant method can be either faster or slower than bisection. You may also observe convergence failures: either convergence to a value that is not near a root or convergence to a value substantially less accurate than expected. Regarding the bisection roots as accurate, are there any examples of convergence to a value that is not near a root in the table? If so, which? Are there any examples of inaccurate roots? If so, which?

2.8   The Regula Falsi method

The regula falsi algorithm is discussed, for example, in [Quarteroni *et al.* (2007)] in Section 6.2.2 and is described below. It can be thought of as a cross between the bisection and secant algorithms, since the interval $[a, b]$ begins as a change-of-sign interval and subsequent reduced intervals are also change-of-sign intervals.

Algorithm Regula(f,a,b,x,ϵ)

(1) Set

$$x = b - \frac{b-a}{f(b) - f(a)} f(b)$$

(2) If $|f(x)| \leq \epsilon$, then accept x as the approximate root, and exit.
(3) If sign($f(a)$) * sign($f(x)$) ≥ 0, then set $a = b$, to keep the change-of-sign interval.
(4) Set $b = x$.
(5) Return to step 1.

Remark 2.5. It is critical to observe that Step 3 maintains the interval $[a, b]$ as a change-of-sign interval that is a subinterval of the initial change-of-sign interval. Thus, regula falsi, unlike the secant method, *must* converge, although convergence might take a long time.

Exercise 2.9.

(1) Starting from your bisect0.m file, write an m-file named regula.m to carry out the regula falsi algorithm.

(2) Test your work by finding the root of `f0(x)=x-1` on the interval `[-1,2]`. You should get the exact answer in a single iteration.

(3) Repeat the experiments from the previous exercise but using `regula.m` and fill in the following table. Pay attention: the last line is for `f3`, not `f5`.

Name	Formula	Interval	Regula approxRoot	Regula Steps	Secant steps	Bisection Steps
f1	x^2-9	[0,5]				
f2	x^5-x-1	[1,2]				
f3	x*exp(-x)	[-1,2]				
f4	2*cos(3*x) -exp(x)	[0,6]				
f5	(x-1)^5	[0,3]				

You should observe both faster and slower convergence, compared with bisection and secant. You should not observe lack of convergence or convergence to an incorrect solution.

(4) Function `f5`, `(x-1)^5` turns out to be very difficult for regula falsi! Loosen your convergence criterion to a tolerance of 10^{-6} and increase the maximum allowable number of iterations to 500,000 and fill in the following additional line of the table. (Be sure that any "`disp`" statements are commented out.)

Name	Formula	Interval	Regula approxRoot	Regula No. Steps
f5	(x-1)^5	[0,3]		

You should observe convergence, but it takes a very large number of iterations.

(5) `regula.m` and `bisect.m` both keep the current iterate, `x` in a change-of-sign interval. Why would it be wrong to use the same convergence criterion in `regula.m` as was used in `bisect.m`?

2.9 Muller's method

In the secant method, the function is approximated by its secant (the linear function passing through the two endpoints) and then the next iterate is taken to be the root of this linear function. Muller's method goes one better and approximates the function by the *quadratic* function passing through the two endpoints and the current iterate. The next iterate is then taken to be the root of this quadratic that is closest to the current iterate.

In more detail, write a function with signature similar to our functions above:

```
[result, itCount] = muller( func, a, b)
```

(You may find it convenient to cut-and-paste some lines from the file `muller.txt`.)

(1) Choose a convergence criterion $\epsilon = 10^{-10}$ and a maximum number of iterations
ITMAX = 100

(2) Choose a third point, not quite at the center of the change-of-sign interval [a,b]

```
x0 = a;
x2 = b;
x1 = 0.51*x0 + 0.49*x2;
```

(3) Evaluate y0, y1, and y2 as values of func at x0, x1, and x2.

(4) Determine coefficients A, B, and C of the polynomial passing through the three points (x0,y0), (x1,y1) and (x2,y2), $y(x) = A(x - x_2)^2 + B(x - x_2) + C$:

```
A = ( (y0 - y2) * (x1 - x2) - (y1 - y2) * (x0 - x2) ) / ...
      ( (x0 - x2) * ( x1- x2) * (x0 - x1) );
B = ( (y1 - y2) * (x0 - x2)^2 - (y0 - y2) * (x1 - x2)^2 ) / ...
      ( (x0 - x2) * (x1 - x2) * (x0 - x1) );
C = y2;
```

(5) If the polynomial has real roots, find them and choose the one closer to x2, otherwise choose something reasonable:

```
if A ~= 0

   disc = B*B - 4.0*A*C;
   disc = max( disc, 0.0 );

   q1 = (B + sqrt(disc) );
   q2 = (B - sqrt(disc) );

   if abs(q1) < abs(q2)
     dx = -2.0*C/q2;
   else
     dx = -2.0*C/q1;
   end

elseif B ~= 0
   dx = -C/B;
else
   error(['muller: algorithm broke down at itCount=',
   num2str(itCount)])
end
```

(6) Discard the point (x0, y0) and add the new point

```
x0 = x1;
y0 = y1;
```

```
x1 = x2;
y1 = y2;


x2 = x1 + dx;
y2 = func(x2);
```

(7) Exit the loop and return `result` = `x2` if the residual (`y2`) is smaller than ϵ.

(8) Go back to Step 4 unless `ITMAX` is exceeded, in which case print an error message.

Exercise 2.10.

(1) Following the outline above, create a file `muller.m` that carries out Muller's method.

(2) Test `muller.m` on the simple linear function $f_0(x) = x - 1$ starting with the interval `[0,2]`. You should observe convergence in a single iteration. If you do not observe convergence in a single iteration, you probably have a bug. Fix it before continuing. Explain why (one sentence, please) it should require only a single iteration.

(3) Test `muller.m` on the function `f1` on the interval `[0, 5]`. Again, you should observe convergence in a single iteration. If you do not, you probably have a bug. Fix it before continuing. Explain why (one sentence, please) it should require only a single iteration.

(4) Repeat the experiments from the previous exercise but using `muller.m` and fill in the following table.

Name	Interval	Muller `approxRoot`	Muller Steps	Regula Steps	Secant Steps	Bisection Steps
f1	[0,5]					
f2	[1,2]					
f3	[-1,2]					
f4	[0,6]					
f5	[0,3]					

You should observe that, since the interval at each iteration need not be a change-of-sign interval, it is possible for the method to fail, as with `secant`. It is possible to combine several methods in such a way that the speed of `secant` or `muller` is partially retained but failure is avoided by maintaining a change-of-sign interval at each iteration. One such method is called Brent's method. See, for example, [Brent's Method], [Brent (1970)], or [Atkinson (1978)], Section 2.8.

Chapter 3

The Newton–Raphson method

3.1 Introduction

In Chapter 2, you saw that the bisection method (Section 2.4) is very reliable, but it can be relatively slow, and it does not generalize easily to more than one dimension. The secant (Section 2.7) and Muller's (Section 2.9) methods are faster, but Newton's method (also known as the Newton–Raphson method) is by far the most commonly used.

In this chapter you will see Newton's method for finding roots of functions. Newton's method naturally generalizes to multiple dimensions and can be much faster than methods such as bisection. On the negative side, it requires a formula for the derivative as well as the function, and it can easily fail. Nonetheless, it is a workhorse method in numerical analysis.

Exercises 3.1 through 3.5 form a sequence to introduce Newton's method. Exercise 3.9 shows that Newton's method works just as well for complex numbers as for real numbers, and is used as an introduction to the method in two dimensions in Chapter 4. The other exercises in this chapter address individual issues.

3.2 Stopping tests

Root finding routines check after each step to test whether the current result is good enough. The tests that are made are called "termination conditions," "convergence conditions" or "stopping tests". Common tests include:

Residual size $|f(x)| < \epsilon$
Increment size $|x_{\text{new}} - x_{\text{old}}| < \epsilon$
Number of iterations: `itCount > ITMAX`

The size of the residual looks like a good choice because the residual is zero at the solution; however, it turns out to be a poor choice because the residual can be small even if the iterate is far from the true solution. You saw this situation when you found the root of the function $f_5(x) = (x-1)^5$ in Chapter 2. The size of the increment is a reasonable choice because Newton's method (usually) converges quadratically, and when it does the increment is an excellent approximation of the true error. The third stopping criterion, when the number of iterations exceeds a maximum, is a safety measure to assure that the iteration will always terminate in finite time.

It should be emphasized that the stopping criteria are based on *estimated errors*. In this and later chapters, you will see many expressions for estimating errors. You should not confuse the estimated error with the *true error*, $|x_{\text{approx}} - x_{\text{exact}}|$. The true error is not included among the stopping tests because you would need to know the exact solution to use it.

3.3　Failure

You have seen that the bisection method is very reliable and rarely fails but always takes a (sometimes large) fixed number of steps. Newton's method works rapidly and correctly a good deal of the time, but does fail. And Newton's method works in more than one dimension. One objective of this chapter is to see how Newton can fail and get a flavor of what might be done about it.

Although you will seem to spend a lot of time looking at failures, you should still expect that Newton's method will converge most of the time. It is just that when it converges there is nothing special to say, but when it fails to converge there is always the interesting questions of, "Why?" and "How can I remedy it?"

3.4　Introduction to Newton's Method

You will find the definition of Newton's method in, for example, [Quarteroni *et al.* (2007)] pp. 255f, with convergence discussed on pp. 263f and systems of equations on pp. 286ff, in [Atkinson (1978)] (Section 2.2), or in [Ralston and Rabinowitz (2001)] (pp. 361ff). The idea of Newton's method is that, starting from a guessed point x_0, find the equation of the *straight* line that passes through the point $(x_0, f(x_0))$ and has slope $f'(x_0)$. The next iterate, x_1, is simply the root of this linear equation, *i.e.*, the location that the straight line intersects the x-axis. Newton's method converges for a reasonably large class of problems and is a workhorse solution method for nonlinear equations.

A "quick and dirty" derivation of the formula can be taken from the definition of the derivative of a function. Assuming you are at the point $x^{(k)\dagger}$, the equation of the tangent line to f can be written as

$$\frac{f(x^{(k)} + \Delta x) - f(x^{(k)})}{\Delta x} = f'(x^{(k)}). \tag{3.1}$$

If f were linear, then f would be the same as its tangent line, and $(x^{(k)} + \Delta x)$ would be its root. Since f might not be linear, this equality merely defines the next

---

$\dagger$Superscripts in parentheses are used to denote iteration number to reduce confusion among iteration number, exponentiation and vector components.

iterate. This yields

$$\frac{-f(x^{(k)})}{\Delta x} \approx f'(x^{(k)}),$$

or

$$\Delta x = x^{(k+1)} - x^{(k)} = -\frac{f(x^{(k)})}{f'(x^{(k)})}. \qquad (3.2)$$

Newton's method also will work for vectors, so long as the derivative is properly handled. Assume that $\mathbf{x}$ and $\mathbf{f}$ are n-vectors. Then the Jacobian matrix is the matrix $\mathbf{J}$ whose elements are

$$\mathbf{J}_{ij} = \frac{\partial f_i}{\partial x_j}.$$

(Here, $i = 1$ for the first row of $\mathbf{J}$, $i = 2$ for the second row of $\mathbf{J}$, *etc.*, so that the MATLAB matrix subscripts correspond to the usual ones.) In this case, $\mathbf{J}$ replaces f', as in (3.5) below.

Convergence properties of Newton's method are given in the following theorem.

Theorem 3.1 (Newton's method).

(a) *Suppose that $f(x)$ is a differentiable function with a root r, so that $f(r) = 0$. If $f'(r) \neq 0$ and x^0 is some value near r, then the sequence defined by*

$$x^{(k+1)} = x^{(k)} - f(x^{(k)})/f'(x^{(k)}) \qquad (3.3)$$

converges to r. Furthermore,

$$|x^{(k+1)} - r| \leq C|x^{(k)} - r|^2 \qquad (3.4)$$

for k large enough and a constant C.

(b) *Suppose that $\mathbf{f}(\mathbf{x})$ is a differentiable vector-valued function of a vector with a root $\mathbf{r}$, so that $f(\mathbf{r}) = \mathbf{0}$. If the Jacobian derivative is denoted $\mathbf{J}$ and if $\mathbf{J}(\mathbf{r})$ is an invertible matrix, and if $\mathbf{x}^{(0)}$ is some value near enough to $\mathbf{r}$, then the sequence defined by*

$$\mathbf{x}^{(k+1)} - \mathbf{x}^{(k)} = -\mathbf{J}^{-1}(\mathbf{x}^{(k)})^{-1}\mathbf{f}(\mathbf{x}^{(k)}) \qquad (3.5)$$

converges to $\mathbf{r}$. Furthermore,

$$\|\mathbf{x}^{(k+1)} - \mathbf{r}\| \leq C\|\mathbf{x}^{(k)} - \mathbf{r}\|^2 \qquad (3.6)$$

for k large enough and a constant C.

Remark 3.1. In practice, the inverse operation in (3.5) is *never* used! Instead, the MATLAB backslash (solution) operator is used instead because it is about three times faster and often requires substantially less storage.

Remark 3.2. The quadratic convergence property (3.4) and (3.6) is remarkable. Once you have achieved the first decimal place of accuracy, one more iteration will provide at least one more decimal place accuracy! For smaller problems, each subsequent iteration will *double* the number of decimal places of accuracy, although for larger problems roundoff errors will often limit the increase in accuracy. As a rule of thumb, once convergence starts, you should observe at least one or two additional decimal places of accuracy per iteration.

3.5 Writing MATLAB code for functions

Newton's method requires *both* the function value and its derivative, unlike the bisection method that requires only the function value. You have seen how MATLAB functions can return several results (the root and the number of iterations, for example). In this chapter, you will use this same method to return both the function value and its derivative. For example, the `f1.m` file from Chapter 2. can be modified in the following way to return both the value (`y`) and its derivative (`yprime`).

```
function [y,yprime]=f1(x)
  % [y,yprime]=f1(x) computes y=x^2-9 and its derivative, yprime=2*x

  % your name and the date
  if numel(x)>1  % check that x is a scalar
    error('f1: x must be a scalar!')
  end

  y=x^2-9;
  yprime=2*x;
end
```

Remark 3.3. Conveniently, this modified version could still be used for the bisection method because MATLAB returns only the first value (`y`) unless the second is explicitly requested.

Remark 3.4. Although the syntax `[value,derivative]` appears similar to a vector with two components, it is not that at all! You will see in the next chapter cases where `value` is a vector and `derivative` is a matrix.

In the following exercise, you will write four more function m-files similar to `f1.m` to provide several functions to test Newton's method. These functions are the same as ones written for an earlier chapter, but they include the derivative.

For this exercise, you will compute derivatives by hand and include them in the same function m-file as the function itself. There are several other possible programming strategies for computing derivatives.

- You could compute the derivative value in a new m-file with a new name such as df0, df1, *etc.*
- You could compute the derivative using some sort of divided difference formula. This method is useful for very complicated functions.
- You could use the symbolic capabilities in MATLAB symbolic capabilities to symbolically differentiate the functions. This method seems attractive, but it would greatly increase the time spent in function evaluation.
- There are automated ways to discover the formula for a derivative from the m-file or symbolic formula defining the function, and these can be used to generate the m-file for the derivative automatically.

These strategies will not be used in this chapter.

Exercise 3.1.

(1) Construct five function m-files f0.m, f1.m, f2.m, f3.m and f4.m, to return both the function value and that of its derivative. Find the formulæ by hand and include them in the mfles as is done above for f1.m. The functions are:

f0	y=x-1	yprime=
f1	y=x^2-9	yprime=
f2	y=x^5-x-1	yprime=
f3	y=x*exp(-x)	yprime=
f4	y=2*cos(3*x)-exp(x)	yprime=

(2) Include the code for f2.m.
(3) Use the help f0 command to confirm that your comments include at least the formula for f0, and similarly for f1, f2, f3 and f4.
(4) What are the values of the function and derivative at $x = -1$ for each of the five functions?

3.6   Newton's method: MATLAB code

In the next exercise, you will get down to the task of writing Newton's method as a function m-file. In this m-file, you will see how to use a *variable* number of arguments in a function to simplify later calls. The maximum number of iterations will be an optional argument. When this argument is omitted, it takes on its default value.

The m-file you are about to write is shorter than bisect.m and the following instructions take you through it step by step. These instructions include testing, checking correctness, finding errors and correcting your work, all of which are essential for writing correct code.

The theoretical convergence rate of Newton's method is quadratic. In your work, you will use this fact to help ensure that your code is correct. This rate depends on an accurate computation of the derivative. Even a seemingly minor error in the derivative can yield a method that does not converge very quickly or that diverges. In the table at the end of the exercise, none of the cases should take as many as 25 iterations. If they do, check your derivative formula. (As a general rule, if you apply Newton's method and it takes hundreds or thousands of iterations but it does converge, check your derivative calculation very carefully. In the overwhelming majority of cases you will find a bug there.)

Exercise 3.2.

(1) Writing `newton.m`.

(a) Open a new, empty m-file named `newton.m`. You can use either the menus or the command `edit newton.m`.

(b) Start off the function with its signature line and follow that with comments. The comments should repeat the signature as a comment, contain a line or two explaining what the function does, explain what each of the parameters mean, and include your name and the date. You may find it convenient to download the outline `newton.txt`.

```
function [x,numIts]=newton(func,x,maxIts)
  % [x,numIts]=newton(func,x,maxIts)
  % func is a function handle with signature [y,yprime]=func(x)
  % on input, x is the initial guess
  % on output, x is the final solution
  % EPSILON is _____
  % maxIts is _____
  % maxIts is an optional argument
  % the default value of maxIts is 100
  % numIts is _____
  % Newton's method is used to find x so that func(x)=0

  % your name and the date

  % check that x is a scalar
  if numel(x) > 1  % numel gives the number of elements
    error('newton: x must be a scalar')
  end
```

(You may have to complete this exercise before coming back to explain all of the parameters.)

(c) From a programming standpoint, the iteration should be limited to a fixed (large) number of steps. The reason for this is that a loop that never terminates appears to you as if MATLAB remains "busy" forever.

In the following code, the special MATLAB variable `nargin` gives a count of *the number of input arguments* that were used in the function call. This allows the argument `maxIts` to be omitted when its default value is satisfactory.

```
if nargin < 3
  maxIts=100;      % default value if maxIts is omitted
end
```

(d) Define the convergence criterion

```
EPSILON = 5.0e-5;
```

(e) Some people like to use `while` loops so that the stopping criterion appears at the beginning of the loop. This author's opinion is that if you are going to count the iterations you may as well use a `for` statement. Start off the loop with

```
for numIts=1:maxIts
```

(f) Evaluate the function and its derivative at the current point x, much as was done in `bisect.m`. Remember that it is customary to indent the statements within a loop. You may choose the variable names as you please. Recall that the syntax for using two values from a single function call is

```
[value,derivative]=func(x);
```

(g) Define a MATLAB variable `increment` as the negative ratio of the function value divided by the value of its derivative. This is the right side of Equation (3.3).

(h) Complete Equation (3.3) with the statement

```
x = x + increment;
```

(i) Finish the loop and the m-file with the following statements

```
errorEstimate = abs(increment);

disp(strcat(num2str(numIts), ' x=', num2str(x), ...
  ' error estimate=', num2str(errorEstimate)));

if errorEstimate<EPSILON
    return;
  end
end
% if get here, the Newton iteration has failed!
error('newton: maximum number of iterations exceeded.')
end
```

The `return` statement causes the function to return to its caller before all `maxIts` iterations are complete. If the error estimate is not satisfied within `maxIts` iterations, then the MATLAB `error` function will cause the calculation to terminate with a red error message. It is a good idea to include the name of the function as part of the error message so you can find where the error occurred.

The `disp` statement prints out some useful numbers during the iteration. For the first few exercises in this chapter, leave this printing in the file, but when the function is being used as part of a calculation, it should not print extraneous information.

(j) Complete the comments at the beginning of the function by replacing the empty lines.

(2) Testing your code.

(a) Use the `help newton` command to confirm your comments are correct.

(b) Recall the Newton iteration should take only a single iteration when the function is linear. Test your `newton` function on the linear function `f0` that you wrote in the previous exercise. Start from an initial guess of `x=10`. The default value of `maxIts` (100) is satisfactory, so you can omit it and use the command

```
[approxRoot numIts]=newton(@f0,10)
```

The correct solution, of course, is `x=1` and it should take a single iteration. It actually takes a second iteration in order to recognize that it has converged, so `numIts` should be 2. If either the solution or the number of iterations is wrong, you have a bug: fix your code. **Hint:** The mistake might be either in the derivative from `f0.m` or in the m-file `newton.m`.

(3) Checking your work

(a) Test your `newton` function on the quadratic function `f1` that you wrote in the previous exercise. Start from an initial guess of `x=0.1`. The default value of `maxIts` (100) is satisfactory, so you can omit it. The correct solution, of course, is `x=3`. How many iterations are needed?

(b) As you know, the theoretical convergence rate for Newton is quadratic. This means that if you compute the quantity

$$r_2^{(k)\dagger} = \frac{|\Delta x^{(k)}|}{|\Delta x^{(k-1)}|^2}$$

then $r_2^{(k)}$ should approach a non-zero constant as $k \to \infty$. You have only taken a few iterations, but you can look at the sequence of values of `errorEstimate` values. Does it appear that the ratio $r_2^{(k)}$ is approaching a non-zero limit for this case of Newton applied to `f1`? (If not, you have a bug: fix your code. The mistake could be either in `f1.m` or in `newton.m`.) Based on the iterations you have, what do you judge is the value of the constant?

(c) What is the true error in your approximate solution, $|x - 3|$? Is it roughly the same size as `EPSILON`?

---

†Superscripts in parentheses are used to denote iteration number to reduce confusion among iteration number, exponentiation and vector components.

(d) Try again, starting at x=-0.1, you should get x=-3.

(4) Using your code, fill in the following table:

Name	Formula	guess	Approx. Root	No. Steps
f0	x-1	10		
f1	x^2-9	0.1		
f2	x^5-x-1	10		
f3	x*exp(-x)	0.1		
f4	2*cos(3*x)-exp(x)	0.1		
f4	2*cos(3*x)-exp(x)	1.5		

Remark 3.5. The theoretical convergence rate of Newton's method is quadratic. In your work in Exercise 3.2, you used this fact to help ensure that your code is correct. This rate depends on an accurate computation of the derivative. Even a seemingly minor error in the derivative can yield a method that does not converge very quickly or that diverges. In the table in Exercise 3.2, none of the cases should have taken so many as 20 iterations. If they did, check your derivative formula.

3.7 Non-quadratic convergence

The proofs of convergence of Newton's method show that quadratic convergence depends on the ratio f''/f' being finite at the solution. In most cases, this means that $f' \neq 0$, but it can also mean that both f'' and f' are zero with their ratio remaining finite. When f''/f' is not finite at the desired solution, the convergence rate deteriorates to linear. You will see this illustrated in the following exercises.

Exercise 3.3.

(1) Write a function m-file for the function f6=(x-4)^2, returning both the function and its derivative.

(2) You recall using newton to find a root of f1=x^2-9 starting from x=0.1. How many iterations did it take?

(3) Now try finding the root (x=4) of f6 using Newton, again starting at x=0.1. How many iterations does it take? Since the exact answer is x=4, the true error is abs(x-4). Is the true error larger than EPSILON or smaller?

(4) Now write a new function m-file for f7=(x-4)^20 and try finding the root (x=4) of f7 using Newton, again starting at x=0.1. For this case, increase maxIts to some large number such as 1000 (recall maxIts is the optional third argument to newton). How many iterations does it take? Is the true error larger than EPSILON or smaller? In this case, you see that convergence rate can deteriorate substantially.

(5) Look at the final two iterations and compute the values of the ratios

$$r_1 = \frac{|\Delta x^{(k+1)}|}{|\Delta x^{(k)}|}, \text{ and} \tag{3.7}$$

$$r_2 = \frac{|\Delta x^{(k+1)}|}{|\Delta x^{(k)}|^2}. \tag{3.8}$$

In this case, you should notice that r_1 is *not* nearly zero, but instead is a number not far from 1.0, and r_2 has become large. This is characteristic of a linear convergence rate, and not a quadratic convergence rate.

In practice, if you don't know the root, how can you know whether or not the derivative is zero at the root? You can't, really, but you can tell if the convergence rate is deteriorating. If the convergence is quadratic, r_2 will remain bounded and r_1 will approach zero. If the convergence deteriorates to linear, r_2 will become unbounded and r_1 will remain larger than zero. In the following exercise, you will use this fact to improve your newton.m function to better handle the case of linear convergence.

Exercise 3.4.

(1) Coding changes

   (a) You are now going to add some code to newton.m. Because newton.m is working code, it is a good idea to keep it around so if you make errors in your changes then you can start over. Make a copy of newton.m called newton_backup.m, and change the name of the function inside the file from newton to newton_backup. If you need to, you can always return to a working version of Newton's method in newton_backup.m. (You should never have two different files containing functions with the same name because MATLAB may get "confused" about which of the two to use.)
   Important: The following changes should be made to the newton.m file, leaving newton_backup.m as a true copy of your original.

   (b) Just before the beginning of the loop in newton.m, insert the following statement

   ```
   increment=1;  % this is an arbitrary value
   ```

   Just before setting the value of increment inside the loop, save increment in oldIncrement:

   ```
   oldIncrement=increment;
   ```

   (c) After the statement x=x+increment;, compute r1 and r2 according to (3.7) and (3.8) respectively.

   (d) Comment out the existing disp statement and add another disp statement to display the values of r1 and r2.

(2) Use your revised `newton` to fill in the following table, using the final values at convergence, starting from `x=1.0`. Enter your estimate of the limiting values of `r1` and `r2` and use the term "unbounded" when you think that the ratio has no bound. As in the previous exercise, use `maxIts=1000` for these tests.

Function	numIts	r1	r2	True error	True err smaller than est?
f1=x^2-9					
f6=(x-4)^2					
f7=(x-4)^20					

Exercise 3.4 shows that quadratic convergence sometimes fails, usually resulting in a linear convergence rate, and you can estimate the rate. This is not always a bad situation — it *is* converging after all — but now the stopping criterion is not so good. In practice, it is usually as important to know that you are reliably close to the answer as it is to get the answer in the first place.

In the cases of $(x-4)^2$ and $(x-4)^{20}$, you saw that r_1 turned out to be approaching a constant, not merely bounded above and below. If r_1 were exactly a constant, then

$$x^{(\infty)} = x^{(n)} + \sum_{k=n}^{\infty} \Delta x^{(k)}$$

because it converges and the sum collapses. The exact root here is denoted by $x^{(\infty)}$. Because r_1 is constant, for $k > n$, $\Delta x^{(k)} = r_1^{k-n} \Delta x^{(n)}$. Hence,

$$x^{(\infty)} = x^{(n)} + \Delta x^{(n)} \sum_{k=n}^{\infty} r_1^{k-n}$$

$$= x^{(n)} + \frac{\Delta x^{(n)}}{1 - r_1}.$$

This equation indicates that a good error estimate would be

$$|x^{(\infty)} - x^{(n)}| \approx \frac{|\Delta x^{(n)}|}{1 - r_1}.$$

Exercise 3.5.

(1) Coding changes

   (a) Comment out the `disp` statement displaying r_1 and r_2 in `newton.m` since it will become a distraction when large numbers of iterations are needed.

   (b) Conveniently, `r1` either goes to zero or remains bounded. If the sequence converges, `r1` should remain below 1, or at least its average should remain below 1. Replace the if-test for stopping in `newton` to

```
       if errorEstimate < EPSILON*(1-r1)
         return;
       end
```

Note: This code is mathematically equivalent to

```
errorEstimate < EPSILON/(1-r1),
```

but has been multiplied through by $(1 - r_1)$ to avoid the problems that occur when $r_1 \geq 1$. Additionally, convergence will never be indicated when $r_1 \geq 1$ because `errorEstimate` is non-negative.

(2) Use `newton` starting from $x = 0.1$ to fill in the following table, where you can compute the absolute value of the true error because you can easily guess the exact solution. (Continue using `maxIts=1000` for this exercise.)

Function	numIts	True err	True err smaller than est?
f1=x^2-9			
f6=(x-4)^2			
f7=(x-4)^20			

Compare this table with the one from Exercise 3.4. You should see that the modified convergence does not harm quadratic convergence (about the same number of iterations required) and greatly improves the estimated error and stopping criterion in the linear case. A reliable error estimate is almost as important as the correct solution — if you don't know how accurate a solution is, what can you do with it?

Remark 3.6. This modification of the stopping criterion is very nice when r_1 settles down to a constant value quickly. In real problems, a great deal of care must be taken because r_1 can cycle among several values, some larger than 1, or it can take a long time to settle down.

Remark 3.7. In the rest of this chapter, you should continue using `newton.m` with the convergence criterion involving $(1 - r_1)$ that you just worked on.

3.8   Choice of initial guess

The theorems about Newton's method generally start off with the assumption that the initial guess is "close enough" to the solution. Since you don't know the solution when you start, how do you know when it is "close enough?" In one-dimensional problems, the answer is basically that if you stay away from places where the derivative is zero, then any initial guess is OK. More to the point, if you know that the solution lies in some interval and $f'(x) \neq 0$ on that interval, then the Newton iteration will converge to the solution, starting from any point in the interval.

When there are zeros of the derivative nearby, Newton's method can display highly erratic behavior and may or may not converge. In the next exercise, you will see an example of this erratic behavior.

Exercise 3.6. In this and the following exercise, you will be interested in the sequence of iterates, not just the final result. Re-enable the `disp` statement displaying the values of the iterates in `newton.m`.

(1) Write a function m-file for the `cosmx` function used in Chapter 2 ($f(x) = \cos x - x$). Be sure to calculate both the function and its derivative, as you did for `f1`, `f2`, `f3` and `f4`.

(2) Use `newton` to find the root of `cosmx` starting from the initial value `x=0.5`. What is the solution and how many iterations did it take? (If it took more than ten iterations, go back and be sure your formula for the derivative is correct.)

(3) Again, use `newton` to find the root of `cosmx`, but start from the initial value `x=12`. Note that $3\pi < 12 < 4\pi$, so there are several zeros of the derivative between the initial guess and the root. You should observe that it takes the maximum number of iterations and seems not to converge.

(4) Try the same initial value `x=12`, but also use `maxIts=5000`. (To do this, include 5000 in the call: `newton('cosmx',12,5000)`.) (This is going to cause a large number of lines of printed information, but you are going to look at some of those lines.) Does it locate a solution in fewer than 5000 iterations? How many iterations does it take? Does it get the same root as before?

(5) Look at the sequence of values of `x` that Newton's method chose when starting from `x=12`. There is no real pattern to the numbers, and it is pure chance that finally put the iteration near the root. Once it is "near enough," of course, it finds the root quickly as you can see from the estimated errors. Is the final estimated error smaller than the square of the immediately preceding estimated error?

You have just observed a common behavior: that the iterations seem to jump about without any real pattern until, seemingly by chance, an iterate lands inside the circle of convergence and they converge rapidly. This has been described as "wandering around looking for a good initial guess." It is even more striking in multidimensional problems where the chance of eventually landing inside the ball of convergence can be very small.

3.9   No root

Sometimes people become so confident of their computational tools that they attempt the impossible. What would happen if you attempted to find a root of a function that had no roots? Basically, the same kind of behavior occurs as when there are zeros of the derivative between the initial guess and the root.

Exercise 3.7.

(1) Write the usual function m-file for f8=x^2+9.
(2) Apply **newton** to the function f8=x^2+9, starting from x=0.1. Describe what happens.
(3) *Intermediate prints will no longer be needed in this chapter.* Comment out the disp statements in **newton.m**. Leave the error statement intact.

3.10   Complex functions

But the function $x^2 + 9$ *does* have roots! The roots are complex, but MATLAB knows how to do complex arithmetic. (Actually the roots are imaginary, but MATLAB does not distinguish between imaginary and complex.) All MATLAB needs is to be "reminded" to use complex arithmetic.

Warning: If you have used the letter i as a variable in your MATLAB session, its special value as the square root of -1 has been obscured. To eliminate the values you might have given it and return to its special value, use the command clear i. It is always safe to use this command, so it is a good idea to use it just before starting to use complex numbers. For those people who prefer to use j for the imaginary unit, MATLAB understands that one, too.

Remark: For complex constants, MATLAB will accept the syntax 2+3i to mean the complex number whose real part is 2 and whose imaginary part is 3. It will also accept 2+3*i to mean the same thing and it is necessary to use the multiplication symbol when the imaginary part is a variable as in x+y*i.

Exercise 3.8.

(1) Apply **newton** to f8=x^2+9, starting from the initial guess x=1+1i. You should get the result 0+3i and it should take fewer than 10 iterations. Recall that if you call the **newton** function in the form

```
[approxRoot,numIts]=newton(@f8,1+1i)
```

then the root and number of iterations will be printed even though there are no intermediate messages.

(2) Fill in the following table by using **newton** to find a root of **f8** from various initial guesses. The exact root, as you can easily see, is $\pm 3i$, so you can compute the true error (typically a small number), as you did in Exercise 3.1.

Initial guess	numIts	true error
1+1i		
1-1i		
10+5i		
10+eps*i		

(Recall that **eps** is the smallest number you can add to 1.0 and still change it.)

Look carefully at what happened with the last case in Exercise 3.8. The derivative is $f_8'(x) = 2x$ and is not zero near $x = (10+(\text{eps})i) \approx 10$. In fact, the derivative is zero only at the origin. The origin is not near the initial guess nor either root. It turns out that the complex plane is just that: a plane. While Newton's method works in two or more dimensions, it is harder to see when it is going to have problems and when not. This will be elaborated a bit in a later section and also in the next chapter.

3.11 Unpredictable convergence

The earlier, one-dimensional cases presented in this chapter might lead you to think that there is some theory relating the initial guess and the final root found using Newton's method. For example, it is natural to expect that Newton's method will converge to the root nearest the initial guess. *This is not true in general!* In the exercise below, you will see that it is not possible, in general, to predict which of several roots will arise starting from a particular initial guess.

Consider the function $f_9(z) = z^3 + 1$, where $z = x + iy$ is a complex number with real part x and imaginary part y. This function has the following three roots.

$$\omega_1 = -1$$
$$\omega_2 = (1 + i\sqrt{3})/2$$
$$\omega_3 = (1 - i\sqrt{3})/2.$$

In the following exercise, you will choose a number of regularly-spaced complex points $z(k,j)$ in the square $[-2,2] \times [-2,2] \in \mathbb{C}$. Then, using each of these points as initial guess, you will use your **newton.m** function to solve $f(z) = 0$. You will then define an array **whichRoots(k,j)** that equals 1 if the root found by **newton.m** is ω_1, that equals 2 if the root found by **newton.m** is ω_2, and that equals 3 if the root found by **newton.m** is ω_3.

You will then construct an image from the array `whichRoots` by coloring each starting point in the square according to which of the three roots was found when starting from the given point. The surprise comes in seeing that nearby initial guesses do not necessarily result in the same root at convergence.

It turns out that the set you will plot is a Julia set (fractal). If you wish to learn more about "Newton Fractals," there is a good [Newton Fractal] article or a very comprehensive paper [Hubbard *et al.* (2001)]. In addition, the informally written and informative article [Tathan (2017)] contains interesting information and plots.

Remark 3.8. The MATLAB functions `min` and `max` have more advanced versions than you have seen so far. The format is

```
[value, index] = min( vector )
```

The returned quantity `value` is the smallest value found in the array `vector`, and the returned quantity `index` is the index of that value so that `value=vector(index)`.

Exercise 3.9.

(1) Writing the code

    (a) Be sure your `newton.m` file does not have any active `disp` statements in it.

    (b) Create a function m-file named `f9.m` that computes the function $f_9(z) = z^3 + 1$ as well as its derivative.

    (c) Download the file `myfractal.txt`. Copy the following portion of that file to a script m-file named `myfractal.m`.

```
NPTS=100;
clear i
clear whichRoots
x=linspace(-2,2,NPTS);
y=linspace(-2,2,NPTS);
omega(1)= -1;
omega(2)= (1+sqrt(3)*i)/2;
omega(3)= (1-sqrt(3)*i)/2;

close   %causes current plot to close, if there is one
hold on
for k=1:NPTS
  for j=1:NPTS
    z=x(k)+i*y(j);
    plot(z,'.k') % plot as black points
  end
end
hold off
```

Be sure to add comments that identify the purpose of the file and your name.

(d) Execute this m-file to generate a plot of 10000 black points that fill up a square in the complex plane.

(e) What is the purpose of the statement `clear i`?

(f) What would happen if the two statements `hold on` and `hold off` were to be omitted?

(g) Next, discard the two `hold` statements and replace the `plot` statement with the following statements, that can also be found in the file `myfractal.txt`.

```
root=newton(@f9,z,500);
[difference,whichRoot]=min(abs(root-omega));
if difference>1.e-2
   whichRoot=4;
end
whichRoots(k,j)=whichRoot;
```

(h) If the variable `root` happened to take on the value `root=0.50-i*0.86`, what would the values of `difference` and `whichRoot` be? (Recall that `(sqrt(3)/2)=.866025....`)

(i) If the function `f9` were not properly programmed, it might be possible for the variable `root` to take on the value `root=0.51-i*0.88`. If this happened, what would the values of `difference` and `whichRoot` be?

(j) The array `whichRoots` contains integers between 1 and 4 but the value 4 should never occur. MATLAB has a "colormap" called "flag" that maps the numbers 1, 2, 3, and 4 into red, white, blue and black respectively. Add the following instructions (that can be found in the file `myfractal.txt`) after the loop to plot the array `whichRoots` and the three roots as black asterisks.

```
imghandle=image(x,y,which');
colormap('flag') % red=1, white=2, blue=3, black=4
axis square
set(get(imghandle,'Parent'),'YDir','normal')

% plot the three roots as blacK asterisks
hold on
plot(real(omega),imag(omega),'k*')
hold off
```

Remark 3.9. The `image` command does not plot the array `whichRoots` in the usual way. In order to get the picture to display in the usual way, the transpose of `whichRoots` is plotted and "handle graphics" is used to restore the y-axis to its usual orientation. For more information about "handle graphics" search for "graphics object programming" in the MATLAB help documentation.

(2) Using the code

    (a) Execute your modified m-file.

> **Remark 3.10.** The position (z=x+i*y) on this image corresponds to the initial guess and the color values 1 (red), 2 (white) and 3 (blue) correspond to roots 1, 2 and 3 to which the Newton iteration converged starting from that initial guess. This image illustrates that, for example, some initial guesses with large positive real parts converge to the root $\omega_1 = -1$ despite being closer to both of the other roots.

    (b) (**Optional**) It is characteristic of fractal images that they appear qualitatively similar at finer and finer levels. To see this effect, modify your script m-file so that instead of working on the square $[-2, 2] \times [-2, 2]$, it works on the square $[-0.5, 0.5] \times [-0.5, 0.5]$.

3.12   Newton-like methods without analytic derivative

One disadvantage of Newton's method is that you have to supply not only the function, but also a derivative. In the following exercise, you will see how to make life a little easier by numerically approximating the derivative of the function instead of finding its formula. One way would be to consider a nearby point, evaluate the function there, and then approximate the derivative by

$$y' \approx \frac{f(x + \Delta x) - f(x)}{\Delta x}. \tag{3.9}$$

There are many possible ways to pick step sizes Δx that change as the iteration proceeds. One way is considered in Exercise 3.11 but here you will look at simply choosing a fixed value in the following two exercises.

Exercise 3.10.

(1) Writing your code

    (a) Make a copy of `newton.m` and call it `newtonfd.m`. The convergence criterion should involve the factor `(1-r1)` in `newtonfd.m` because using an approximate derivative usually leads to slower linear convergence.

    (b) Change the signature (first line) to be

       `function [x,numIts]=newtonfd(f,x,dx,maxIts)`

    and change the comment lines to reflect the new function.

    (c) Since `maxIts` is now the fourth variable, change the test involving `nargin` to indicate the new number of variables.

    (d) Replace the evaluation of the variable `derivative` through a function call with the expression described in Equation (3.9) using the constant stepsize `dx`.

(2) Test your code using the function `f1(x)=x^2-9`, starting from the point `x = 0.1`, using an increment `dx=0.00001`. You should observe essentially the same converged value in the same number (±1) of iterations as using `newton`.

(3) For the two functions `f2(x)=x^5-x-1` and `f6(x)=(x-4)^2`, and the starting point `x = 5.0`, compute the number of steps taken to converge using the true Newton method, and `newtonfd` with different stepsizes `dx`. You should observe convergence to a solution in all cases, but you will have to increase the value of `maxIts` by a very great deal to get `f6` to converge for the cases of large `dx`. In all cases, the converged `newtonfd` solution should be close to the `newton` solution.

	Number of steps f2	Number of steps f6
using newton		
dx = 0.00001		
dx = 0.0001		
dx = 0.001		
dx = 0.01		
dx = 0.1		
dx = 0.5		

As you can see, the choice of `dx` is critical. It is also quite difficult in more complicated problems.

Remark 3.11. As mentioned earlier, Newton's method generally converges quite quickly. If you write a Newton solver and observe poor or no convergence, the first thing you should look for is an error in the derivative. One way of finding such an error is to apply a method like `newtonfd` and print both the estimated derivative and the analytical derivative computed inside your function. They should be close. This strategy is especially useful when you are using Newton's method in many dimensions, where the Jacobian can have some correctly-computed components and some components with errors in them.

You saw the secant method in Section 2.7 and it had the update formula

$$x = b - \left(\frac{b - a}{f(b) - f(a)} \right) f(b). \tag{3.10}$$

Exercise 3.11.

(1) Compare (3.10) with (3.3) and show that by properly identifying the variables x, a, and b then (3.10) can be interpreted as a Newton update using an approximation for $f'(x^{(k)})$. Explain your reasoning.

(2) Starting from your `newton.m`, copy it to a new m-file named `newsecant.m` and modify it to carry out the secant method. The secant method requires an extra

value for the iterations to start, but there is only the value x in the signature line. Take b=x and a=x-0.1 *only for the first iteration.* Do not use `secant.m` from Chapter 2 because it uses a different convergence criterion.

(3) Fill in the following table.

Function	start	Newton numIts	Newton solution	newsecant numIts	newsecant solution
f1=x^2-9	0.1				
f3=x*exp(-x)	0.1				
f6=(x-4)^2	0.1				
f7=(x-4)^20	0.1				

You should observe that both methods converge to the same solutions but that the secant method takes more iterations, but not nearly so many as `newtonfd` with a larger *fixed* `dx`. This is because the theoretical convergence rate is quadratic for Newton and $(1 + \sqrt{5})/2 \approx 1.62$ for secant because of the approximation to the derivative.

3.13 Square roots

Some computer systems compute square roots in hardware using an algorithm often taught in secondary school and similar to long division. Other computer systems leave square roots to be computed in software. Some systems, however, use iterative methods to compute square roots. With these systems, great care must be taken to be sure that a good initial guess is chosen and that the stopping criterion is reliable. You can find a very technical discussion of these issues in a paper by [Wang and Schulte (2005)].

Exercise 3.12. One well-known[1] iterative method for computing the square root of the positive number a is

$$x^{(k+1)} = .5 \left(x^{(k)} + \frac{a}{x^{(k)}} \right). \tag{3.11}$$

This iteration can be started with, for example, $x^{(0)} = .5(a + 1)$. This initial guess is reasonable because $\sqrt{a}$ is always between a and 1.

(1) Explain why (3.11) is the result of applying Newton's method to the function $f(x) = x^2 - a$.

(2) In the case $a > 0$, explain why convergence cannot deteriorate from quadratic. For this case, it makes sense to use a relative error estimate to stop iterating when

$$|x^{(k+1)} - x^{(k)}| \le 3(\mathsf{eps}) |x^{(k+1)}|.$$

---

[1]A parlor trick to compute square roots "in your head" is based on this iterative method.

The value 3(eps) is used instead of just eps to avoid the possibility of a one-bit error causing convergence failure.

(3) Write a MATLAB function m-file named newton_sqrt.m to compute the square root using (3.11). Write this function directly from (3.11) — do not call your previously-written newton function.

(4) Fill in the following table using your newton_sqrt.m. (The true error is the difference between the result of your function and the result from the Matlab sqrt function.)

Value	square root	true error	no. iterations
a=9			
a=1000			
a=12345678			
a=0.000003			

Chapter 4

Multidimensional Newton's method

Chapter 3 discusses Newton's method for nonlinear equations in one real or complex variable. In this chapter, the discussion is extended to two or more dimensions. One example is solution of water flow through a simple piping network. Another of the examples includes a common application of Newton's method, *viz.*, nonlinear least squares fitting. This application is known to have a very small radius of convergence, and some variations on Newton's method when it is hard to find a good initial guess are presented. Quasi-Newton methods are also discussed. Since code for the Jacobian matrix is a common source of errors, use of finite difference approximations of derivatives is illustrated to check the consistency of the Jacobian and function.

Exercises 4.1 and 4.2 involve modifications to `newton.m` from Chapter 3 to allow vector functions and their Jacobian matrices. As a test of those changes, Exercise 4.3 presents a two-dimensional case. Exercises 4.4 and 4.5 illustrate the use of MATLAB to investigate a case involving of slowly-convergent iterations. Exercises 4.6 and 4.7 illustrate a well-behaved Newton solution of a nonlinear flow network. Exercises 4.8 and 4.9 present a so-called parameter identification problem that has a very small radius of convergence. Exercises 4.10 and 4.11 present softening approaches to improving the radius of convergence, and Exercise 4.12 presents continuation methods as a more successful approach to enlarging the radius of convergence. Exercises 4.13 and 4.14 introduce quasi-Newton methods as an approach to speeding up Newton's method for large systems.

4.1 Introduction

Suppose $\mathbf{x}$ is a vector in $\mathbb{R}^N$, $\mathbf{x} = (x_1, x_2, \ldots, x_N)$, and $\mathbf{f}(\mathbf{x})$ a differentiable vector function,

$$\mathbf{f} = (f_m(x_1, x_2, \ldots, x_N)), \text{ for } m = 1, 2, \ldots, N,$$

with Jacobian matrix J,

$$J_{mn} = \frac{\partial f_m}{\partial x_n}.$$

Recall that Newton's method can be written

$$\mathbf{x}^{(k+1)} - \mathbf{x}^{(k)} = - \left(J(\mathbf{x}^{(k)}) \right)^{-1} \mathbf{f}(\mathbf{x}^{(k)})^{\dagger}. \tag{4.1}$$

As an implementation note, the inverse of J should not be computed. Instead, the system

$$J(\mathbf{x}^{(k)})(\mathbf{x}^{(k+1)} - \mathbf{x}^{(k)}) = -\mathbf{f}(\mathbf{x}^{(k)})$$

should be solved. As you will see later in this workbook, this will save about a factor of three in computing time over computing the inverse. MATLAB provides a special, division-like symbol for this solution operation: the backslash (\) operator. If you wish to solve the system

$$J\boldsymbol{\Delta}\mathbf{x} = -\mathbf{f}$$

then the solution can be written as

$$\boldsymbol{\Delta}\mathbf{x} = -J\backslash\mathbf{f}.$$

Note that the divisor is written so that it appears "underneath" the backslash. Mathematically speaking, this expression is equivalent to

$$\boldsymbol{\Delta}\mathbf{x} = -J^{-1}\mathbf{f}$$

but using \ is about three times faster.

You might wonder why MATLAB needs a new symbol for division when there is a perfectly good division symbol already. The reason is that matrix multiplication is not commutative, so order is important. Multiplying a matrix times a (column) vector requires the vector to be on the right. If you wanted to "divide" the vector $\mathbf{f}$ by the matrix J using the ordinary division symbol, you would have to write $\mathbf{f}/J$, but this would seem to imply that $\mathbf{f}$ is a *row* vector because it is to the left of the matrix. If $\mathbf{f}$ is a column vector, you would need to write $\mathbf{f}^T/J$, but this expression is awkward, and does not really mean what you want because the result would be a row vector. The MATLAB method:

$$J\backslash\mathbf{f}$$

is better.

4.2   Modifications to `newton.m` for vector functions

Exercise 4.1.

(1) Start from your version of `newton.m` from Chapter 3. Change its name to `vnewton.m` and change its signature and check on vector variables to the following. You may find it convenient to download the file `vnewton.txt` and copy from it.

---

†Superscripts in parentheses are used to denote iteration number to reduce confusion among iteration number, exponentiation and vector components.

```
function [x,numIts]=vnewton(func,x,maxIts)
  % comments

  % check that x is a column vector (all rows, only 1 column)
  [rows,cols]=size(x);
  if cols>1
    error('vnewton: x must be a column vector')
  end
```

and modify it so that it:

(a) Uses the MATLAB \ operator instead of scalar division for `increment`;
(b) Uses the `norm` of the increment and `norm` of the old increment instead of `abs` to determine `r1` and the error estimate;
(c) Has no `disp` statements; and,
(d) Has appropriately modified comments.

(Leaving the `disp` statements in will cause syntax errors when vector arguments are present, so be sure to eliminate them.)

(2) Test your modifications by comparing the value of the root and the number of iterations required for the complex scalar (not vector) case from last time $(f_8(z) = z^2 + 9)$. This shows that `vnewton` will still work for scalars. Your results should agree with those from `newton`.

(3) Fill in the following table (similar to Exercise 3.9). Recall that the column `abs(error)` refers to the *true* error, the absolute value of the difference between the computed solution and the true solution.

Initial guess	numIts	abs(error)	numIts(newton)
1+i			
1-i			
10+5i			
10+eps*i			

4.3 A complex function revisited

It is possible to rewrite a complex function of a complex variable as a *vector* function of a *vector* variable. This is done by equating the real part of a complex number with the first component of a vector and the imaginary part of a complex number with the second component of a vector.

Remark 4.1. Recall that you denote subscripts in MATLAB using parentheses. For example, the 13th element of a vector f would be denoted f_{13} mathematically and in MATLAB by `f(13)`.

Consider the function $f_8(z) = z^2 + 9$. Write $z = x_1 + x_2 i$ and $f(z) = f_1 + f_2 i$. Plugging these variables in yields

$$f_1 + f_2 i = (x_1^2 - x_2^2 + 9) + (2x_1 x_2)i.$$

This can be written in an equivalent matrix form as

$$\begin{bmatrix} f_1(x_1, x_2) \\ f_2(x_1, x_2) \end{bmatrix} = \begin{bmatrix} x_1^2 - x_2^2 + 9 \\ 2x_1 x_2 \end{bmatrix}. \tag{4.2}$$

Exercise 4.2. Finding roots from the *vector* form of f8, (4.2).

(1) Write a function m-file f8v.m to compute the vector function described above in Equation (4.2) and its Jacobian. It should be in the form needed by **vnewton** and with the outline

```
function [f,J]=f8v(x)
  % comments

  % your name and the date

  << your code >>

end
```

where f is the two-dimensional column vector from Equation (4.2) and J is its Jacobian matrix. **Hint:** Compute, by hand, the formulas for df1dx2 ($= \frac{\partial f_1}{\partial x_2}$), df1dx1, df2dx1, and df2dx2. Then set

```
J=[df1dx1 df1dx2
   df2dx1 df2dx2];
```

(2) Test your **vnewton.m** using f8v.m starting from the column vector [1;1] and comparing with the results of **newton.m** using f8.m and starting from 1+1i in Exercise 4.1. Both solution and number of iterations should agree exactly. If they do not, use the debugger to compare results after 1, 2, *etc.* iterations.

(3) Fill in the table below. Note that the norm of the error here should be the same as the absolute value of the error for f_8 in Exercise 4.1, and the correct solutions are $\begin{bmatrix} 0 \\ \pm 3 \end{bmatrix}$.

Initial guess	numIts	norm(error)
[1;1]		
[1;-1]		
[10;5]		
[10;eps]		

At this point, you should be confident that `vnewton` is correctly programmed and yields the same results as `newton` from Chapter 3. In the following exercise, you will apply `vnewton` to a simple nonlinear problem.

Exercise 4.3. Use Newton's method to find the two intersection points of the parabola $x_2 = x_1^2 - x_1$ and the ellipse $x_1^2/16 + x_2^2 = 1$.

(1) Plot both the parabola and the ellipse on the same plot, showing the intersection points. Be sure to show the whole ellipse so you can be sure there are exactly two intersection points.

(2) Write a MATLAB function m-file `f3v.m` to compute the vector function that is satisfied at the intersection points. The outline of the function should be the following. You may find it convenient to download the file `f3v.txt` to start from.

```
function [f,J]=f3v(x)
% [f,J]=f3v(x)
% f and x are both 2-dimensional column vectors
% J is a 2 X 2 matrix
% more comments

% your name and the date

% check that x is a 2-dimensional column vector
[rows,cols]=size(x);
if cols>1 | rows ~= 2
  error('f3v: x must be a 2-dimensional column vector')
end

<< your code >>

end
```

Hint: The vector `f=[f1;f2]`, where `f1` and `f2` are each zero at the intersection points.

(3) Starting from the initial column vector guess `[2;1]`, what is the intersection point that `vnewton` finds, and how many iterations does it take?

(4) Find the other intersection point by choosing a different initial guess. What initial guess did you use, what is the intersection point, and how many iterations did it take?

4.4   Slow to get started

In Exercise 4.2, you saw that the guess [10;eps] resulted in a relatively large number of iterations. While not a failure, a large number of iterations is unusual. In the next exercise, you will investigate this phenomenon using MATLAB as an investigative tool.

Exercise 4.4. In this exercise, you will look more carefully at the iterates in the poorly-converging case from Exercise 4.2. Because you are interested in the sequence of iterates, you will generate a special version of vnewton.m that returns the full sequence of iterates. It will return the iterates as a matrix whose columns are the iterates, with as many columns as the number of iterates. (One would not normally return so much data from a function call, but you have a special need for this exercise.)

(1) (a) Make a copy of vnewton.m and call it vnewton1.m. Change its signature line to the following

     `function [x, numIts, iterates]=vnewton1(func,x,maxIts)`

   (b) Just *before* the loop, add the following statement

     `iterates=x;`

     to initialize the variable iterates.

   (c) Add the following statement *just after* the new value of x has been computed.

     `iterates=[iterates,x];`

     Since iterates is a matrix and x is a column vector, this MATLAB statement causes one column to be appended to the matrix iterates for each iteration.

   (d) Replace the **error** function call at the end of the function with a disp function call. The reason for this change is that you will be using vnewton1 later in this lab to see how Newton's method can fail.

(2) Test your modified function vnewton1 using the same f3v.m from above, starting from the vector [2;1]. You should get the same results as before for solution and number of iterations. Check that the size of the matrix is 2×(number of iterations+1), *i.e.*, it has 2 rows and (numIts+1) columns.

The study of the poorly-converging case continues with the following exercise. The objective is to try to understand what is happening. The point of the exercise is two-fold: on the one hand to learn something about Newton iterations, and on other hand to learn how one might use MATLAB as an investigative tool.

Exercise 4.5.

(1) Apply `vnewton1.m` to the function `f8v.m` starting from the column vector `[10;eps]`. This is the poorly-converging case.
(2) Plot the iterates on the plane using the command

```
plot(iterates(1,:),iterates(2,:),'*-')
```

It is very hard to interpret this plot, but it is a place to start. Note that most of the iterates lie along the x-axis, with some quite large.
(3) Use the zoom feature of plots in MATLAB (the magnifying glass icon with a + sign) to look at region with horizontal extent around $[-20, 20]$. It is clear that most of the iterates appear to be on the x-axis.
(4) Look at the formula you used for the Jacobian in `f8v.m`. Explain why, if the initial guess starts with $x_2 = 0$ exactly, then all subsequent iterates also have $x_2 = 0$.

Remark 4.2. You have used MATLAB to help formulate a hypothesis that can be proved.

(5) You should be able to see what is happening. The initial guess has $x_2 \approx 0$, so subsequent iterates stay near the x-axis. Since the root is at $x_2 = 3$, it takes many iterates before the x_2 component of the iterate can increase sufficiently to enter the region of quadratic convergence.
(6) Use a semilog plot to see how the iterates grow in the vertical direction:

```
semilogy( abs( iterates(2,:) ) )
```

The plot shows the x_2 component seems to grow exponentially (linearly on the semilog plot) with seemingly random jumps superimposed.

You now have a rough idea of how this problem is behaving. It converges slowly because the initial guess is not close enough to the root for quadratic convergence to be exhibited. The x_2 component grows exponentially, however, and eventually the iterates become close enough to the root to exhibit quadratic convergence. It is possible to prove these observations, although the proof is beyond the scope of this chapter.

4.5   A nonlinear flow network

Another example of the use of Newton's method to solve a nonlinear system comes from computing water flow in the following network of pipes and junctions.

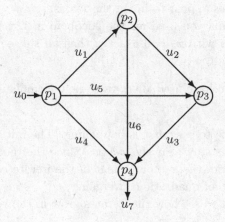

In this network, water is imagined to be flowing in pipes along the arrows, and connections among the pipes are denoted by circles. The velocity of the water in each pipe is denoted u_i, for $i = 0, \ldots, 7$, (positive in the direction of the arrow) and the pressure of the water at each of the four connections is denoted p_i, $i = 1, 2, 3, 4$. There is no connection between the pipes that appear to cross at the center of the network. The velocity of the incoming flow, u_0 is specified to be $u_0 = 1$.

The relationship between the pressures at the two ends of a pipe and the velocity of water through the pipe is quadratic, $p_{\text{out}} - p_{\text{in}} = Ku|u|$ where K is the pipe resistance, a constant. This constitutive law is observed in each of the six pipes with connections at either end.

Since water is conserved, the sum of all the water flowing through each connection must be zero.

As a consequence of the physical law of conservation and the constitutive law relating pressure and velocity through a pipe, the following equations can be written.

$$f_1 = u_0 - u_1 - u_4 - u_5 = 0$$
$$f_2 = u_1 - u_2 - u_6 = 0$$
$$f_3 = u_2 - u_3 + u_5 = 0$$
$$f_4 = u_3 + u_4 + u_6 - u_7 = 0$$
$$f_5 = p_2 - p_1 - K_1 u_1 |u_1| = 0$$
$$f_6 = p_3 - p_2 - K_2 u_2 |u_2| = 0 \qquad (4.3)$$
$$f_7 = p_4 - p_3 - K_3 u_3 |u_3| = 0$$
$$f_8 = p_4 - p_1 - K_4 u_4 |u_4| = 0$$
$$f_9 = p_3 - p_1 - K_5 u_5 |u_5| = 0$$
$$f_{10} = p_4 - p_2 - K_6 u_6 |u_6| = 0$$
$$f_{11} = p_1 + p_2 + p_3 + p_4 = 0.$$

The last of these equations has been added because, as a quick glance should tell you, the same constant can be added to all pressures without changing the rest of the equations. This final equation makes the system uniquely solvable.

The system of equations (4.3) has eleven equations in eleven unknowns (u_k for $k = 1, \ldots, 7$ and p_j for $j = 1, \ldots, 4$) with $u_0 = 1$ as a given boundary condition. In the following pair of exercises, you will see how the solution to this system of nonlinear equations can be found using Newton's method.

Exercise 4.6. In this exercise you will be writing a MATLAB function to evaluate the equations in the system (4.3). A MATLAB vector f with eleven components will be used to represent the eleven f_i in (4.3), and a MATLAB vector x with eleven components will represent the seven u_k and four p_j so that x(k)=u_k for $k = 1, \ldots, 7$ and x(k) = p_{k-7} for $k = 8, \ldots, 11$.

(1) Writing flow.m. You may find it convenient to download the file flow.txt to work from.

(a) Begin a MATLAB function m-file named flow.m with the outline

```
function [f,J]=flow(x)
% [f,J]=flow(x)
% comments

% your name and the date

<< your code >>

end
```

with comments describing the signature of the function, the meaning of the variables, and the purpose of the function.

(b) Define the flow resistances as

```
K=[1,3,5,7,9,11];
```

(c) Add MATLAB code to evaluate the eleven components of the MATLAB variable f in terms of the components of the MATLAB variable x, according to Equation (4.3). For example, the sixth component is given by

```
f(6)=x(10)-x(9)-K(2)*x(2)*abs(x(2));
```

Hints:

H1 The quantity u_0 is a constant, equal to 1. You might define the variable U0=1; in your code.

H2 One way to be sure that f is a column vector is to set f=zeros(11,1); first.

H3 Another way to make f into a column vector is to assign values to f(6,1) instead of f(6).

(d) The following formula gives the derivative of $u|u|$.

$$\frac{d}{du}(u|u|) = 2|u|.$$

Convince yourself that this formula is true by considering the cases that: (1) $u \leq 0$ and (2) $u \geq 0$, and noting the two values agree at $u = 0$.

(e) Complete your function by computing the Jacobian matrix, J, as an 11×11 matrix whose components are given as

$$J_{mn} = \frac{\partial f_m}{\partial x_n}.$$

For example, the first and sixth rows of J are given as

```
J(1,:)=[-1,0,0,-1,-1,0,0,0,0,0,0];
J(6,:)=[0,-2*K(2)*abs(x(2)),0,0,0,0,0,0,-1,1,0];
```

(2) The Jacobian matrix can be *approximately* evaluated as the vector of all ones using a finite difference formula. Copy the code to a script m-file named checkJ.m (or download the file) and use it to check your work

```
u=ones(11,1);  % choose a vector
[f,J]=flow(u);
delta=.001;
% make sure approximate J starts fresh
clear approxJ
for k=1:11
  uplus=u;
  uplus(k)=u(k)+delta;
  [fplus]=flow(uplus);
  uminus=u;
  uminus(k)=u(k)-delta;
```

```
    [fminus]=flow(uminus);
    approxJ(:,k)=(fplus-fminus)/(2*delta);
end
```

This code should yield an approximate Jacobian matrix that is close to the matrix J computed by the function `flow`. If it does not, go back and fix your error.

Exercise 4.7. In this exercise, you will be solving the flow network using your `vnewton.m` code, and you will at the same time be checking that your code in `flow.m` is reasonable.

(1) You know Newton's method should converge quadratically. *Temporarily* print `norm(increment)` inside the loop inside `vnewton.m` and apply it to `flow`, starting from the column vector of all ones. Once convergence starts, you should see the norm of `increment` decline *very rapidly* to zero. What are the final three values of `norm(increment)`? This case should take fewer than ten iterations.

(2) Remove the temporary print from `vnewton.m`, and modify `flow.m` so that `K=[1,1,1,1,1,1]`, and solve the resulting system. If you look back at the system of equations (4.3), you can easily prove that $u_7 = u_0$. Does your solution satisfy this relationship?

(3) Modify `flow.m` so that `K(4)=1` and all the rest of the values are `1.e5`. This effectively "blocks" all the pipes except number 4, so most of the flow should run through pipe 4. What is the value of u_4?

(4) Modify `flow.m` so that all the components of K equal 1 except `K(5)=1e5` and `K(6)=1e5`, blocking those pipes. Your solution should have $u_1 \approx u_2 \approx u_3$. What are these three values?

4.6 Nonlinear least squares

A common application of Newton's method for vector functions is to nonlinear curve fitting by least squares. This application falls into the general topic of "optimization" wherein the extremum of some function $F : \mathbb{R}^n \to \mathbb{R}$ is sought. If F is differentiable, then its extrema are given by the solution of the system of equations $\partial F / \partial x_k = 0$ for $k = 1, 2, \ldots, n$, and the solution can be found using Newton's method.

The differential equation describing the motion of a weight attached to a damped spring without forcing is

$$m \frac{d^2 v}{dt^2} + c \frac{dv}{dt} + kv = 0,$$

where v is the displacement of the weight from equilibrium, m is the mass of the weight, c is a constant related to the damping of the spring, and k is the spring stiffness constant. The physics of the situation indicate that m, c and k should be positive. Solutions to this differential equation are of the form

$$v(t) = e^{-x_1 t}(x_3 \sin x_2 t + x_4 \cos x_2 t),$$

where $x_1 = c/(2m)$, $x_2 = \sqrt{k - c^2/(4m^2)}$, and x_3 and x_4 are values depending on the position and velocity of the weight at $t = 0$. One common practical problem (called the "parameter identification" problem) is to estimate the values $x_1 \ldots x_4$ by observing the motion of the spring at many instants of time.

After the observations have progressed for some time, you have a large number of pairs of values (t_n, v_n) for $n = 1, \ldots, N$. The question to be answered is, "What values of x_k for $k = 1, 2, 3, 4$ would best reproduce the observations?" In other words, find the values of $\mathbf{x} = [x_1, x_2, x_3, x_4]^T$ that minimize the norm of the differences between the formula and observations. Define

$$F(\mathbf{x}) = \sum_{n=1}^{N}(v_n - e^{-x_1 t_n}(x_3 \sin x_2 t_n + x_4 \cos x_2 t_n))^2 \qquad (4.4)$$

and the minimum of F is sought.

Remark 4.3. The particular problem as stated can be reformulated as a linear problem, resulting in reduced numerical difficulty. However, it is quite common to solve the problem in this form and in more difficult cases it is not possible to reduce the problem to a linear one.

In order to solve this problem, it is best to note that when the minimum is achieved, the gradient $\mathbf{f} = \nabla F$ must be zero. The components f_k of the gradient $\mathbf{f}$ can be written as

$$f_1 = \frac{\partial F}{\partial x_1} = 2\sum_{k=1}^{K}\Big[(v_k - e^{-x_1 t_k}(x_3 \sin x_2 t_k + x_4 \cos x_2 t_k))t_k e^{-x_1 t_k}$$

$$\times (x_3 \sin x_2 t_k + x_4 \cos x_2 t_k)\Big]$$

$$f_2 = \frac{\partial F}{\partial x_2} = -2\sum_{k=1}^{K}\Big[(v_k - e^{-x_1 t_k}(x_3 \sin x_2 t_k + x_4 \cos x_2 t_k))e^{-x_1 t_k}$$

$$\times (x_3 t_k \cos x_2 t_k - x_4 t_k \sin x_2 t_k)\Big] \qquad (4.5)$$

$$f_3 = \frac{\partial F}{\partial x_3} = -2\sum_{k=1}^{K}(v_k - e^{-x_1 t_k}(x_3 \sin x_2 t_k + x_4 \cos x_2 t_k))e^{-x_1 t_k}\sin(x_2 t_k)$$

$$f_4 = \frac{\partial F}{\partial x_4} = -2\sum_{k=1}^{K}(v_k - e^{-x_1 t_k}(x_3 \sin x_2 t_k + x_4 \cos x_2 t_k))e^{-x_1 t_k}\cos(x_2 t_k).$$

(One rarely does this kind of calculation by hand any more. The MATLAB symbolic toolbox is one example of a computer algebra package that can greatly reduce the manipulative chore.)

To apply Newton's method to $\mathbf{f}$ as defined in (4.5), the sixteen components of the Jacobian matrix are also needed. These are obtained from f_i above by differentiating with respect to x_j for $j = 1, 2, 3, 4$.

Remark 4.4. The function F is a real-valued function of a vector. Its gradient, $\mathbf{f} = \nabla F$, is a vector-valued function. The gradient of $\mathbf{f}$ is a matrix-valued function. The gradient of $\mathbf{f}$, called the "Jacobian" matrix in the above discussion, is the second derivative of F, and it is sometimes called the "Hession" matrix. In that case, the term "Jacobian" is reserved for the gradient. This latter usage is particularly common in the context of optimization.

In the following exercise, you will see that Newton's method applied to this system can require the convergence neighborhood to be quite small.

Exercise 4.8.

(1) Download code for the least squares objective function F, its gradient $\mathbf{f}$, and the Jacobian matrix J. The file is called `objective.m` because a function to be minimized is often called an "objective function." (The objective is to minimize the objective function.)
(2) Use the command `help objective` to see how to use it.
(3) Compute $\mathbf{f}$ at the solution x=[0.15;2.0;1.0;3] to be sure that the function is zero there.
(4) Compute the determinant of J at the solution x=[0.15;2.0;1.0;3] to see that it is nonsingular.
(5) Since this particular problem seeks the minimum of a quadratic functional, the matrix J must be positive definite and symmetric. Check that J is symmetric and then use the MATLAB `eig` function to find the four eigenvalues of J to see they are each positive.

Exercise 4.9. The point of this exercise is to see how sensitive Newton's method can be when the initial guess is changed. Fill in the following table with either the number of iterations required or the word "failed," using `vnewton.m` and the indicated initial guess. Note that the first line is the correct solution. *For this exercise, restrict the number of iterations to be no more than 100.*

Initial guess	Number of iterations
[0.15; 2.0; 1.0; 3]	
[0.15; 2.0; 0.9; 3]	
[0.15; 2.0; 0.0; 3]	
[0.15; 2.0;-0.1; 3]	
[0.15; 2.0;-0.3; 3]	
[0.15; 2.0;-0.5; 3]	
[0.15; 2.0; 1.0; 4]	
[0.15; 2.0; 1.0; 5]	
[0.15; 2.0; 1.0; 6]	
[0.15; 2.0; 1.0; 7]	
[0.15; 1.99; 1.0; 3]	
[0.15; 1.97; 1.0; 3]	
[0.15; 1.95; 1.0; 3]	
[0.15; 1.93; 1.0; 3]	
[0.15; 1.91; 1.0; 3]	
[0.17; 2.0; 1.0; 3]	
[0.19; 2.0; 1.0; 3]	
[0.20; 2.0; 1.0; 3]	
[0.21; 2.0; 1.0; 3]	

You can see from the previous exercise that Newton can require a precise guess before it will converge. Sometimes some iterate is not far from the ball of convergence, but the Newton step is so large that the next iterate is ridiculous. In cases where the Newton step is too large, reducing the size of the step might make it possible to get inside the ball of convergence, even with initial guesses far from the exact solution. This is the strategy examined in the following section.

4.7   Softening (damping)

Exercise 4.9 suggests that Newton gets in trouble when its increment is too large. One way to mitigate this problem is to "soften" or "dampen" the iteration by putting a fractional factor on the iterate

$$\mathbf{x}^{(n+1)} = \mathbf{x}^{(k)} - \alpha J(\mathbf{x}^{(k)})^{-1}\mathbf{f}(\mathbf{x}^{(k)}) \tag{4.6}$$

where α is a number smaller than one. It should be clear from the convergence proofs you have seen for Newton's method that introducing the softening factor α destroys the quadratic convergence of the method. This raises the question of stopping. In the current version of `vnewton.m`, you stop when `norm(increment)` gets small enough, but if `increment` has been multiplied by `alpha`, then convergence could happen immediately if `alpha` is very small. It is important to make sure `norm(increment)` is not multiplied by `alpha` before the test is done.

Try softening in the following exercise.

Exercise 4.10.

(1) Starting from your `vnewton.m` file, copy it to a new file named `snewton0.m` (for "softened Newton"), change its signature to

```
function [x,numIts]=snewton0(f,x,maxIts)
```

and change it to conform with Equation (4.6) with the fixed value $\alpha = 1/2$. Don't forget to change the comments and the convergence criterion.
Warning: The backslash operator does not observe the "operator precedence" you might expect, so you need parentheses. For example, `3*2\4=0.6667`, but `3*(2\4)=6`.

(2) Returning to Exercise 4.9, to the nearest 0.01, how large can the first component of the initial guess get before the iteration diverges? (Leave the other three values at their correct values.)

Softening by a constant factor can improve the initial behavior of the iterates, but it destroys the quadratic convergence of the method. Further, it is hard to guess what the softening factor should be. There are tricks to soften the iteration in such a way that when it starts to converge, the softening goes away ($\alpha \to 1$). One such trick is to compute

$$\Delta x = -J(\mathbf{x}^{(k)})^{-1}\mathbf{f}(\mathbf{x}^{(k)})$$
$$\alpha = \frac{1}{1 + \beta\|\Delta x\|}$$
$$\mathbf{x}^{(k+1)} = \mathbf{x}^{(k)} + \alpha\Delta x$$

(4.7)

where the value $\beta = 10$ is a conveniently chosen constant. You should be able to see how you might prove that this softening strategy does not destroy the quadratic convergence rate, or, at least, allows a superlinear rate.

Remark 4.5. The expression in (4.7) is designed to keep the largest step below one tenth of the Newton step. This is a very conservative strategy. Note also that another quantity could be placed in the denominator, such as $\|\mathbf{f}\|$, so long as it becomes zero at the solution.

Exercise 4.11.

(1) Starting from your `snewton0.m` file, copy it to a new file named `snewton1.m`, change its signature to

```
function [x,numIts]=snewton1(f,x,maxIts)
```

and change it so that α is not the constant $1/2$, but is determined from Equation (4.7). Don't forget to change the comments.

(2) How many iterations are required to converge, starting from x(1)=0.20 and the other components equal to their converged values? How many iterations were required by snewton0? You should see that snewton1 has a slightly larger ball of convergence than snewton0, and converges much faster.

(3) Using snewton1.m, to the nearest 0.01, how large can the first component of the initial guess get before the iteration diverges? (Leave the other three values at their correct values.)

4.8   Continuation or homotopy methods

As can be seen in the previous few exercises, there are ways to improve the radius of convergence of Newton's method. For some problems, such as the curve-fitting problem above, they just don't help enough. There is a class of methods called continuation or homotopy methods (or Davidenko's method, [Ralston and Rabinowitz (2001)], page 363f) that can be used to find good initial guesses for Newton's method. Some other references include [Verschelde (1999)], [Davidenko (originator)], [Davidenko (originator)], and [Ortega and Rheinboldt (1970)] (pages 230–234).

The previous section is concerned with solving for the minimum value of an objective function $F(x)$. Suppose there is another, much simpler, objective function $\Phi(x)$, whose minimum (different from that of F) is easy to find using Newton's method. One possible choice would be $\Phi(x) = \|x - x_0\|^2$, for some choice of x_0. For $0 \le p \le 1$, consider the new objective function

$$G(x, p) = pF(x) + (1 - p)\Phi(x). \tag{4.8}$$

When $p = 0$, G reduces to Φ and is easy to minimize (to a result that you already know) but when $p = 1$, G is equal to F and its minimum is the desired minimum. All you need to do is minimize $G(x, p)$ for the sequence $0 = p_1 < p_2 < \cdots < p_{n-1} < p_n = 1$ where you use the solution x_k from parameter p_k as the initial guess for the p_{k+1} case. For properly chosen sequences, the result p_k from step k will be within the radius of convergence for the next step p_{k+1} and the final minimum will be the desired one. The trick is to find a "properly chosen sequence," and there is considerable mathematics involved in doing so. In this chapter, you will simply take a uniform sequence of values. This method can work quite nicely, as you see in the following exercise.

Exercise 4.12.

(1) Suppose that x_0 is a fixed four-dimensional vector and x is a four-dimensional variable. Define

$$\Phi(x) = \sum_{k=1}^{4} (x_k - (x_0)_k)^2.$$

Its gradient is

$$\phi_k = \frac{\partial \Phi}{\partial x_k}$$

and the Jacobian matrix is given by

$$J_{k,\ell} = \frac{\partial \phi_k}{\partial x_\ell}.$$

The following code outline computes Φ and its derivatives in a manner similar to `objective.m`. You may find it convenient to download the outline file `easy_objective.txt`.

```
function [f,J,F]=easy_objective(x,x0)
 % [f,J,F]=easy_objective(x-x0)
 % more comments

 % your name and the date

 if norm(size(x)-size(x0)) ~= 0
   error('easy_objective: x and x0 must be compatible.')
 end
 F=sum((x-x0).^2);
 % f(k)=derivative of F with respect to x(k)
 f=zeros(4,1);
 f= ???
 % J(k,ell)=derivative of f(k) with respect to x(ell)
 J=diag([2,2,2,2]);
end
```

Copy it into a file named `easy_objective.m` and complete the expression for the vector `f`.

Remark 4.6. The chosen value for `x0` is problem-related and amounts to a vague approximation to the final solution.

(2) What are the values of `f`, `J` and `F` for `x0=[0;2;1;2]` and `x=[0;0;0;0]` and also for `x0=[0;2;1;2]` and `x=[1;-1;1;-1]`? You should be able to confirm that these values are correct by an easy calculation.

(3) Copy the following code to a file named `homotopy.m`, and complete the code. You may find it convenient to download the file `homotopy.txt`.

```
function [f,J,F]=homotopy(x,p,x0)
  % [f,J,F]=homotopy(x,p,x0)
  % computes the homotopy or Davidenko objective function
  % for 0<=p<=1

  [f1,J1,F1]=objective(x);
  [f2,J2,F2]=easy_objective(x,x0);
  f=p*f1+(1-p)*f2;
  J=???
  F=???


end
```

(4) Place the following code into a MATLAB script file named **dvdnko.m** or download it.

```
x0=[0.24; 2; 1; 3];
x=ones(4,1);
STEPS=1000;
MAX_ITERS=100;
p=0;
% print out table headings
fprintf('   p    n    x(1)    x(2)    x(3)    x(4)\n');
for k=0:STEPS
  p=k/STEPS;
  [x,n]=vnewton(@(xx) homotopy(xx,p,x0),x,MAX_ITERS);
  % the fprintf statement is more sophisticated than disp
  if n>3 || k==STEPS || mod(k,20)==0
    fprintf('%6.4f %2d %7.4f %7.4f %7.4f %7.4f\n',p,n,x);
  end
end
```

Try this code with x0=[0.24; 2; 1; 3]. Does it successfully reach the value p=1? Does it get the same solution values for x as in Exercises 4.9 through 4.11?

(5) Explain what the expression

```
@(xx) homotopy(xx,p,x0)
```

means in the context of **dvdnko.m**. Why is this construct used instead of simply using @homotopy?

(6) Test x0=[.25;2;1;3]; in easy_objective. Does the **dvdnko** script successfully reach the value p=1? To the nearest 0.05, how far can you increase the first component and still have it successfully reach p=1?

(7) Success of the method depends strongly on the sizes of the steps taken in moving from the simple objective to the true objective. Change STEPS from 1000 to 750, thus increasing the size of the steps. Starting from x0=[.25;2;1;3]; in easy_objective, to the nearest 0.05, how far can you increase the first component and still have it successfully reach p=1?

(8) Returning to STEPS=1000, can you start from x0=[0;2;1;3] and reach the solution? How about x0=[-0.5;2;1;3]?

As you can see, the idea behind continuation methods is powerful. The best that could be done in Exercise 4.11 is deviate from the exact answer by a few percent, while in Exercise 4.12 you more than tripled the first component and still achieved a correct solution. Nonetheless, some care must be taken in choosing x0 and STEPS. Nothing is free, however, and you probably noticed that 1000 steps takes a bit of time to complete.

4.9 Quasi-Newton methods

The term "quasi-Newton" method basically means a Newton method using an approximate Jacobian instead of an exact one. You saw in Chapter 3 that approximating the Jacobian can result in a linear convergence rate instead of the usual quadratic rate, so quasi-Newton methods can take more iterations than true Newton methods will take. On the other hand, inverting an $N \times N$ matrix takes time proportional to N^3, while solving a matrix (after inverting it) and constructing the matrix in the first place take time proportional to N^2.

For very large linear systems, then, enough time could be saved by solving an approximate Jacobian system to make up for the extra iterations required because true Newton converges faster. In the exercise below, the inverse of the Jacobian matrix will be saved from iteration to iteration, and under proper circumstances, re-used in later iterations because it is "close enough" to the inverse of the true Jacobian matrix.

Remark 4.7. You will see in the next part of this book that one almost never constructs the inverse of a matrix because simply solving a linear system takes much less time than constructing the inverse and multiplying by it. Furthermore, solving a linear system by a direct method involves constructing two matrices, one lower triangular and one upper triangular. These two matrices can be solved efficiently, and the two matrices can be saved and used again. For this chapter, however, you will be constructing the inverse matrix and saving it for future use because it is conceptually simpler.

One choice of nonlinear vector function can be given by the following expression for its components, for $k = 1, \ldots, N$,

$$(f_{14}(x))_k = (d_k + \varepsilon)x_k^n - \sum_{j=k+1}^{N} \frac{x_j^n}{j^2} - \sum_{j=1}^{k-1} \frac{x_{k-j}^n}{j^2} - \frac{k}{N(1+k)} \qquad (4.9)$$

where $n = 2$, $N = 3000$, $\epsilon = 10^{-5}$ and

$$d_k = \sum_{j \neq k} \frac{1}{j^2}.$$

Note that $d_k < \sum_{k=1}^{\infty} 1/k^2 = \pi^2/6 = B_1$, the first Bernoulli number.

This function is an example of a nonlinear function whose Jacobian matrix is full and for which there is no simple pattern to its non-zero entries. Nonetheless, it can be expressed in a relatively compact form (4.9).

Exercise 4.13.

(1) For the case that $N = 3$, compute by hand (using symbols for x, n, ε, and d_k) the column vector f of length 3 and the 3×3 matrix J from (4.9). You should recognize that the matrix J is not symmetric and has no zero components, and these facts are true no matter how large N is chosen.
(2) Copy the following code to a function m-file f13.m and fill in values for the Jacobian matrix J by taking derivatives "in your head" and using the resulting formulæ in place of ???. You may find it convenient to download the file f13.txt.

```
function [f,J]=f13(x)
  % [f,J]=f13(x)
  % large function to test quasi-Newton methods

  % your name and the date

  [N,M]=size(x);
  if M ~= 1
    error(['f13: x must be a column vector'])
  end

  n=2;  % exponent in function definition (4.9)
  epsilon=1.e-5; % epsilon in (4.9)
  f=zeros(N,1);
  J=zeros(N,N);

  j=(1:N)';
  jn=j.^n;
  dd=sum(1./jn);
```

```
xn=x.^n;
for k=1:N
  f(k)=(dd-1/k^2+epsilon)*xn(k) - sum(xn(k+1:N)./jn(k+1:N)) ...
        -sum(xn(k-1:-1:1)./jn(1:k-1))-k/(N*(k+1));
end


if nargout>1 % if need Jacobian too, compute it
  J=zeros(N,N);
  for k=1:N
    J(k,k)=???
    J(k,k-1:-1:1)=-1./jn(1:k-1)*n.*x(k-1:-1:1).^(n-1);
    J(k,k+1:N)=???
  end
end
end
```

Remark 4.8. Recall that the `nargout` function returns the number of requested output variables. For `[f,J]=f13(x)`, `nargout=2` and for `f=f13(x)`, `nargout=1`.

(3) Compare your hand work with the function for the case that `x=[1;1;1]`. They should agree.

(4) It is extremely important for this exercise that the Jacobian matrix is correctly computed. The terms in $f_{13}(x)$ are all quadratic in x, and for any quadratic function $g(x)$,

$$\frac{dg}{dx} = \frac{g(x + \Delta x) - g(x - \Delta x)}{2\Delta x} \qquad (4.10)$$

exactly for any reasonable value of Δx.

Choose `x=[1;2;3]`, $\Delta x = 0.1$, and `N=3`, and use (4.10) to compute the finite difference Jacobian

$$(\texttt{Jfd})_{k,j} = \frac{d(f_{13})_k}{dx_j}, \qquad \text{for } k, j = 1, \ldots, 3.$$

Compare `Jfd` with `J` from `f13` by showing the (matrix) norm of the difference is zero or roundoff (`norm(J - Jfd)`). If they do not agree, be sure to find your error.

Hint: You can do this in three steps instead of nine by choosing the three column vectors $\Delta x = [0.1; 0; 0]$, $\Delta x = [0; 0.1; 0]$ and $\Delta x = [0; 0; 0.1]$,

---

You have seen that when your `vnewton` function is converging quadratically, the ratio `r1` becomes small. One way to improve speed might be to stop using the current Jacobian matrix when `r1` is small, and just use the previous one. If you save the inverse Jacobians from step to step, this can improve speed. In the next exercise, you will construct a quasi-Newton method that does just this.

Exercise 4.14.

(1) Writing `qnewton.m`:

    (a) Make a copy of your `vnewton.m` file called `qnewton.m` and change the outline to read

```
function [x,numIts]=qnewton(func,x,maxIts)
  % [x,numIts]=qnewton(func,x,maxIts)
  % more comments

  % your name and the date

  << your code >>

end
```

       You may find it convenient to download the file `qnewton.txt` and copy from it.

    (b) Just before the start of the loop, initialize a variable

```
skipNext = false;
```

    (c) Replace the two lines defining `oldIncrement` and `increment` with the following lines

```
    if ~skipNext
      tim=clock;
      Jinv=inv(derivative);
      inversionTime=etime(clock,tim);
    else
      inversionTime=0;
    end
    oldIncrement=increment;
    increment = -Jinv*value;
```

    (d) Add the following lines just after the computation of `r1`.

```
% Use r1 to determine if the NEXT iteration should be skipped
skipNext = r1 < 0.2;
fprintf('it=%d, r1=%e, inversion time=%e.\n',numIts,r1, ...
    inversionTime)
```

       Make sure there are no other print statements inside `qnewton.m`.

(2) In addition, make a copy of your completed `qnewton.m` to a new file `qnewtonNoskip.m` (or use "save as"). Change the name inside the file, and replace The line

```
    skipNext = r1 < 0.2;
```

with

```
    skipNext = false;
```

This will give you a file for comparing times for true Newton against those for quasi-Newton.

(3) The following command both solves a moderately large system and gives the time it takes:

```
tic;[v,its]=qnewton(@(x) f13(x),linspace(1,10,3000)');toc
```

How long does it take? How many iterations? How many iterations were skipped (took no time)? What is the total time for inversion as a percentage of the total time taken?

(4) Repeat the same experiment but use `qnewtonNoskip`.

```
tic;[v0,its0]=qnewtonNoskip(@(x) f13(x),linspace(1,10,3000)');toc
```

How long does it take? How many iterations? How far apart are the solutions (`norm(v-v0)/norm(v)`?

(5) For this case, which is faster (takes less total time) qnewton or qnewtonNoskip?

Remark 4.9. On the author's computer, `qnewton` is a few seconds faster. If you have a faster computer, you may observe a smaller difference, or even that `qnewton` is slower. If so, you can try the same comparison with N=4000 or larger.

Remark 4.10. It turns out that `qnewtonNoskip` is somewhat slower than `vnewton` because the inverse matrix is constructed explicitly. Instead of computing and saving `Jinv`, you could compute and save the factors of the matrix. A version of `qnewtonNoskip` programmed this way would take about the same amount of time as `vnewton`, and `qnewton` would be faster yet.

Remark 4.11. The choice `skip = (r1 < 0.2)` is very much problem-dependent. There are more reliable ways of deciding when to skip inverting the Jacobian, but they are beyond the scope of this chapter.

Remark 4.12. For large matrix systems, especially ones arising from partial differential equations, it is faster to solve the system using iterative methods. In this case, there are two iterations: the nonlinear Newton iteration and the linear solution iteration. For these systems, it can be more efficient to stop the linear solution iteration before full convergence is reached, saving the time for unnecessary iterations. This approach is also called a "quasi-Newton method."

Remark 4.13. When a sequence of similar problems is being solved, such as in Davidenko's method or in time-dependent partial differential equations, quasi-Newton methods can save considerable time in the solution at each step because it is often true that the Jacobian changes relatively slowly.

Chapter 5

Interpolation on evenly-spaced Points

5.1 Introduction

This chapter is concerned with interpolating data with polynomials and with trigonometric functions. There is a difference between interpolating and approximating. Interpolating means that the function *passes through* all the given points with no regard to the points between them. Approximating means that the function is close to all the given points and and has some additional mathematical properties often making it close to nearby points as well. It might seem that interpolating is always better than approximating, but there are many situations in which approximation is the better choice.

Polynomials are commonly used to interpolate data. There are several methods for defining, constructing, and evaluating the polynomial of a given degree that passes through data values. This chapter will present solving for the coefficients of the polynomial by setting up a linear system called *Vandermonde's equation*, and representing the polynomial as the sum of simple *Lagrange polynomials* and interpolation using trigonometric polynomials.

You will see examples showing that interpolation does not necessarily mean good approximation and that one way that a polynomial interpolant can fail to approximate is because of a bad case of "the wiggles." Trigonometric polynomial interpolation can do better, but can also break down.

In this chapter, both the Vandermonde matrix and Lagrange polynomials are used to generate polynomial interpolating functions. Exercises 5.1 and 5.2 use the Vandermonde matrix approach, with 5.3 presenting a nontrivial example. Exercise 5.4 presents the Runge example of a function that is not well approximated by its interpolant because of the "wiggles," and Exercise 5.5 suggests why the problem occurs. Exercises 5.6 and 5.7 introduce Lagrange polynomials to generate the same polynomial interpolating functions as the Vandermonde matrix approach. Exercises 5.8 through 5.11 are a sequence discussing trigonometric interpolation. The final exercise, 5.12, discusses two-dimensional interpolation as used in finite element theory and is a much more difficult exercise than the rest.

5.1.1 *Notation*

This chapter is focussed on finding functions that pass through certain specified points. In customary notation, you are given a set of points $\{(x_k, y_k) | k = 1, 2, \ldots, n\}$ and you want to find a function $f(x)$ that passes through each point ($f(x_k) = y_k$ for $k = 1, 2, \ldots, n$). That is, for each of the abscissæ, x_k, the function value $f(x_k)$ agrees with the ordinate y_k. The given points are variously called the "given" points, the "specified" points, the "data" points, *etc.* In this chapter, the term "data points" will be used.

Written in the customary notation, it is easy to see that the quantities x and x_k are essentially different. MATLAB has no kerning or font differentiation, so it can be difficult to keep the various quantities straight. This discussion will use the name xval to denote the values of x and the names xdata and ydata to denote the data x_k and y_k. The variable names fval or yval are used for the value $y = f(x)$ to emphasize that it is an interpolated value.

Generally, one thinks of the values xval as not equal to any of the data values xdata although, in this chapter, you will often set xval to one or more of the xdata values in order to test that your interpolation is correct.

5.1.2 MATLAB *Tips*

You will sometimes need to determine if an integer m is not equal to an integer n. The MATLAB syntax for this is

```
if m ~= n
```

Many people enclose the logical expression in parentheses, but this is not required. Numbers with decimal parts should *never* be tested for equality! Instead, the absolute value of the difference should be tested to see if it is small. The reason is that numbers are rarely known to full 16-digit precision and, in addition, small errors are usually made in representing them in binary instead of decimal form. To check if a number x is equal to y, you should use

```
if abs(x-y) <= TOLERANCE
```

where TOLERANCE represents some reasonable value chosen based on the problem.

MATLAB represents a polynomial as the vector of its coefficients. MATLAB has several commands for dealing with polynomials:

c=poly(r) Finds the coefficients of the polynomial whose roots are given by the vector r.

r=roots(c) Finds the roots of the polynomial whose coefficients are given by the vector c.

c=polyfit(xdata, ydata, n) Finds the coefficients of the polynomial of degree n passing through, or approaching as closely as possible, the points xdata(k),ydata(k), for k = 1, 2, ..., K, where $K - 1$ need not be the same as n.

yval=polyval(c, xval) Evaluates a polynomial with coefficients given by the vector c at the values xval(k) for $k = 1, 2, ..., K$ for $K \geq 1$.

When there are more data values than the minimum, the polyfit function returns the coefficients of a polynomial that "best fits" the given values in the sense of least squares. This polynomial approximates, but does not necessarily interpolate, the data. In this chapter, you will be writing m-files with functions similar to polyfit but that generate polynomials of the precise degree determined by the number of data points. (N=numel(xdata)-1).

The coefficients c_k of a polynomial in MATLAB are, by convention, defined as

$$p(x) = \sum_{k=1}^{N} c_k x^{N-k} \tag{5.1}$$

and in MATLAB are represented as a vector c. With this definition of the vector c of coefficients of a polynomial:

- N=numel(c) is one higher than the degree of $p(x)$.
- The subscripts run *backwards*! c(1) is the coefficient of the term with degree N-1, and the constant term is c(N).

In this and some later chapters, you will be writing m-files with functions analogous to polyfit and polyval, using several different methods. Rather than following the MATLAB naming convention, functions with the prefix coef_ will generate a vector of **coefficients**, as by polyfit, and functions with the prefix eval_ will **eval**uate a polynomial (or other function) at values of xval, as with polyval.

In the this chapter and often in later chapters you will be using MATLAB functions to construct a known polynomial and using it to generate "data" values. Then the interpolation functions recover the original, known, polynomial. This strategy is a powerful tool for illustrative and debugging purposes, but practical use of interpolation starts from arbitrary data, not contrived data.

Remark 5.1. Newton's method (Chapter 3) can be derived by determining a linear polynomial (degree 1) that passes through the point (a, f_a) with derivative f'_a. That is $p(a) = f_a$ and $p'(a) = dp/dx(a) = f'_a$. One way of looking at this is that an interpolating function, in this case a linear polynomial, is being constructed that explains all the currently known data. Examining the graph of the polynomial, evaluating it at other points, determining its integral or derivative, or doing other things with the interpolate become possible as surrogates for the original function. This is a reason that polynomial interpolation is important even when the functions in are not polynomials.

5.2   Vandermonde's Equation

Here is one way to see how to organize the computation of the polynomial that passes through a set of data.

Suppose you want to determine the linear polynomial $p(x) = c_1 x + c_2$ that passes through the data points (x_1, y_1) and (x_2, y_2). You can solve a set of linear equations for c_1 and c_2 constructed by plugging in the two data points into the general linear polynomial. These equations are

$$c_1 x_1 + c_2 = y_1$$
$$c_1 x_2 + c_2 = y_2$$

or, equivalently,

$$\begin{bmatrix} x_1 & 1 \\ x_2 & 1 \end{bmatrix} \begin{bmatrix} c_1 \\ c_2 \end{bmatrix} = \begin{bmatrix} y_1 \\ y_2 \end{bmatrix}$$

which (usually) has the solution

$$c_1 = (y_2 - y_1)/(x_2 - x_1)$$
$$c_2 = (x_2 y_1 - x_1 y_2)/(x_2 - x_1).$$

Compare that situation with the case determining the quadratic polynomial $p(x) = c_1 x^2 + c_2 x + c_3$ that passes through three sets of data values. Then the following set of three linear equations for the polynomial coefficients c must be solved.

$$\begin{bmatrix} x_1^2 & x_1 & 1 \\ x_2^2 & x_2 & 1 \\ x_3^2 & x_3 & 1 \end{bmatrix} \begin{bmatrix} c_1 \\ c_2 \\ c_3 \end{bmatrix} = \begin{bmatrix} y_1 \\ y_2 \\ y_3 \end{bmatrix}.$$

These are examples of second- and third-order *Vandermonde Equations*. It is characterized by the fact that for each row (sometimes column) of the coefficient matrix, the successive entries are generated by a decreasing (sometimes increasing) set of powers of a set of variables.

You should be able to see that, for *any* collection of abscissæ and ordinates, it is possible to define a linear system that should be satisfied by the (unknown) polynomial coefficients. Solving the system, and solving it accurately, is one way to determine the interpolating polynomial.

Using MATLAB to construct and solve the Vandermonde equation involves setting up the coefficient matrix A. Use the MATLAB variables xdata and ydata to represent the quantities x_k and y_k, and assume them to be row vectors of length (numel) N.

```
for j = 1:N
  for k = 1:N
    A(j,k) = xdata(j)^(N-k) ;
  end
end
```

The right hand side comes from the ordinates ydata, assumed to be a row vector. After setting it up, solving the linear system is easy. Just write:

```
c = A \ ydata';
```

Recall that the backslash symbol means to solve the system with matrix A and right side ydata'. Notice that ydata' is the transpose of the row vector ydata in this equation in order to make it a column vector. (By default, MATLAB constructs row vectors unless told to do otherwise.)

Remark 5.2. A better way to be sure that the backslash operator is operating on a *column* vector is to use the **reshape** function, because it will work whether its argument is a row vector or a column vector.

Exercise 5.1. The MATLAB built-in function **polyfit** finds the coefficients of a polynomial through a set of points. You will write your own using the Vandermonde matrix. (This is the way that the MATLAB function **polyfit** works.) You can find some of the code and data below in the file **exer1.txt**.

(1) Write a MATLAB function m-file, **coef_vander.m** with signature and outline

```
function c = coef_vander ( xdata, ydata )
% c = coef_vander ( xdata, ydata )
% xdata= ???
% ydata= ???
% c= ???
% other comments

% your name and the date

<< your code >>

end
```

that accepts a pair of row vectors **xdata** and **ydata** of arbitrary but equal length, and returns the coefficient vector **c** of the polynomial that passes through that data. Be sure to complete the comments with question marks in them.
Warning: Think carefully about what to use for N.

(2) Test your function by computing the coefficients of the polynomial through the following data points. (This polynomial is $y = x^2$, so you can check your coefficient vector "by inspection.")

```
xdata= [ 0   1   2 ]
ydata= [ 0   1   4 ]
```

(3) Test your function by computing the coefficients of the polynomial that passes through the following points

```
xdata= [  -3  -2 -1  0  1   2    3]
ydata= [1636 247 28  7  4  31  412]
```

(4) Confirm using `polyval` that your polynomial passes through these data points.

(5) Double-check your work by comparing with results from the MATLAB `polyfit` function. Please include both the full polyfit command and its result.

In the following exercise you will construct a polynomial using `coef_vander` to interpolate data points and then you will see what happens *between* the interpolation points.

Exercise 5.2.

(1) Consider the polynomial whose roots are $r = [-2 - 1 1 2 3]$. Use the MATLAB `poly` function to find its coefficients. Call these coefficients `cTrue`.

(2) This polynomial obviously passes through zero at each of these five "data" points. You want to check if your `coef_vander` function can reproduce it. To use `coef_vander`, you need a sixth point. You can "read off" the value of the polynomial at `x=0` from its coefficients `cTrue`. What is this value?

(3) Use the `coef_vander` function to find the coefficients of the polynomial passing through the "data" points

```
xdata=[ -2 -1 0  1 2 3];
ydata=[  0  0 ?? 0 0 0];
```

Call these coefficients `cVander`.

(4) Use `coef_vander` to find the coefficients using only the five roots as `xdata`. Name these coefficients something other than `cVander`. Explain your results.

(5) Use the following code. (You may find it convenient to download `exer2.txt`.) to compute and plot the values of the true and interpolant polynomials on the interval [-3,2]. If you look at the last line of the code, you will see an estimate of the difference between the two curves. How big is this difference? You will be using essentially this same code in several following exercises. You should be sure you understand what it does. Copy the code to a script m-file named `exer2.m` and complete it by replacing the question marks.

```
% construct many test points (for plotting)
xval=linspace(-2,3,4001);
% construct the true test point values, for reference
yvalTrue=polyval(cTrue,xval);

% use Vandermonde polynomial interpolation coefficients
% to evaluate the interpolant at the test points
yval=polyval(cVander, ???);
```

```
% plot reference values in thick green
plot(xval,yvalTrue,'g','linewidth',4);
hold on
% plot interpolation data points as black plus signs
plot(xdata,ydata,'k+');
% plot interpolant in thin black
plot(xval,yval,'k');
hold off

% estimate the relative approximation error of the interpolant
approximationError=max(abs(yvalTrue-yval))/max(abs(yvalTrue))
```

You should observe the two curves are the same. The second curve appears as a thin black curve overlaying the thick green curve.

Remark 5.3. Relative error is used here so that the case where `yvalTrue` are very large or very small numbers does not cause difficulty.

5.3   Interpolating a function that is not a polynomial

Interpolating functions that are polynomials and using polynomials to do it is cheating a little bit. You have seen that interpolating polynomials can result in interpolants that are essentially identical to the original polynomial. Results can be much less satisfying when polynomials are used to interpolate functions that are not themselves polynomials. At the interpolation points, the function and its interpolant agree exactly, so you need to examine the behavior *between* the interpolation points. In the following exercise, you will see that some non-polynomial functions can be interpolated quite well, and in the subsequent exercise you will see one that cannot be interpolated well. The example used here is due in part to C. Runge, [Runge (1901)].

Exercise 5.3. In this exercise you will construct interpolants for the hyperbolic sine function $\sinh(x)$ and see that it and its polynomial interpolant are quite close.

(1) Interpolate the function $y = \sinh(x)$ on the interval $[-\pi, \pi]$, so copy `exer2.m` to `exer3.m` and modify it to examine the behavior of the polynomial interpolant to the hyperbolic sine function for five evenly-spaced points using the following steps.

(a) Add the following code at the beginning:

```
% construct N=5 data points
```

```
N=5;
xdata=linspace(-pi,pi,N);
ydata=sinh(xdata);
```

    (b) Correct the calculations of xval and yvalTrue.

    (c) Add the following code to compute cVander before it is used

```
% compute Vandermonde polynomial interpolation coefficients
cVander=coef_vander(xdata,ydata);
```

(2) Execute exer3.m. By zooming, *etc.*, confirm visually that the exponential and its interpolant agree at the interpolation points.

(3) Using more data points gives higher degree interpolation polynomials. Fill in the following table using Lagrange interpolation with increasing numbers of data points.

N	Approximation error
5	
11	
21	

You should have observed in Exercise 5.3 that the approximation error becomes quite small. The hyperbolic sine function is entire, as are polynomials, so they share one essential feature. In the following exercise, you will see that attempts to interpolate functions that are not entire can give poor results.

Exercise 5.4.

(1) Construct a function m-file for the Runge example function $y = \frac{1}{1+x^2}$. Name the file runge.m and give it the signature and outline

```
function y=runge(x)
  % y=runge(x)
  % comments

  % your name and the date

  << your code >>

end
```

Use componentwise (vector) division and exponentiation (./ and .^).

(2) Copy exer3.m to exer4.m and modify it to use the Runge example function.

(3) Confirm visually that the Runge example function and its interpolant agree at the interpolation points, but not necessarily between them.

(4) Using more data points gives higher degree interpolation polynomials. Fill in the following table using Lagrange interpolation with increasing numbers of data points.

N	Approximation error
5	
11	
21	

(5) Are you surprised to see that the errors do not decrease?

Many people expect that an interpolating polynomial $p(x)$ gives a good approximation to the function $f(x)$ everywhere, no matter what function you choose. If the approximation is not good, it is expected it to get better if the number of data points is increased. These expectations will be fulfilled only when the function does not exhibit some "essentially non-polynomial" behavior. You will see why the Runge example function cannot be approximated well by polynomials in the following exercise.

Exercise 5.5. The Runge example function has Taylor series

$$\frac{1}{1+x^2} = 1 - x^2 + x^4 - x^6 + \dots \tag{5.2}$$

as you can easily prove. This series has a radius of convergence of 1 in the complex plane. Polynomials, on the other hand, are entire functions, *i.e.*, their Taylor series converge everywhere in the complex plane. No finite sum of polynomials can be anything but entire, but no entire function can interpolate the Runge example function on a disk with radius larger than one about the origin in the complex plane. If there were one, it would have to agree with the series (5.2) inside the unit disk, but the series diverges at $x = i$ and an entire function cannot have an infinite value (a pole).

(1) Using `exer4.m`, look at the nontrivial coefficients (c_k) of the interpolating polynomials and fill in the following table.

N	Coefficients			
5	$c_5 =$	$c_3 =$	$c_1 =$	
11	$c_{11} =$	$c_9 =$	$c_7 =$	$c_5 =$
21	$c_{21} =$	$c_{19} =$	$c_{17} =$	$c_{15} =$
limit	+1	-1	+1	-1

(2) Look at the trivial coefficients (c_k) of the interpolating polynomials by filling in the following table. (Look carefully at what the colon notation does.)

N	max(abs(c(2:2:end)))
5	
11	
21	

You should see that the interpolating polynomials are "trying" to reproduce the Taylor series (5.2). These polynomials cannot agree with the Taylor series at all points, though, because the Taylor series does not converge at all points.

5.4   Lagrange Polynomials

Imagine a set of N distinct abscissæ x_k, $k = 1, \ldots, N$ and think about the problem of constructing a polynomial that has (not yet specified) values y_k at these points. Now suppose there is a polynomial $\ell_7(x)$ whose value is zero at each x_k, $k \neq 7$, and is 1 at x_7. Then the intermediate polynomial $y_7\ell_7(x)$ would have the value y_7 at x_7, and be 0 at all the other x_i. Doing the same for each abscissa and adding the intermediate polynomials together results in the polynomial that interpolates the data without solving any equations!

In fact, the *Lagrange polynomials* ℓ_k are easily constructed for any set of abscissae. Each Lagrange polynomial will be of degree $N - 1$. There will be N Lagrange polynomials, one per abscissa, and the kth polynomial $\ell_k(x)$ will have a special relationship with the abscissa x_k, namely, it will be 1 there, and 0 at the other abscissæ.

In terms of Lagrange polynomials, then, the interpolating polynomial has the form:

$$p(x) = y_1\ell_1(x) + y_2\ell_2(x) + \ldots + y_N\ell_N(x) \tag{5.3}$$

$$= \sum_{k=1}^{N} y_k\ell_k(x)$$

Assuming you can determine these *magical* polynomials, this is a *second* way to define the interpolating polynomial to a set of data.

Remark 5.4. The strategy of finding a function that equals 1 at a distinguished point and zero at all other points in a set is very powerful. If $y_k = 1$ in (5.3), then $p(x) \equiv 1$, so the $\ell_k(x)$ are an example of a "partition of unity." One of the places you will meet it again is when you study the construction of finite elements for solving partial differential equations.

In the next two exercises, you will be constructing polynomials through the same points as in the previous exercise. Since there is only one nontrivial polynomial of degree $(N - 1)$ through N data points, the resulting interpolating polynomials are the same in these and the previous exercise.

In the next exercise, you will construct the Lagrange polynomials associated with the data points, and in the following exercise you will use these Lagrange polynomials to construct the interpolating polynomial.

Exercise 5.6. In this exercise you will construct Lagrange polynomials based on given data points. Recall the data set for $y = x^2$:

```
k :    1   2   3
xdata= [ 0   1   2   ]
ydata= [ 0   1   4   ]
```

(Actually, ydata is immaterial for construction of $\ell_k(x)$.) In general, the formula for $\ell_k(x)$ can be written as:

$$\ell_k(x) = (f_1(x))(f_2(x)) \cdots (f_{k-1}(x)) \cdot (f_{k+1}(x)) \cdots (f_N(x)) \qquad (5.4)$$

(skipping the kth factor), where each factor has the form

$$f_j(x) = \frac{(x - x_j)}{(x_k - x_j)}. \qquad (5.5)$$

(1) Explain in a sentence or two why the formula (5.4) using the factors (5.5) yields a function

$$\ell_k(x_j) = \begin{cases} 1 & j = k \\ 0 & j \neq k \end{cases}. \qquad (5.6)$$

(2) Write a MATLAB function m-file called `lagrangep.m` that computes the Lagrange polynomials (5.4) for any `k`. (One of the MATLAB toolboxes has a function named "`lagrange`", so this one is named "`lagrangep`".)

(a) The signature and outline should be

```
function pval = lagrangep( k , xdata, xval )
% pval = lagrangep( k , xdata, xval )
% comments
% k= ???
% xdata= ???
% xval= ???
% pval= ???

% your name and the date

<< your code >>

end
```

(b) Add code to evaluate the kth Lagrange polynomial for the abscissæ `xdata` at the point `xval`.
    Hint: You can implement the general formula using code like the following.

```
pval = 1;
for j = 1 : ???
  if j ~= k
    pval = pval .* ???   % elementwise multiplication
  end
end
```

Remark 5.5. If xval is a vector of values, then pval will be the vector of corresponding values, so that an elementwise multiplication (.*) is being performed.

(c) If you have not done so already, add the final **end** statement to close out the function.

(3) Using (5.6), determine the values of lagrangep(1, xdata, xval) for xval=xdata(1), xval=xdata(2) and xval=xdata(3).

(4) Does lagrangep give the correct values for lagrangep(1, xdata, xdata)? For lagrangep(2, xdata, xdata)? For lagrangep(3, xdata, xdata)?

───────────────

Exercise 5.7. The hard part is done. Now you can use your lagrangep routine as a helper for a second replacement for the polyfit-polyval pair, called eval_lagr, that implements Equation (5.3). Unlike coef_vander, the coefficient vector of the polynomial does not need to be generated separately because it is so easy, and that is why eval_lagr both fits and evaluates the Lagrange interpolating polynomial.

(1) Write a MATLAB function m-file called eval_lagr.m with the signature and outline

```
function yval = eval_lagr ( xdata, ydata, xval )
  % yval = eval_lagr ( xdata, ydata, xval )
  %   comments

  % your name and the date

  << your code >>

end
```

This function should take the data values xdata and ydata, and compute the value of the interpolating polynomial at xval according to (5.3), using your lagrangep function for the Lagrange polynomials. Be sure to include comments to that effect. The file exer7.txt contains the above outline and also the test data used below.

(2) Test `eval_lagr` on the simplest data set you have been using.

```
    k :     1   2   3
 xdata= [   0   1   2 ]
 ydata= [   0   1   4 ]
```

by evaluating it at `xval=xdata`. You should get `ydata` back.

(3) Test your function by interpolating the polynomial that passes through the following points, again by evaluating it at `xval=xdata`.

```
xdata= [   -3 -2 -1  0  1   2    3]
ydata= [1636 247 28  7  4  31  412]
```

(4) Repeat Exercise 5.2 using Lagrange interpolation.

(a) Return to the polynomial constructed in Exercise 5.2 with roots `r=[-2 -1 1 2 3]`. and coefficients `cTrue`.

(b) Reconstruct (using `polyval`) or recall the `ydata` values associated with `xdata=[-2 -1 0 1 2 3]`.

(c) Copy `exer2.m` to `exer7.m` and modify it to use Lagrange interpolation, plot `yvalLag` and `yvalTrue`, and compute the error between them.

5.5 Trigonometric polynomial interpolation

Interpolation by trigonometric functions is discussed by [Quarteroni *et al.* (2007)] in Section 10.1, [Atkinson (1978)] in Section 3.8, and [Ralston and Rabinowitz (2001)] on page 271, among others. Trigonometric interpolation is closely related to approximation by Fourier series, but the focus is on interpolation in this chapter and approximation will be discussed in Chapter 8.

The basic expression for trigonometric interpolation using $(2N + 1)$ functions over the interval $[-\pi, \pi]$ is

$$f(x) = \sum_{k=1}^{2N+1} a_k e^{i(k-N-1)x}. \tag{5.7}$$

(There are $(2N + 1)$ points because the focus here is to use real functions and real interpolants. Thus, one trigonometric function $e^{i(k-N-1)x}$ is the constant term when $k = N + 1$, and all the rest come in complex conjugate pairs.)

Using the $2N + 1$ evenly-spaced points in the interval $(-\pi, \pi)$ given by

$$x_j = \frac{2\pi(j - N - 1)}{2N + 1}, \text{ for } j = 1, 2, \ldots, (2N + 1), \tag{5.8}$$

the coefficients can be determined from

$$a_k = \frac{1}{2N + 1} \sum_{j=1}^{2N+1} e^{-i(k-N-1)x_j} f(x_j), \text{ for } k = 1, 2, \ldots, (2N + 1). \tag{5.9}$$

These equations can be found directly by multiplying (5.7) through by each of the functions $e^{-i(k-N-1)x}$ in turn, applying it to the values x_k, and solving the resulting system for a_j, taking advantage of the properties of the complex exponential to simplify the equations.

Remark 5.6. The trigonometric coefficients a_k in (5.9) play the same role as the polynomial coefficients c_k in (5.1).

In the following exercises, you are going to write functions `coef_trig` to be analogous to `polyfit` and `coef_vander`, and also `eval_trig` to be analogous to `polyval` and `eval_lagr`.

Exercise 5.8.

(1) Construct a function m-file named `coef_trig.m` with the signature and outline

```
function a=coef_trig(func,N)
  % a=coef_trig(func,N)
  % func=???
  % N=???
  % a=???
  % comments

  % your name and the date

  << your code >>

end
```

This function should evaluate the trigonometric coefficients a_k according to Equation (5.9). Use Equation (5.8) to determine the points x_k. Remember that the length (`numel`) of a should be `2*N+1`. It is more efficient to use vector (componentwise) notation and the MATLAB **sum** function, but if you cannot see how to do that, just use `for` loops. **Hint:** You can use the following code to generate the points x_k without writing a `for` loop.

```
xdata = 2*pi*(-N:N)/(2*N+1);
```

(2) Test your function by applying it to the function $f(t) = e^{ix}$, for N=10. (You can either write an m-file for e^{ix} or use the @ command to define an "anonymous function.") By examining Equation (5.7), you should be able to see that $a_k = 1$ for $k = N + 2$ and zero otherwise.

(3) Test your function again by applying it to $f(x) = \sin 4x$ with N=10. You should see that a_k is zero for all but two subscripts. What subscripts (k) have non-zero a_k and what are the values?

Exercise 5.9.

(1) Construct a function m-file named `eval_trig.m` with the signature and outline

```
function fval=eval_trig(a,xval)
% fval=eval_trig(a,xval)
  % a=???
  % xval=???
  % fval=???
  % comments

  % your name and the date

  << your code >>

end
```

This function should evaluate Equation (5.7) at an arbitrary collection of points, xval.

(2) Test your function by first using `coef_trig` to find the coefficients for the function $\sin 4x$ with `n=10` (you did this in the previous exercise), and then applying `eval_trig` to those coefficients at 4001 equally-spaced points in the interval $[-\pi, \pi]$. What is the approximation error between the interpolated values and $\sin 4x$? Plot the function and interpolant on the same plot: the lines should overlap. You can modify the m-files `exer2.m` or `exer7.m` and call your modified m-file `exer9.m`. **Warning:** Your interpolated values may appear to MATLAB to be complex, and MATLAB may not plot them the way you expect. In that case, verify that their imaginary parts (MATLAB function `imag`) are zero and plot only the real (MATLAB function `real`) parts.

Trigonometric polynomial interpolation does well even when the functions are not themselves trigonometric polynomials.

Exercise 5.10.

(1) Modify `exer9.m` to use the function $y = x(\pi^2 - x^2)$, and call the modified file `exer10.m`. If you have not done so already, modify your script file to include plotting plus signs at each interpolation point. Plot the function and its interpolant for N=5, 10, 15 and N=20. Fill in the following table.

N	y = x.*(pi^2-x.^2) approximation error
5	
10	
15	
20	

(2) Do the same thing for the Runge example function and fill in the following table.

N	Runge approximation error
2	
4	
5	

(3) For the N=2 case, and the Runge example function, what are the five points at which the trigonometric interpolation should *interpolate* the Runge example function? Check that `runge` and `eval_trig` agree (up to roundoff) at those points.

The previous exercise shows that trigonometric polynomial interpolation does well for some functions. It does much less well when the function is not continuous or not periodic in $[-\pi, \pi]$. Consider, for example, the function

$$f(x) = \begin{cases} x + \pi & \text{for } x < 0 \\ x - \pi & \text{for } x \geq 0 \end{cases}.$$

Exercise 5.11.

(1) Copy the following code into a function m-file named `sawshape5.m`, or download the file `sawshape5.txt` and modify it. Be sure to add the usual comments.

```
function y=sawshape5(x)
  % y=sawshape5(x)
  % x=???
  % y=???
  % comments

  % your name and the date

  kless=find(x<0);
  kgreater=find(x>=0);
  y(kless)=x(kless)+pi;
  y(kgreater)=x(kgreater)-pi;
end
```

(2) The MATLAB `find` command is a *very* useful function. Use the `help find` command or the Help facility to see what it does.

(3) If the letters a through j represent a sequence of ten increasing values, and if `x=[a b c d e f g h i j i h g]`, what is the result of the function calls `find(x==c)`, `find(x==h)`, and `find(x>=g)`?

(4) Use a script file based on `exer9.m` to fill in the following table by interpolating the `sawshape5` function using trigonometric interpolation.

N	sawshape5 approximation error
5	
10	
100	

You can see that the approximation error is not shrinking to zero, but if you look at the plots you will see that the error really is shrinking everywhere except near $x = 0$, where the oscillations get more rapid but do not get smaller. This behavior is typical and is called Gibbs's phenomenon. Please send me the plots for N=5 and N=100.

(5) Examine the plot for the N=5 case. Are the plus signs in the proper places to indicate the interpolation points?

Increasing the order of polynomial interpolation can lead to divergence of approximation because of "the wiggles." Trigonometric interpolation does not diverge as $n \to \infty$, but Gibbs's phenomenon slows convergence drastically at discontinuities.

5.6 Interpolation over triangles

In this exercise, you will see how to find a quadratic interpolating polynomial for a function given on an arbitrary triangle in the plane. The method used will be a two-stage one:

(1) Map the given triangle into a "reference" triangle; and,
(2) Use the two-dimensional Lagrange interpolating polynomials over the reference triangle to interpolate the function.

This two-stage approach is used because it is more convenient to generate the Lagrange polynomials once over a standard triangle than to generate Lagrange polynomials for an arbitrary triangle. A second reason is that each of the stages is easily programmed and can be used repeatedly. Computer programs using finite element methods may involve thousands or millions of triangles.

Consider the following reference triangle, $T_{\text{ref}} = \{(\xi, \eta) \,|\, 0 \leq \xi \leq 1, \ 0 \leq \eta \leq (1 - \xi)\}$.

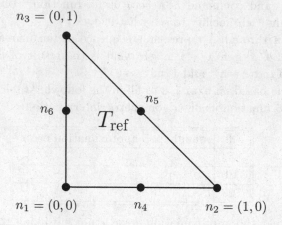

In this triangle, the three nodes n_4, n_5, and n_6 are midway along the indicated sides.

There are three linear Lagrange polynomials defined on T_{ref}. It should be clear that any linear polynomial can be written as a linear combination of $\ell_1 \ldots \ell_3$.

$$\ell_1(\xi, \eta) = 1 - \xi - \eta$$
$$\ell_2(\xi, \eta) = \xi$$
$$\ell_3(\xi, \eta) = \eta.$$

There are six quadratic Lagrange polynomials defined on T_{ref}. It should be clear that any quadratic polynomial can be written as a linear combination of $q_1 \ldots q_6$.

$$q_1(\xi, \eta) = 2(1 - \xi - \eta)(.5 - \xi - \eta)$$
$$q_2(\xi, \eta) = 2\xi(\xi - .5)$$
$$q_3(\xi, \eta) = 2\eta(\eta - .5)$$
$$q_4(\xi, \eta) = 4\xi(1 - \xi - \eta)$$
$$q_5(\xi, \eta) = 4\xi\eta$$
$$q_6(\xi, \eta) = 4\eta(1 - \xi - \eta).$$

In the following exercise, you will be using these polynomials to interpolate functions on the following triangle.

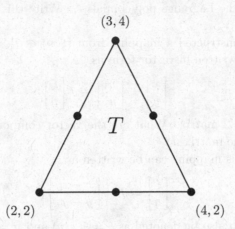

The relationship between the reference triangle T_{ref} expressed in (ξ, η) coordinates and the triangle T expressed in (x, y) coordinates, as well as the function $p(x, y) = e^{0.1xy}$, can be summarized in the following table.

n	(x, y)	(ξ, η)	p
1	$(2, 2)$	$(0, 0)$	$e^{0.4}$
2	$(4, 2)$	$(1, 0)$	$e^{0.8}$
3	$(3, 4)$	$(0, 1)$	$e^{1.2}$
4	$(3, 2)$	$(0.5, 0)$	$e^{0.6}$
5	$(3.5, 3)$	$(0.5, 0.5)$	$e^{1.05}$
6	$(2.5, 3)$	$(0, 0.5)$	$e^{0.75}$

Exercise 5.12. Note: It is possible to do this exercise without writing any MATLAB code, doing your work by hand. If you wish to use the symbolic toolbox for this lab, it would be a good learning exercise.

(1) Confirm that ℓ_i are Lagrange polynomials by directly computing the values $\ell_i(\xi, \eta)$, $i = 1, \ldots, 3$, for the nodal points n_j $j = 1, \ldots, 3$. Each ℓ_i should be 1 at n_i and 0 at n_j for $j \neq i$.

(2) Confirm that q_i are Lagrange polynomials by directly computing the values $q_i(\xi, \eta)$, $i = 1, \ldots, 6$, for the nodal points n_j $j = 1, \ldots, 6$. Each q_i should be 1 at n_i and 0 at n_j for $j \neq i$.

(3) Find a linear function $x = x(\xi, \eta)$ so that $x(n_1) = 2$, $x(n_2) = 4$ and $x(n_3) = 3$. Note that 2, 4, and 3 are the x-coordinates of the three vertices of T. You can use the Lagrange polynomials ℓ_i. Write this as a formula for x in terms of ξ and η.

(4) Similarly, find a linear function $y = y(\xi, \eta)$ so that $y(n_1) = 2$, $y(n_2) = 2$ and $y(n_3) = 4$. Note that 2, 2, and 4 are the y-coordinates of the three vertices of

T. You can use the Lagrange polynomials ℓ_i. Write this as a formula for y in terms of ξ and η.

(5) You have just constructed a mapping from $(\xi, \eta) \in T_{\text{ref}}$ to $(x, y) \in T$. This mapping can be written in vector form as

$$\begin{bmatrix} x \\ y \end{bmatrix} = \begin{bmatrix} x_0 \\ y_0 \end{bmatrix} + J \begin{bmatrix} \xi \\ \eta \end{bmatrix}$$

where J is a 2×2 matrix. What are the vector components x_0, y_0, and the components of the matrix J?

(6) The inverse of this mapping can be written as

$$\begin{bmatrix} \xi \\ \eta \end{bmatrix} = J^{-1} \begin{bmatrix} x - x_0 \\ y - y_0 \end{bmatrix}.$$

This mapping can also be denoted as $\xi = \xi(x, y)$ and $\eta = \eta(x, y)$.

(7) The quadratic polynomial interpolant $p(x, y)$ of the function $e^{0.1xy}$ over the triangle T can be written as

$$p(x, y) = \sum_{k=1}^{6} a_k q_k(\xi(x, y), \eta(x, y)) \tag{5.10}$$

(also see the table at the beginning of this exercise). What are the values a_k for $k = 1, \ldots, 6$? **Hint:** The top of the triangle is the point $x = 3$ and $y = 4$. What are the right and left sides of (5.10) at that point?

(8) What is the value of $p(3.5, 2.5)$?

Chapter 6

Polynomial and piecewise linear interpolation

6.1 Introduction

You saw in Chapter 5 that the interpolating polynomial could get *worse* (in the sense that values at intermediate points are far from the function) as its degree increased. This means that your strategy of using equally spaced data for high degree polynomial interpolation is *a bad idea*. It turns out that equidistant spacing must always result in poor asymptotic convergence rates! See the excellent article [Platte *et al.* (2011)].

In this chapter, you will test other possible strategies for spacing the data, including the optimal Chebyshev strategy. These test will rely on polynomial interpolation functions from Chapter 5 and the Runge test function. You will also investigate piecewise linear interpolation and piecewise constant interpolation. Unlike polynomial interpolation, piecewise polynomial interpolation comes with a guarantee that the error will get smaller when more points are taken. Because the accuracy comparisons involve similar steps in each case, the approach used in Chapter 5 will first be codified in a general experimental framework.

Exercise 6.1 standardizes a way of comparing different choices of spacing strategies, with application in Exercises 6.2, 6.3, 6.7 and 6.8. Exercises 6.4 and 6.5 are purely programming exercises intended to introduce recursive MATLAB functions because students are not presumed to be familiar with recursive functions. Exercise 6.6 discusses Chebyshev polynomials and lay some groundwork for Exercises 6.7 and 6.8, although these two exercises can be completed without relying on results from Exercise 6.6. Exercises 6.9–6.12 form a sequence culminating in piecewise linear approximation for continuous functions. Exercise 6.13 presents two ways of estimating convergence rates of piecewise polynomial interpolants. Exercise 6.14 discusses piecewise constant interpolation for continuous functions, and Exercise 6.15 discusses convergence of the derivative of a piecewise linear interpolant to the derivative of a differentiable function. This latter exercise is intended for more advanced students.

6.2 An experimental framework

In the previous chapter, you compared several different methods of polynomial interpolation, and you used a common approach that standardized the comparisons. Part of this chapter will be to generate polynomial interpolants for a few different functions on different sets of points. You will be comparing the accuracy of the interpolating polynomials, just as you did last chapter. Instead of repeating basically the same code all the time, it is more convenient to automate the process in an m-file. The utility function in the next exercise is designed to test interpolation for different functions on different sets of points.

Remark 6.1. It can be very helpful to build a utility such as the one below to standardize the comparisons as well as to make them quick and easy. It is both error-prone and bothersome to repeat the same sequence of commands again and again.

Exercise 6.1. In this exercise, you will construct a utility function m-file that will take as input a function (handle) to interpolate and a set of points `xdata` on which to perform the interpolation. In the previous chapter, the points `xdata` were uniformly distributed but in this chapter you will investigate non-uniform distributions. The approximation error *between interpolation points* will be computed by picking a much larger number of test points than are used for interpolation and computing the maximum error on this larger set of test points. The function uses `eval_lagr.m` and `lagrangep.m` from Chapter 5. (It could equally well use `coef_vander.m` and `polyval`.) An outline of this function follows, and can also be found in the file `test_poly_interpolate.txt`.

```
function max_error=test_poly_interpolate(func,xdata )
  % max_error=test_poly_interpolate(func,xdata )
  % utility function used for testing polynomial interpolation
  % func is the function to be interpolated
  % xdata are abscissae at which interpolation takes place
  % max_error is the maximum difference between the function
  %   and its interpolant

  % your name and the date

  % Choose the number of the test points and generate them
  % Use 4001 because it is odd, capturing the interval midpoint
  NTEST=4001;
  % construct NTEST points evenly spaced so that
  % they cover the interpolation interval in a standard way, i.e.,
  % xval(1)=xdata(1) and xval(NTEST)=xdata(end)
  xval= ???
```

```
% we need the results of func at the points xdata to do the
% interpolation WARNING: this is a vector statement
% In a real problem, ydata would be "given" somehow, and
% a function would not be available
ydata=func(xdata);

% use Lagrange interpolation from Chapter 5 to do the interpolation
% WARNING: these use componentwise (vector) statements.
% Generate yval as interpolated values corresponding to xval
yval=eval_lagr( ???

% we will be comparing yval with the exact results of func at xval
% In a real problem, the exact results would not be available.
yexact= ???

% plot the exact and interpolated results on the same plot
% this gives assurance that everything is reasonable
plot( ???

% compute the error in a standard way.
max_error=max(abs(yexact-yval))/max(abs(yexact));
end
```

(1) Look at the code in `test_poly_interpolate.m` and complete the statements that have question marks in them.

(2) Test your `test_poly_interpolate.m` function by getting the same approximation values for the Runge example function $f(x) = 1/(1 + x^2)$ on the interval $[-\pi, \pi]$ with uniformly spaced points as you should have seen in last chapter and are reproduced here. Recover your `runge.m` file from last chapter or rewrite it. Do not forget to use componentwise (vector) syntax.

Runge function, evenly spaced points $[-\pi, \pi]$

ndata	Max Error
5	0.31327
11	0.58457
21	3.8607

(3) Construct `xdata` containing `ndata=5` evenly spaced points between -5 and $+5$ (*not* the same interval) and use the utility `test_poly_interpolate` to plot and evaluate the error between the interpolated polynomial and the exact values of the Runge example function, and fill in the following table.

Runge function, evenly spaced points $[-5, 5]$

ndata	Max Error
5	
11	
21	

After looking at the plotted comparisons between the Runge example function and its polynomial fit, you might guess that the poor fit is because the points are evenly spaced. Maybe they should be concentrated near the endpoints (so they don't oscillate wildly there) or near the center (maybe the oscillations are caused by too many points near the endpoints). Let's examine these two hypotheses.

The strategy used in the next exercise is to use a function to change the distribution of points. That is, pick a nonlinear function f that maps the interval $[-1, 1]$ into itself (you will use x^2 and $x^{1/2}$) and also pick an affine function g that maps the given interval, $[-5, 5]$ into $[-1, 1]$. Then use the points $x_k = g^{-1}(f(g(\tilde{x}_k))) = g^{-1} \circ f \circ g(\tilde{x}_k)$, where $\tilde{x}_k$ are uniformly distributed.

Exercise 6.2. You just were looking at evenly-spaced points in $[-5, 5]$. One way to concentrate these points near zero would be to recall that, for numbers less than one, $|x|^2 < |x|$. Hence, if `xdata` represents a sequence of numbers uniformly distributed between -5 and 5, then the expression `5*( sign(xdata).*abs(xdata./5).^2 )` yields a similar sequence of numbers concentrated near 0. (The `sign` and `abs` functions are used to get the signs correct for negative values.)

(1) Construct 11 points uniformly distributed between $[-5, 5]$ in a vector called `xdata`. Use the transformation

```
xdata=5*( sign(xdata).*abs(xdata./5).^2 )
```

Look at the concentrated distribution of points using the command

```
plot(xdata,zeros(size(xdata)),'*')
```

to plot the redistributed points along the x-axis. Your points should be concentrated near the center of the interval $[-5, 5]$ but with endpoints -5 and $+5$. Be sure that the endpoints are correct.

(2) Use this transformation and the `test_poly_interpolate` utility to fill in the following table for the Runge example function with `ndata` points concentrated near 0.

Runge function, points concentrated near 0

ndata	Max Error
5	
11	
21	

You should observe that the error is rising dramatically faster than in the uniformly distributed case.

Exercise 6.3. Points can be concentrated near the endpoints in a similar manner, by forcing them away from zero using the $|x|^{1/2} = \sqrt{|x|}$ function.

(1) Use a transformation similar to the one in the previous exercise but with $|x|^{1/2}$ with 11 points. Take care to get the signs of negative xdata correct.
(2) Plot your points using the following command.

```
plot(xdata,zeros(size(xdata)),'*')
```

(3) Use this transformation and the test_poly_interpolate utility to fill in the following table for the Runge example function with ndata points concentrated near the endpoints.

Runge function, points concentrated near endpoints

ndata	Max Error
5	
11	
21	

You should observe that the error is shrinking, in contrast with the uniformly distributed case.

You could ask why x^2 and $x^{1/2}$ were chosen, out of the many possibilities. Basically, it is a guess. It turns out, though, that there is a systematic way to pick optimally-distributed (Chebyshev) points. It is also true that the Chebyshev points are closely related to trigonometric interpolation (see Chapter 5).

6.3 Chebyshev polynomials

6.3.1 *Recursive functions*

In this section, a brief introduction to programming recursive functions is offered. As you will see, recursive functions can be very short and elegant when explicit formulas or loops are awkward and difficult to understand.

Consider the factorial function. Its mathematical definition can be written as

$$0! = 1$$
$$n! = n\left((n-1)!\right) \qquad \text{for } n \geq 1.$$

A function defined this way is termed "recursive" because it is defined in terms of itself. You have seen such definitions often in your theoretical work, but they can be confusing when you try to program them.

Here, you will see how to write recursive functions in MATLAB m-files. It can be shown that any function written in a recursive manner can be rewritten using loops; however, the recursive form of the function is generally shorter and easier to understand.

Exercise 6.4. The following code implements the factorial function. It will be named `rfactorial` because MATLAB already has the factorial function and it is called `factorial`.

(1) Copy or download the following code to a file named `rfactorial.m`.

```
function f=rfactorial(n)
  % f=rfactorial(n) computes n! recursively

  % your name and the date

  if n<0
    error('rfactorial:
    cannot compute factorial of negative integer.');
  elseif n==0  %double equals sign for logical testing
    f=1;
  else
    f=n*rfactorial(n-1);
  end
end
```

(2) Confirm that `rfactorial` gets the same result for 5! as the MATLAB `factorial` function.

(3) Suppose you are computing `rfactorial(3)`. When `rfactorial` starts up, the first thing it does is check if `n<0`. Since `n=3`, it goes on and discovers that before it can continue, it must start up a *new copy* of `rfactorial` and compute `rfactorial(2)`. In turn, this new copy checks that `2>1` and starts up *another new copy* of `rfactorial` so that there are now two copies of `rfactorial` waiting and one copy active. In your own words, explain how the computation continues to get a result of 6.

(4) If you attempt to use `rfactorial` for a negative integer, it will print an error message and stop because of the `error` function. What would have happened if the code did not test if `n<0`?

The factorial function is really very simple, and recursion is, perhaps, too powerful a method to employ. After all, the factorial can be written as a simple loop or, even more simply, as `prod(1:n)`. Consider the Fibonacci series

$$f_1 = 1$$
$$f_2 = 1$$
$$f_n = f_{n-1} + f_{n-2} \qquad \text{for } n > 2.$$
$$\text{(6.1)}$$

Values of f_n can also be computed using Binet's formula

$$f_n = \frac{(1 + \sqrt{5})^n - (1 - \sqrt{5})^n}{2^n \sqrt{5}}.$$

You can find more information about the Fibonacci series from the Wolfram (Mathematica) web site [Chandra and Weisstein (2020)].

Exercise 6.5.

(1) Write a MATLAB function named `fibonacci` that accepts a value of n and returns the element f_n of the Fibonacci series according to the prescription given in (6.1).
(2) Confirm that your function correctly computes the first few Fibonacci numbers: 1,1,2,3,5,8,13.
(3) Confirm that Binet's formula gives the approximately same value when `n=13` as your `fibonacci` gives.

The message you should be getting is that recursion is a powerful programming technique that allows you to do some things more elegantly than using loops (*e.g.*, `fibonacci`), and to do some things with essentially the same effort as using loops (*e.g.*, `factorial`). Although not presented in this chapter, there are examples where a recursive program is simply the wrong approach.

It is also true that some recursive programs (such as `factorial`) run just about as fast as their loop counterparts while others (such as `fibonacci`) can be *much* slower than the same thing written as a loop.

The Chebyshev polynomial of degree n is given by

$$T_n(x) = \cos(n \cos^{-1} x). \qquad \text{(6.2)}$$

This formula doesn't look like it generates a polynomial, but the trigonometric formulæ for sums of angles and the relation that $\sin^2 \theta + \cos^2 \theta = 1$ can be used to show that it does generate a polynomial.

These polynomials satisfy an orthogonality relationship and a three-term recurrence relationship

$$T_n = 2x T_{n-1}(x) - T_{n-2}(x), \text{ with } T_0(x) = 1 \text{ and } T_1(x) = x. \qquad \text{(6.3)}$$

The facts that make Chebyshev polynomials important for interpolation are

- The peaks and valleys of T_n are smallest of all polynomials of degree n over $[-1, 1]$ with $T_n(1) = 1$. (See [Quarteroni *et al.* (2007)], Property 10.2, [Atkinson (1978)], Theorem 4.09, page 222, or [Chebyshev].)
- On the interval $[-1, 1]$, each polynomial oscillates about zero with peaks and valleys all of equal magnitude. (See [Quarteroni *et al.* (2007)], Property 10.1, [Atkinson (1978)], Theorem 4.10, page 224.)

Thus, when $\{x_1, x_2, \ldots, x_n\}$ in (6.4) are chosen to be the roots of $T_n(x)$, then the bracketed expression in (6.4) is proportional to T_n, and the bracketed expression is minimized over all polynomials.

Before going on to use the Chebyshev points, it is a good idea to confirm that the two expressions (6.2) and (6.3) do, in fact, refer to the same polynomials. Mathematically speaking, the best way to approach the problem is to use the symbolic toolbox, which can produce the identity that can become the central portion of a proof. Instead, you will approach the problem numerically. This approach will yield a convincing example, but no proof.

Exercise 6.6.

(1) Write a MATLAB function with the signature and outline,

```
function tval=cheby_trig(xval,degree)
  % tval=cheby_trig(xval,degree)

  % your name and the date

  if nargin==1
    degree=7;
  end

  << your code >>

end
```

to evaluate $T_n(x)$ as defined using trigonometric functions in (6.2) above. If the argument `degree` is omitted, it defaults to 7, a fact that will be used below.

(2) Write a recursive MATLAB function with the signature and outline,

```
function tval=cheby_recurs(xval,degree)
  % tval=cheby_recurs(xval,degree)

  % your name and the date

  if nargin==1
    degree=7;
```

```
    end

    << your code >>

end
```

to evaluate $T_n(x)$ as defined recursively in (6.3) above. If the argument `degree` is omitted, it defaults to 7, a fact that will be used below.

(3) Show that `cheby_trig` and `cheby_recurs` agree for degree=4 (T_4) at the points x=[0,1,2,3,4] by computing the largest absolute value of the differences at those points. What is this value?

Remark 6.2. If two polynomials of degree 4 agree at 5 points, they agree everywhere. Hence *if (6.2) defines a polynomial,* then (6.2) and (6.3) produce identical values for T_4. This is why only five test points are required.

(4) Plot values of `cheby_trig` and `cheby_recurs` on the interval $[-1.1, 1.1]$ for degree=7 (T_7). Use at least 100 points for this plot in order to see the details. The two lines should lie on top of one another. You should also observe visually that the peaks and valleys of T_7 are of equal absolute value.
Suggestion: You can plot the first curve with a wider line than the second, as you did in Chapter 5, to see both lines.

(5) Visually choose change-of-sign intervals around the largest and second-largest roots in your plot of T_7. Using your `bisect.m` file from Chapter 2 (or use the MATLAB `fzero` function), find the largest and second-largest roots of $T_7(x) = 0$ on the interval $[-1.1, 1.1]$.

6.4 Chebyshev points

If you have no idea what function is generating the data, but you are allowed to pick the locations of the data points beforehand, the Chebyshev points are the smartest choice. (There is more on Chebyshev points and interpolation in [Quarteroni *et al.* (2007)] in Sections 10.1–10.3, and in [Atkinson (1978)] on page 228.)

The approximation error between a function f and its polynomial interpolant p at any point x is given by an expression of the form:

$$f(x) - p(x) = \left[\frac{(x - x_1)(x - x_2) \ldots (x - x_n)}{n!} \right] f^n(\xi) \qquad (6.4)$$

where ξ is an unknown nearby point. This is something like an error term for a Taylor's series.

You can't do a thing about the expression $f^n(\xi)$, because f is an arbitrary (smooth enough) function, but you *can* affect the magnitude of the bracketed expression. For instance, if you are only using a single data point ($n = 1$) in the interval [10,20], then the very best choice for the data point would be $x_1 = 15$, because then the maximum absolute value of the expression

$$\left[\frac{(x - x_1)}{1!}\right]$$

would be 5. The worst value would be to choose x at one of the endpoints, in which case the maximum value would be doubled. This doesn't guarantee better results, but it improves the chance.

The Chebyshev points are the zeros of the Chebyshev polynomials. They can be found from (6.2):

$$\cos(n \cos^{-1} x) = 0,$$
$$n \cos^{-1} x = (2k - 1)\pi/2,$$
$$\cos^{-1} x = (2k - 1)\pi/(2n),$$
$$x = \cos\left(\frac{(2k - 1)\pi}{2n}\right).$$

For a given number n of data points, then, the Chebyshev points on the interval $[a, b]$ can be constructed in the following way.

(1) Pick equally-spaced angles $\theta_k = (2k - 1)\pi/(2n)$, for $k = 1, 2, \ldots, n$.
(2) The Chebyshev points on the interval $[a, b]$ are given as

$$x_k = 0.5(a + b + (a - b) \cos \theta_k),$$

for $k = 1, 2, \ldots, n$.

Exercise 6.7.

(1) Write a MATLAB function m-file called `cheby_points.m` that returns in the vector `xdata` the values of the `ndata` Chebyshev points in the interval `[a,b]`. You can do this in 3 lines using vector notation. You can find the following outline in the file `cheby_points.txt`.

```
function xdata = cheby_points ( a, b, ndata )
  % xdata = cheby_points ( a, b, ndata )
  % more comments

  % your name and the date

  k = (1:ndata);  %vector, NOT A LOOP
  theta = << vector expression involving k >>
  xdata = << vector expression involving theta >>
end
```

or you can use a `for` loop. (The vector notation is more compact and runs faster, but is a little harder to understand.)

(2) To check your work, use `cheby_points` to find the Chebyshev points on the interval $[-1, 1]$ for `ndata=7`. Do the largest and second largest roots agree with your roots of T_7 computed above?

(3) What are the five Chebyshev points for `ndata=5` and `[a,b]=[-5,5]`?

(4) You should observe that the Chebyshev points are not uniformly distributed on the interval but they are *symmetrical about its center*. It is easy to see from (6.2) that this fact is true in general.

(5) Repeat the comparisons in Exercises 6.2 and 6.3, but with Chebyshev points on $[-5, 5]$. Fill in the table. (Note the extra row!) You should see smaller errors than in either of the previous exercises, especially for larger `ndata`.

Runge function, Chebyshev points

ndata	Max Error
5	
11	
21	
41	

In the following exercise, you will *numerically* examine the hypothesis that the Chebyshev points are the best possible set of interpolation points. You will do this by running many tests, each one using a (pseudo-) randomly-generated set of interpolation points. If the theory is correct, none of these tests should result in an error smaller than the error obtained using the Chebyshev points.

Exercise 6.8.

(1) Because you will be running a large number of cases, remove or comment out the plotting statements in `test_poly_interpolate`.

(2) The MATLAB `rand` function generates a matrix filled with (pseudo-) randomly generated numbers uniformly distributed in the interval $[0, 1]$. What does the following statement do?

```
xdata=[-5,sort(10*(rand(1,19)-.5)),5];
```

(3) If you execute the previous command twice, will you get the same values for `xdata`?

(4) Use the following loop to test a number of different sets of 21 interpolation points. This loop should take much less than a minute. (You can copy this code from `exer8.txt`.)

```
for k=1:500
  xdata=[-5, sort(10*(rand(1,19)-.5)), 5];
  err(k)=test_poly_interpolate(@runge,xdata);
end
```

(5) What are the largest and smallest observed values of `err`? How do they compare with the error you found in the previous exercise for 21 Chebyshev points?

Remark 6.3. It is, in fact, possible for a randomly-generated set of points to yield a smaller error than the Chebyshev points when computed using `test_poly_interpolate`. The Chebyshev points are not necessarily optimal for the Runge example function–they are optimal over the set of all smooth functions. It is also possible because `test_poly_interpolate` computes the error at only 4001 points.

Remark 6.4. The larger the number of trials, the more rigorous this kind of testing becomes. It is never, of course, a proof, but it can be a way of discovering a counterexample.

The topic now switches from interpolation using a single polynomial to interpolation using different polynomials on different subintervals. This "piecewise" interpolation is a much better strategy than using single polynomials for most applications.

6.5   Piecewise linear interpolation

6.5.1   *Bracketing*

The following section presents interpolation using *piecewise linear* functions, instead of polynomial functions. That means that, in order to evaluate the interpolating function you first must know in which of the intervals (pieces) x lies, and then compute the value of the function depending on the endpoints of the interval. This section addresses the issue of how to find the left and right end points of the subinterval on which x lies.

You will write a *utility routine* to perform this task. This utility routine must be correct, because you need to be able to rely on it. As usual, it's one of those things that's very easy to describe, and not quite so easy to program.

Suppose you are given a set of N abscissæ, in increasing order

$$x_1 < x_2 < \cdots < x_N$$

with functional values y_n for $n = 1, 2, \ldots, N$ that together define a piecewise linear function $\ell(x)$. In order to evaluate $\ell(x)$ for a given x, you need to know the index, n, so that $x \in [x_n, x_{n+1}]$. This situation is illustrated below, with $N = 4$, and $x \in [x_2, x_3]$. so $n = 2$.

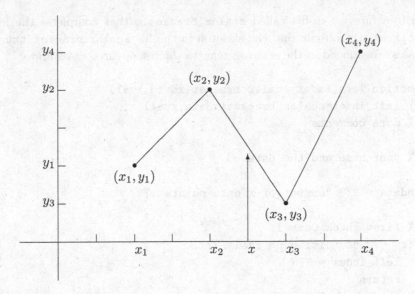

In MATLAB notation, this amounts to the following discussion. Denote N by **ndata** and the x_n by the MATLAB vector **xdata**. Assume that the components of **xdata** are strictly increasing. Now suppose that you are given a value **xval** and you are asked to find an integer named **left_index** so that the subinterval [**xdata(left_index)**, **xdata(left_index+1)**] contains **xval**. The index **left_index** is that of the left end point of the subinterval, and the index (**left_index+1**) is that of its right end point. (The term *index* in a computer program generally means the same thing as *subscript* in a mathematical expression.)

The following clarifications are needed. You seek an integer value **left_index** so that one of the following cases holds.

Case 1 If **xval** is less than **xdata(1)**, then regard **xval** as in the first interval and **left_index=1**; or

Case 2 If **xval** is greater than or equal to **xdata(ndata)**, then regard **xval** as in the final interval and **left_index=ndata-1** *(read that value carefully)*; or,

Case 3 If **xval** satisfies **xdata(k)<=xval** and **xval<xdata(k+1)**, then **left_index=k**.

It would be desirable to write a MATLAB function to perform this task even when **xval** is a vector of values, but it is not easy to do it efficiently. Instead, in the following exercise, you will write a function **scalar_bracket** to do it for the case **xval** is a scalar, not a vector. In Exercise 6.10, you will see how it can be written for vectors.

Exercise 6.9. In this exercise, you will be writing a MATLAB function m-file called **scalar_bracket.m** and test it thoroughly. Later, you will see how to write it for vectors.

(1) Write a function m-file called `scalar_bracket.m` that completes the following skeleton. You can find the skeleton in the file `scalar_bracket.txt`. The "cases" mentioned in the skeleton refer to the list of three cases above.

```
function left_index=scalar_bracket(xdata,xval)
  % left_index=scalar_bracket(xdata,xval)
  % more comments

  % your name and the date

  ndata = ??? "number of x data points" ???

  % first check Case 1
  if  ??? "condition on xval" ???
    left_index = ???
    return
  % then check Case 2
  elseif ??? "condition on xval" ???
    left_index = ???
    return
  % finally check Case 3
  else
    for k = 1:ndata-1
      if ??? "condition on xval" ???
        left_index = ???
        return
      end
    end
    error('Scalar_bracket: this cannot happen!')
  end
end
```

(2) Make up a set of at least five values for `xdata` (don't forget to make them ascending!), and a set of at least seven test values `xval`, and test `scalar_bracket` with the values you have selected, one at a time. Make the tests as comprehensive as you can. Your tests should include:

- One point to the left of all `xdata`.
- One point to the right of all `xdata`.
- One point equal to one of the `xdata` points in the interior.
- One point not equal to any `xdata` but in the interior.

Remark 6.5. The call to the MATLAB `error` function may seem superfluous to you because, indeed, it can never be executed. It is a powerful debugging tool, however. If you make an error in the cases, it might turn out that the error message *does* get executed. In that case, you are alerted to the error immediately instead of having an incorrect value of `left_index` cause an obscure error hundreds of statements later. Anything that saves debugging time is worthwhile.

Exercise 6.10. In this exercise, you will see how the same task can be accomplished using vector statements. The code presented below is written to be as efficient as possible for the case of very large vectors `xval`.

(1) Copy the following code to a file named `bracket.m`, or download it.

```
function left_indices=bracket(xdata,xval)
  % left_indices=bracket(xdata,xval)
  % more comments

  % your name and the date

  ndata=numel(xdata);

  left_indices=zeros(size(xval));
  % Case 1
  left_indices( find( xval<xdata(2) )  )=1;
  % Case 2
  left_indices( find( xdata(ndata-1)<=xval )  )=ndata-1;
  % Case 3
  for k=2:ndata-2
    left_indices(find( (xdata(k)<=xval) & (xval<xdata(k+1)) ))=k;
  end
  if any(left_indices==0)
    error('bracket: not all indices set!')
  end
end
```

(2) Suppose that the letters `a,b,c,...,z` represent an ascending set of numbers. If `xval=[b,m,f,g,h,a,q]`, what would the value of `indices` be as a result of the following statement?

```
indices = find( (g <= xval) & (xval < p) )
```

(3) What would the value of `left_indices` be after the following statements, assuming as above that `xval=[b,m,f,g,h,a,q]`,

```
left_indices=zeros(size(xval));
left_indices(find( (g<=xval) & (xval<p) ))=4
```

(4) Explain the meaning of the statement beginning "`if any.`" Explain why it indicates that "not all indices set."

(5) The loop in `bracket` has `k=2:ndata-2`, but the loop in `scalar_bracket` has `k=1:ndata-1`. Why do the two functions give the same results?

(6) Add a descriptive set of comments to the code.

(7) Use the *same* sets of values for `xdata` and `xval` as you used for `scalar_bracket.m` in the previous exercise. Summarize your results in a table such as this one.

xdata	xval	left_indices (scalar)	left_indices (vector)
—			
—			

Be sure that you get the same set of values for `left_indices` in the scalar and vector cases.

Now you are ready to consider *piecewise linear interpolation*. The idea is that your interpolating function is not going to be a smooth polynomial defined by a formula. Instead, it will be defined by piecing together linear interpolants that go through each consecutive pair of data points. As such, it will be continuous but not necessarily differentiable. To handle values below the first data point, extend the first linear interpolant all the way to the left. Similarly, the linear function is extended in the last interval all the way to the right.

The graph of a piecewise linear function may not be smooth, but one thing is certain: it will *never* oscillate between the data values. And the interpolation error is probably going to involve the size of the intervals, and the norm of the second derivative of the function, both of which are usually easy to estimate. As the number of points gets bigger, the interval size will get smaller, and the norm of the second derivative won't change, so you have good reason assert a well-known convergence result:

Theorem 6.1. *Given an interval* $[a, b]$ *and a function* $f(x)$ *with a continuous second derivative over that interval, and any sequence* $\lambda_n(x)$ *of linear interpolants to* f *with* *the property that h_{max}, the size of the maximum interval, goes to zero as n increases, then the linear interpolants converge to f both pointwise, and uniformly.*

Proof. If C is a bound for the maximum absolute value of the second derivative of the function over the interval, then Taylor's theorem says that the pointwise interpolation error and the L^∞ norm are bounded by Ch_{max}^2, while the L^1 norm is bounded by $(b-a)Ch_{max}^2$, and L^2 norm is bounded by $(\sqrt{b-a})Ch_{max}^2$. $\qquad\square$

Uniform convergence is convergence in the L^∞ norm, and is a much stronger result than pointwise convergence.

Thus, if the only thing you know is that your function has a bounded second derivative on $[a, b]$, then you are guaranteed that the maximum interpolation error you make decreases to zero as you make your intervals smaller.

Remark 6.6. The convergence result for piecewise linear interpolation is so easy to get. Is it really better than polynomial interpolation? Why are there such problems with polynomial interpolation? One reason is that the error result for polynomial interpolation couldn't be turned into a convergence result. Using 10 data points, the error estimate is in terms of the 10th derivative. Then using 20 points, the error estimate is in terms of the 20th derivative. These two quantities can't be compared easily, and for nonpolynomial functions, successively higher derivatives can easily get successively and endlessly *bigger* in norm. The linear interpolation error bound involved a particular *fixed* derivative and held that fixed, so then it was easy to drive the error to zero because the other factors in the error term depended only on interval size.

To evaluate a piecewise linear interpolant function at a point xval, you need to:

(1) Determine the interval (xdata(left_index),xdata(left_index+1)) containing xval;
(2) Determine the equation of the line that passes through the points (xdata(left_index),ydata(left_index)) and (xdata(left_index+1),ydata(left_index+1));
(3) Evaluate that line at xval.

The bracket.m function from the previous exercise does the first of these tasks, but what of the second and third?

Exercise 6.11. This exercise involves writing a MATLAB function m-file called eval_plin.m (**eval**uate **p**iecewise **lin**ear interpolation) to do steps 2 and 3 above.

(1) Write a file called eval_plin.m:
  (a) Start with the outline

```
function yval = eval_plin ( xdata, ydata, xval )
  % yval = eval_plin ( xdata, ydata, xval )
```

```
% more comments

% your name and the date
end
```

Add appropriate comments.

(b) Add lines that use `bracket.m` to find the vector of indices in `xdata` identifying the intervals that each of the values in `xval` lies. (It is possible to do this in one line of code.)

(c) Write down a general formula for the linear function through an arbitrary pair of points. Include this formula among the comments.

(d) Add lines evaluating this formula on the interval you found in step (b). (It is possible to do this in one line of code.)

(2) Test correctness of `eval_plin.m` on the piecewise linear function $f(x) = 3|x|+1$ on the interval $[-2, 2]$ using the following sequence of commands, based on the `test_poly_interpolate` function. You can find these commands in `exer11.m`

```
xdata=linspace(-1,1,7);
ydata=3*abs(xdata)+1;  % the function y=3*|x|+1
plot(xdata,ydata,'*');
hold on
xval=linspace(-2,2,4001);
plot(xval,eval_plin(xdata,ydata,xval));
hold off
```

Your plot should pass through the five data points and consist of two straight lines forming a "V".

A piecewise linear function was chosen for testing in the last part of this exercise for both theoretical and practical reasons. In the first place, you are fitting a piecewise linear function to a function that is already piecewise linear and both the original and the fitted functions have breaks at $x = 0$, so the two functions will agree. This makes it easy to see that the interpolation results are correct. In the second place, it is unlikely you will get it right for this piecewise linear data but have it wrong for other data.

Exercise 6.12. In this exercise, you will examine convergence of piecewise linear interpolants to the Runge example function you have considered before.

(1) Write a function m-file `test_plin_interpolate.m` to do piecewise linear interpolation by following the steps below.

(a) Copy the file `test_poly_interpolate.m` to a file called `test_plin_interpolate.m`. Change the signature and outline to be

```
function max_error=test_plin_interpolate(func,xdata)
% max_error=test_plin_interpolate(func,xdata)
% more comments

% your name and the date
```

and change from using polynomial interpolation to piecewise linear interpolation using eval_plin. Do not forget to change to comments to reflect your coding changes.

(b) Restore the plot statement, if it is not active.

(2) Consider the same Runge example function you used above. Over the interval $[-5, 5]$, use ndata=5 equally spaced data points, use test_plin_interpolate to plot the interpolant and evaluate the interpolation error.

(3) Repeat the evaluation for progressively finer meshes and fill in the following table. You do not need to send me the plots that are generated. **Warning:** The final few of these tests should take less than a few seconds, but if you have written code that happens to be inefficient they can take much longer.

Runge function, Piecewise linear, uniformly-spaced points	
ndata	Maximum Error
5	
11	
21	
41	
81	
161	
321	
641	

(4) Repeat the above convergence study using Chebyshev points instead of uniformly distributed points. You should observe no particular advantage to using Chebyshev points.

Runge function, Piecewise linear, Chebyshev points	
ndata	Maximum Error
5	
11	
21	
41	
81	
161	
321	
641	

The errors in the table decrease as `ndata` increases, as they should. But decreasing error is only half the story: the *rate* that the errors decrease is also important. You know from the lectures that, generally, errors tend to be bounded by Ch^p where C is a constant, often related to a derivative of the function being interpolated, h is approximately proportional to the subinterval size (in the uniform case, $h = 1/\text{ndata}$), and p is an integer, often a small integer.

One way to estimate the rate errors decrease is to plot log(error) versus $\log h$. From the general bound, $\log(\text{error}) \approx \log C + p \log h$ for small enough h, so log(error) versus $\log h$ should yield a curve that becomes straight as h becomes small. Furthermore, plotting log(error) versus $\log h$ is the same thing as plotting error versus h as a log-log plot.

It is possible to visually estimate the error from a log-log plot. To do so, simply plot lines with known slope q until you see one that is roughly parallel to the error plot. You will try this in the following exercise. You could also pick points on the plot and estimate the slope of the line directly, but it is easier to use visual comparison.

Another way to estimate the rate of error decrease is to successively halve the subinterval lengths. You may note that the `ndata` values chosen above result in subintervals halving in length. If the error is roughly proportional to Ch^p, halving h should result in reducing the error by $(1/2)^p$. Thus, |`Max Error(321)`|/|`Max Error(641)`| should be roughly 2^p for some integer p.

Exercise 6.13. This exercise investigates the behavior of the errors in the table (for uniformly-spaced points) in the previous exercise.

(1) Plot the values of `Max Error` *vs.* $h =$(10/`ndata`) (h is approximately proportional to interval size) in the first table in Exercise 6.12 (uniform points) using a log-log plot (the MATLAB function `loglog` is used just like `plot`, but results in a log-log plot). Your points should be roughly a straight line on the plot, especially for larger values of `ndata`.

(2) Visually estimate the slope of the line in the following manner.

  (a) Choose a value of C_3 so that the line $y = C_3 h^3$ passes through the point `h=10/641`, `y=Max Error(641)`.

  (b) With this value of C_3, plot the line $y = C_3 h^3$ on the same log-log plot as the error plot above (use `hold on`). If you have computed C_3 correctly, the two plots should pass through the same point at `h=10/641`.

  (c) Do the same thing for $y = C_2 h^2$.

  (d) Do the same thing for $y = C_1 h$.

  (e) Which of the values p best approximates the slope of the error curve, p=3, p=2, or p=1?

(3) Compute the ratios `Max Error(5)/Max Error(11)`, `Max Error(11)/Max Error(21)`, `Max Error(21)/Max Error(41)`, *etc.* Do they appear to approach 2^p for some integer p? If so, what is your estimate of the value of p?

Remark 6.7. When computing ratios in order to estimate rates of convergence, it is best to choose the ratios with the larger number in the numerator and the smaller number in the denominator. That way, the ratios approach integers. It easier to recognize numbers such as 15.5 as being nearly $2^4 = 16$ than to recognize 0.06452 as being nearly $2^{-4} = 0.0625$.

6.6 Interpolation with piecewise constants

This exercise involves interpolation by piecewise constant functions. It is important to recall that a piecewise constant function is not continuous! The plot of such a function is a sequence of horizontal lines with jumps between them. While it may seem strange to interpolate continuous functions with discontinuous functions, there is no theoretical reason not to do so, and there may be some advantage. For example, the definition of the integral of a function using rectangles that you saw in your undergraduate years amounts to interpolating the function by a piecewise constant function and then finding the integral of the interpolant as a sum of areas of rectangles.

For simplicity, only interpolating continuous functions will be considered.

Given a continuous function $f(x)$, its piecewise constant interpolant $I(x)$ over the interval $x_\ell \leq x < x_{\ell+1}$ is simply $I(x) = .5(f(x_\ell) + f(x_{\ell+1}))$.

Exercise 6.14.

(1) Write a MATLAB function `eval_pconst.m` to evaluate the piecewise constant interpolant of a given continuous function. This function can be based on `eval_plin.m`.
(2) Write a MATLAB function `test_pconst_interpolate.m` similar to `test_plin_interpolate.m`.
(3) Test your code by plotting the piecewise interpolant of the function $f(x) = x$ over the interval $[-5, 5]$ using 11 evenly-spaced points. You should observe that the piecewise constant interpolant agrees with $f(x)$ at the interval midpoints $-4.5, -3.5, -2.5, \ldots 4.5$.
(4) Perform a convergence study by filling the following table using the Runge example function.

Runge function, Piecewise constant	
ndata	Maximum Error
5	
11	
21	
41	
81	
161	
321	
641	

(5) Using either the plot method (first two parts of Exercise 6.13) or the ratio method (third part of Exercise 6.13), estimate the rate of convergence of the interpolant to Runge example as the number of points becomes large. You should observe a slower convergence rate than for piecewise linear interpolation.

6.7   Approximating the derivative, too

It may seem likely that if you have a good approximation of a differentiable function then you could differentiate it to get a good approximation of its derivative. On second thought, you have seen how the Runge example results in wiggles when it is approximated by polynomials, so polynomials probably cannot be used to approximate both a function and its derivative.

Standard theorems, for example, [Quarteroni *et al.* (2007)], Theorem 8.3, page 347, say that a twice-differentiable function can be approximated to $O(h^2)$ by piecewise linear functions (as you saw above in Exercises 6.12–6.13), *and* the derivative of the approximation approximates the derivative of the function to $O(h)$. In the following exercise, you will confirm this for the case of the Runge example function.

Exercise 6.15.

(1) Modify your eval_plin.m from Exercise 6.12 so that it returns *both* yval and y1val, where y1val approximates the derivative of yval. (yval comes from a piecewise linear expression. y1val should come from the derivative of that piecewise linear expression.)

(2) Write another MATLAB function test_plin1_interpolate.m that is similar to test_plin_interpolate.m, and that interpolates the given function (func), but plots and measures accuracy of the *derivative* of the function.

(3) Test test_plin1_interpolate.m on the function $f(x) = 3|x|+1$ on the interval $[-2, 2]$. (Recall this is a test function used in Exercise 6.11.) Explain why you believe your result is correct.

(4) Using either the plot method (first two parts of Exercise 6.13) or the ratio method (third part of Exercise 6.13), estimate the rate of convergence of the derivative of the interpolant to the derivative of Runge's example function on the interval $[-5, 5]$ as the number of points becomes large. You should observe a slower rate of convergence for the derivative of Runge's example function than for Runge's example function itself (in Exercise 6.12). Explain your methodology.

———————————————

Chapter 7

Higher-order interpolation

7.1 Introduction

Piecewise linear interpolation has many good properties. In particular, if the data come from a continuously differentiable function $f(x)$ and if the data points are suitably spread throughout the closed interval, then the interpolant converges to the function. The resulting interpolant consists of straight line segments and so has flats and kinks in it, making it inappropriate for many applications.

In this chapter you will look at several examples of piecewise polynomial interpolation with continuous derivatives from place to place. You will be looking first at piecewise Hermite cubic interpolation.

In this chapter, Exercises 7.1 through 7.4, introduce parametric interpolation. Exercise 7.1 requires `eval_plin.m` and `bracket.m` and 7.3 requires `test_plin_interpolate.m` from Chapter 6. Exercises 7.5 through 7.9 present an application of parametric interpolation in generating quadrilateral meshes for curved objects. These exercises should be relevant for students interested in numerical partial differential equations for engineering applications. Exercises 7.10 through 7.15 discuss spline interpolation, a more traditional topic. Exercises 7.13 through 7.15 employ only MATLAB interpolation functions, not spline functions the student wrote in earlier exercises.

7.2 Parametric interpolation

You are familiar with a *parameterized* curve, in which a special variable, often called t or s, is used to draw objects for which the relationship between x and y may not be functional. For example, one simple definition of a circle is:

$$\begin{aligned} x &= \cos t \\ y &= \sin t \end{aligned} \tag{7.1}$$

for $0 \leq t \leq 2\pi$. Writing the circle this way involves regarding both x and y as functions of the independent variable t. (There is no way to write y as a single function of x, because taking square roots introduces "+" and "−" options.)

129

Suppose you want to do interpolation of some complicated curve? If you have a drawing of the curve, you could simply mark points along the curve that are roughly evenly spaced, and make tables of `tdata`, `xdata` and `ydata`, where the values of `tdata` could be 1, 2, 3, ...

Now if you want to compute intermediate points on the curve, that's really saying, for a given value of `tdata`, what would the corresponding values of `xdata` and `ydata` be? This means doing two interpolations, one for `xdata` as a function of `tdata` and the second for `ydata` as a function of `tdata`.

In the first exercise, you will be dealing with a slightly more complicated curve than a circle. You may recall that a cardioid is a roughly heart-shaped closed curve, but without a sharp point at the bottom. You can find a nice description of cardioid curves with various ways of defining them in the web page [Weisstein (2020a)]. That article also remarks that a cardioid is a special case of a limaçon, and parametric equations for a limaçon can be given by

$$
\begin{aligned}
r &= r_0 + \cos t \\
x &= r \cos t \\
y &= r \sin t.
\end{aligned}
\tag{7.2}
$$

Exercise 7.1. Try the following method for plotting a limaçon. Use `eval_plin.m` and `bracket.m` from Exercises 6.10, and 6.11. Copy the following code to a script m-file named `exer1.m` and use it to generate the plot. You can find `exer1.m` among the download files for this Chapter.

```
% generate the data points for a limacon
numIntervals = 20;
tdata = linspace ( 0, 2*pi, numIntervals + 1 );
r = 1.15 + cos ( tdata );
xdata = r .* cos ( tdata );
ydata = r .* sin ( tdata );

% interpolate them
tval = linspace ( 0, 2*pi, 10*(numIntervals+1) );
xval = eval_plin ( tdata, xdata, tval );
yval = eval_plin ( tdata, ydata, tval );

% plot them with correct aspect ratio
plot ( xval, yval )
axis equal
```

You can see that the plot of the limaçon in Exercise 1 is not very smooth. This is because the line segments are straight lines that meet at corners. In the following section you will see one way of interpolating curves using curved sections whose derivatives are continuous so that there are no corners.

7.3   Cubic Hermite interpolation

Hermite interpolation is discussed by [Quarteroni *et al.* (2007)] in Section 8.5, [Atkinson (1978)] in Section 3.6, and [Ralston and Rabinowitz (2001)] in Section 3.7.

If you were trying to design, say, the shape of the sheet metal pattern for a car door, kinks and corners would not be acceptable. If that's a problem, then perhaps you could try to make sure that our interpolant is smoother by matching *both* the function value and the derivative at each point. *(This assumes you can get the derivative value.)*

Suppose you have points x_k, $k = 1, 2, \ldots, N$ and a differentiable function $f(x)$, so that the values $y_k = f(x_k)$ and $y'_k = f'(x_k)$ can be regarded as data points.

Now consider the situation in the kth interval, $[x_k, x_{k+1}]$. There is a unique cubic polynomial on this interval that takes on the function and derivative values y_k and y'_k at the two endpoints. You may recall that Lagrange interpolation polynomials could be used to write `eval_plin` in the previous chapter. On the interval $[x_k, x_{k+1}]$ there are two special linear polynomials: $\ell_0(x)$ is 1 at x_k and zero at x_{k+1}, and $\ell_1(x)$ is 1 at x_{k+1} and zero at x_k. This feature makes it easy to construct an arbitrary linear polynomial that matches the values y_k and y_{k+1} at the endpoints as $p_{\text{linear}} = y_k \ell_0(x) + y_{k+1} \ell_1(x)$. For this chapter, matching *both* the function *and* derivative values at the endpoints is important, so cubic Hermite polynomials are needed instead of linear ones.

Consider the four Hermite polynomials

$$h_1(x) = \frac{(x - x_{k+1})^2(3x_k - x_{k+1} - 2x)}{(x_k - x_{k+1})^3}$$

$$h_2(x) = \frac{(x - x_k)(x - x_{k+1})^2}{(x_k - x_{k+1})^2}$$

$$h_3(x) = \frac{(x - x_k)^2(3x_{k+1} - x_k - 2x)}{(x_{k+1} - x_k)^3}$$

$$h_4(x) = \frac{(x - x_{k+1})(x - x_k)^2}{(x_{k+1} - x_k)^2}$$

These four cubic polynomials satisfy the following equalities, that are similar to ones in [Quarteroni *et al.* (2007)], page 349, Equation (3.6.11) in [Atkinson (1978)] and Equations (3.7-7) in [Ralston and Rabinowitz (2001)].

	x_k	x_{k+1}
$h_1(x)$	1	0
$h_1'(x)$	0	0
$h_2(x)$	0	0
$h_2'(x)$	1	0
$h_3(x)$	0	1
$h_3'(x)$	0	0
$h_4(x)$	0	0
$h_4'(x)$	0	1

$$(7.3)$$

(This table means that, for example, $h_2(x_k) = 0$, $h_2'(x_k) = 1$, $h_2(x_{k+1}) = 0$, and $h_2'(x_{k+1}) = 0$.) If you know Mathematica, Maple or the MATLAB symbolic toolbox, you can check these equalities easily enough. For the purpose of this chapter, the easiest one to check is h_2, so check it by hand. (Hint: Do not multiply the factors out! You can check by staring at the formula long enough.)

Given these four polynomials, the unique polynomial with the values

$$y_k = f(x_k)$$
$$y_k' = f'(x_k)$$
$$y_{k+1} = f(x_{k+1})$$
$$y_{k+1}' = f'(x_{k+1})$$

can be written as

$$p(x) = y_k h_1(x) + y_k' h_2(x) + y_{k+1} h_3(x) + y_{k+1}' h_4(x) \qquad (7.4)$$

in the interval $[x_k, x_{k+1}]$[1]. If (7.4) is used at two adjacent intervals $[x_{k-1}, x_k]$ and $[x_k, x_{k+1}]$, It is easy to see that both p and p' are continuous at the points $x = x_k$ so that p is C^1 on the combined interval $[x_{k-1}, x_{k+1}]$.

You are now in a position to write a piecewise Hermite interpolation routine. This routine will be similar to `eval_plin.m`.

Exercise 7.2.

(1) Write `eval_pherm.m` by following the following steps.

    (a) Begin a MATLAB function m-file called `eval_pherm.m` with the signature and outline

```
function yval = eval_pherm ( xdata, ydata, ypdata, xval )
  % yval = eval_pherm ( xdata, ydata, ypdata, xval )
  % comments

  % your name and the date
```

---

[1] The interval given by $x_k \leq x \leq x_{k+1}$.

```
        << your code >>

    end
```

and insert comments. The values in `ypdata` are the derivative values at the points `xdata`.

(b) Use the `bracket` function to find the interval in which `xval` lies. As in `eval_plin.m`, this can be done in a single line.

(c) Evaluate the four Hermite functions, $h_1(x)$, $h_2(x)$, $h_3(x)$, and $h_4(x)$ at `xval`. Use componentwise operations if you can or use loops.

(d) Complete `eval_pherm` with the evaluation of `yval` using the expression (7.4) for the Hermite interpolating polynomial.

(2) Check your work by using `eval_pherm` four times, once for each of the polynomials $y = 1$, $y = x$, $y = x^2$, and $y = x^3$ interpolated for the interval `xdata=[0,2]`. Each of these four interpolants should be checked at four points of your choice. You can choose integers for these points, either inside or outside `[0,2]` because points outside `[0,2]` are evaluated using the same formula as those inside but do not choose the points $x = 0$ or $x = 2$. Recall that four points uniquely determine a cubic polynomial, so if you get agreement within roundoff at four points, you know your code is correct.

If your code is not correct, you can debug in the following way. You only need to do the following steps if you are debugging.

(a) For this test problem, there is only one interval, so the result of `bracket` should be the vector `ones(size(xval))`. You can use the debugger to confirm this fact. This eliminates `bracket.m` as the source of the bug.

(b) If you are using elementwise operations, make sure all your multiplications and (especially) divisions are preceded by a dot.

(c) Try the four polynomials $y = 1$, $y = x$, $y = x^2$, and $y = x^3$ on the interval `xdata=[0,1]` (not `[0,2]`) one at a time. The reason to change the interval to `[0,1]` is that the length of this interval is 1 so the denominators in the definitions of the h_k polynomials all become 1. If all four polynomials work but `eval_pherm` does not, odds are you have an error in the denominators of one or more of the $h_k(x)$.

(d) If you cannot get agreement for the polynomials $y = 1$, $y = x$, *etc.*, on `[0,1]`, then one or more of your $h_k(x)$ is probably wrong. Use `eval_pherm` to reproduce each of the $h_k(x)$ for `xdata=[0,1]`. The table (7.3) above provides `ydata` and `ypdata` values. Check your results at the points $x = 0$ and $x = 1$ first, and then for the points $x = 0.25$ and $x = 0.5$ (you will have to use the formulæ to get the data values. If these all agree but `eval_pherm` still does not work correctly, then your implementation of (7.4) is incorrect.

(e) If you have completed the previous two steps and fixed all bugs you found, but you still cannot interpolate the monomials $y = 1$, $y = x$, *etc.*, on $[0,2]$, then one or more of your $h_k(x)$ is still wrong. Try interpolating each of the four Hermite polynomials on `xdata=[0,2]`. When you are finished you will surely have found your bugs.

Now, you will use `eval_pherm.m` to see how piecewise Hermite converges.

Exercise 7.3.

(1) First, modify your MATLAB m-file `runge.m` so that it returns both the Runge function $y = 1/(1 + x^2)$ and its derivative y'. You should double-check that the derivative expression is correct. The signature and outline of the function should be

```
function [y,yprime]= runge ( x )
 % [y,yprime]= runge ( x )
 % comments

 << your code >>

end
```

and it should have appropriate comments. Do not forget to use componentwise syntax.

(2) Recover your `test_plin_interpolate.m` function from Chapter 6. Rename it to `test_pherm_interpolate.m` and modify it to use `eval_pherm.m`.

(3) Using equally spaced data points from 0 to 5 for `xdata` and the `runge.m` function to compute `ydata` and `ypdata`. Estimate the maximum interpolation error by computing the maximum relative difference (infinity norm) between the exact and interpolated values at 4001 evenly spaced points. Fill in the following table, where the column "ratio" is the error on the preceeding line divided by the error on the current line.

Runge function, Piecewise Hermite Cubic		
ndata	error	ratio
5		—
11		
21		
41		
81		
161		
321		
641		

(4) Estimate the rate of convergence by examining the ratios `Err(41)/Err(81)`, *etc.*, and estimating the nearest power of two. You should observe the theoretical convergence rate. (The rate is larger than 2.)

Now, you are in a position to apply Hermite interpolation to generate a very smooth limaçon.

Exercise 7.4.

(1) Write a new script file by following these steps:

    (a) Copy your script m-file `exer1.m` from Exercise 7.1 and rename it to `exer4.m`.

    (b) Differentiate the equations (7.2) defining the cardioid and add expressions for `xpdata` and `ypdata` to `exer4.m`.

    (c) Replace `eval_plin` with `eval_pherm`.

(2) Run the script to plot the limaçon. If this plot does not appear smooth or if you can see where the boundaries of the pieces that make it up lie, you probably have an error in your derivatives of the equations (7.2). Similarly, wiggles and extraneous loops also indicate errors, probably in the derivatives.

7.4 Two-dimensional Hermite interpolation and mesh generation

In order to numerically solve a partial differential equation, the region over which it is to be solved must be broken into small pieces. These pieces are often called "mesh elements," and the process of generating them for arbitrary regions is called "mesh generation." Meshes can be very complex, especially in engineering applications such as stress analysis of an airplane, where centimeter-sized mesh elements must cover all details of an intercontinental jet airliner including windows, decks, the wings, tail and engines. The process of building the outline of a complicated assembly and then using it to generate a mesh is very labor-intensive and can take weeks. The discussion in this chapter is most relevant to generation of quadrilateral and hexahedral meshes, not triangular or tetrahedral meshes.

The lowest-level activity in generating meshes in many commercial packages is defining "Coons Patches" [Coons (1967)]. These are two- or three-dimensional geometrical entities that can, on the one hand, be combined together smoothly, and, on the other hand, easily be subdivided into mesh elements. These Coons Patches are generally built using bicubic Hermite polynomials in two dimensions, and tricubic Hermite polynomials in three dimensions. This chapter will be focussing on two dimensions.

Coons patches are also used in computer graphics, but that application will not be addressed here.

One approach to generating two-dimensional meshes is to take a drawing of the object and cover it with a modest number of patches, each with four curved sides. These patches generally are chosen to match up (to a reasonable approximation) with the boundaries of the object and fit against each other and are no smaller than necessary for these tasks. Once the object in the drawing is completely covered with patches, the patches themselves are broken into smaller mesh pieces. It is important that the mesh lines do not undergo abrupt changes across patch boundaries and it is also important that the user be able to modify the density of mesh lines and to concentrate them near, for example, one boundary. Commercial programs such as Patran [MSC (2020)] and Ansys [ANSYS (2020)] use this approach as one option.

Definition A two-dimensional bicubic Hermite patch (Coons patch) is a smooth map from the unit square $(s, t) \in [0, 1] \times [0, 1]$ to a region $(x(s, t), y(s, t)) \in \mathbb{R}^2$. This mapping is a cubic polynomial in s for each fixed t and a cubic polynomial in t for each fixed s and can be written as a linear combination of the sixteen terms $s^m t^n$ for $n, m = 0, 1, 2, 3$.

A similar definition holds for a two-dimensional patch on a surface in $\mathbb{R}^3$, with an additional function $z(s, t)$. The extension of the definition of bicubic Hermite patches to tricubic Hermite patches in three dimensions is straightforward.

Remark 7.1. A mesh consisting of curved-sided quadrilaterals can be constructed from the union of several bicubic Hermite patches that are adjacent to one another and whose parameterizations along adjacent edges agree. Dividing the square $[0, 1] \times [0, 1]$ into rows and columns of smaller squares and mapping the smaller squares using the bicubic Hermite mappings results in a computational mesh.

Consider the following sketch.

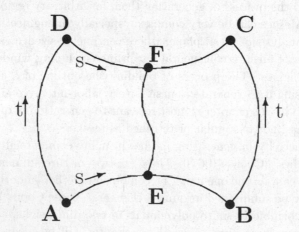

The four outer curves, AB, BC, DC, and AD will form the boundary of the patch.

The two curves AB and DC are parameterized using Hermite cubics in s (with positive direction shown in the figure) and the curves BC, AD and each of the intermediate curves such as EF are parameterized using Hermite cubics in t (with positive direction shown in the figure). Obviously, the coordinates of each of the points A, B, C, and D will be needed. Furthermore, since these are Hermite cubics, the derivatives of $x(s,t)$ and $y(s,t)$ with respect to both s and t will be needed at each of the four corners. Finally, the cross-derivatives $\partial^2 x/\partial s \partial t$ will be needed at each of the four corners. These sixteen data are needed to construct the Coons patch since, in general, a bicubic polynomial is a sum of the sixteen terms $s^m t^n$ for $n, m = 0, 1, 2, 3$.

Among these sixteen data, there is enough information to construct the boundary curve AB parametrically using Hermite polynomials for $x(s, 0)$ and $y(s, 0)$, where s is the independent variable increasing from point A to point B. This interpolation can be done using the `eval_pherm.m` function you wrote earlier. Similarly, the curve DC can be constructed using Hermite polynomials to give $x(s, 1)$ and $y(s, 1)$. Lines of constant s, such as the line EF, can be constructed using the curves AB and DC. The result will be the parametric coordinate functions $x(s, t)$ and $y(s, t)$ for all $(s, t) \in [0, 1] \times [0, 1]$. This is the approach taken in the following exercise.

In the following exercise you will write a script m-file that fills a square with smaller squares. This is a special case of mesh generation. In doing this exercise, we will *not* use any special facts about squares, but will be writing as if the lines were curved. In subsequent exercises, you will see what meshes with curved lines look like.

The four corners, A, B, C, and D, each require data for x and y regarded as functions of s and t. The data are

x	y
$\frac{\partial x}{\partial s}$	$\frac{\partial y}{\partial s}$
$\frac{\partial x}{\partial t}$	$\frac{\partial y}{\partial t}$
$\frac{\partial^2 x}{\partial s \partial t}$	$\frac{\partial^2 y}{\partial s \partial t}$

To see to what these quantities refer, consider the point A. The value of x at point A means its abscissa. The quantity $\frac{\partial x}{\partial s}$ means the derivative of x with respect to the parameter s and is the answer to the question, "How rapidly does x change moving along the curve as s increases from point A?" The quantity $\frac{\partial x}{\partial t}$ is similar, except with respect to the parameter t. Finally, $\frac{\partial^2 x}{\partial s \partial t}$ is the cross derivative. These cross derivatives can be difficult to calculate "in your head" but reasonable meshes can often be generated by taking both $\frac{\partial^2 x}{\partial s \partial t}$ and $\frac{\partial^2 y}{\partial s \partial t}$ to be zero.

Exercise 7.5. Generate a script m-file named `exer5.m` with the commands found below (also found in `exer5.txt`), replacing the ??? with correct expressions. Running this script should produce a square divided into three rectangles horizontally and four rectangles vertically, as shown here

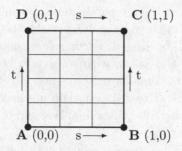

Hints:

- Read through the entire file before making any changes.
- In this simple case, the mapping $(s, t) \mapsto (x(s, t), y(s, t))$ is the identity mapping.
- To see what quantities such as dxdsB should be, answer the question, "As you move along the line from A to B, how is the coordinate x varying as you get to B?"
- If you are having trouble getting it to look right, first plot the outline (the lines with $t = 0$, $t = 1$, $s = 0$, and $s = 1$). Then look at the corner points one at a time. You likely have some of them right and others wrong. Fix the wrong ones, one at a time. Then add the lines with varying s, and finally add the lines with varying t.
- *Never* make more than one change to the script before looking at the resulting plot!

The outline for exer5.m follows:

```
% Generate a mesh based on Hermite bicubic interpolation

% values for point A
xA        = 0;        yA       = 0;
dxdsA     = 1;        dydsA    = 0;
dxdtA     = 0;        dydtA    = 1;
d2xdsdtA  = 0;        d2ydsdtA = 0;

% values for point B
xB        = 1;        yB       = 0;
dxdsB     = ???       dydsB    = ???
dxdtB     = 0;        dydtB    = 1;
d2xdsdtB  = 0;        d2ydsdtB = 0;

% values for point C
xC        = 1;        yC       = 1;
dxdsC     = 1;        dydsC    = 0;
```

```
dxdtC    = ???      ·  dydtC    = ???
d2xdsdtC = 0;          d2ydsdtC = 0;


% values for point D
xD       = 0;          yD       = 1;
dxdsD    = 1;          dydsD    = 0;
dxdtD    = 0;          dydtD    = 1;
d2xdsdtD = ???         d2ydsdtD = ???


% Start off with 4 points horizontally,
% and 5 points vertically
s=linspace(0,1,4);
t=linspace(0,1,5);


% interpolate x along bottom and top, function of s
xAB    =eval_pherm([0,1],[xA,xB]       ,[dxdsA,dxdsB],        s);
dxdtAB=eval_pherm([0,1],[dxdtA,dxdtB],[d2xdsdtA,d2xdsdtB],s);
xDC    =???
dxdtDC=???


% interpolate y along bottom and top, function of s
yAB    =???
dydtAB=???
yDC    =???
dydtDC=eval_pherm([0,1],[dydtD,dydtC],[d2ydsdtD,d2ydsdtC],s);


% interpolate s-interpolations in t-direction
% if variables x and y already exist, they might have
% the wrong dimensions.  Get rid of them before reusing them.
clear x y
for k=1:numel(s)
  x(k,:)=eval_pherm([0,1],[xAB(k),xDC(k)],[dxdtAB(k),dxdtDC(k)],t);
  y(k,:)=???
end


% plot all lines
plot(x(:,1),y(:,1),'b')
hold on
for k=2:numel(t)
  plot(x(:,k),y(:,k),'b')
end
for k=1:numel(s)
```

```
  plot(x(k,:),y(k,:),'b')
end
axis('equal');
hold off
```

It seems natural that s and t would vary linearly along the sides, but that is not necessary. In the following exercise, you will see how nonlinear variation of s can be used to vary the mesh distribution without changing the outline of the figure. So long as s and t vary smoothly and map onto the square $[0, 1] \times [0, 1]$, they are essentially arbitrary.

Exercise 7.6.

(1) Copy your `exer5.m` to `exer6.m` and change it to generate 10 rectangular elements horizontally and 15 vertically (11 points horizontally, 16 points vertically).
(2) Modify $\partial x / \partial s$ at points A and D to be 3.0 and at points B and C to be 0.1. Leave all other values alone. You should see that the mesh elements become much thinner as they get closer to the right side. This is particularly valuable in aeronautical calculations of air flow over a wing because the "boundary layer" is remarkably thin near an aircraft's skin and mesh elements must be very thin there.

Just because you can correctly generate a straight-line mesh does not mean your code is correct. In the next two exercises, you are going to continue debugging your MATLAB programming by testing with two more difficult shapes. The first shape is a quadrilateral with straight sides, and the subsequent exercise uses curved sides.

Exercise 7.7. Copy the file `exer5.m` to a file `exer7.m` (or you can use "Save as" in the File menu). Modify it in the following ways and execute it to generate the plot:

- Change the points so that point A is $(-1, 0)$ and point C is $(1, 0.5)$.
- Change the number of points for s to 20 and the number of points for t to 15.
- Choose the slopes of the lines so that the *boundary* lines are straight with mesh points uniformly distributed along them. For example, both values `dxdtA` and `dydtA` should be 1 because both x and y increase by 1 as t goes from 0 to 1 starting at point A and ending at point D.
- Change the values `d2xdsdt=-1` and `d2ydsdt=-0.5` at all four corner points.

All of the resulting mesh lines (both boundary and interior) should be straight lines with uniformly spaced divisions.

Hint: If you cannot get it to look right, try the following strategy.

(1) Starting from `exer5.m`, make the three changes:

    (a) Change the coordinates of point A to $(-1, 0)$ and of point C to $(1, 0.5)$;

    (b) Change the number of points of `s` and `t` to 20 and 15, respectively; and,

    (c) Change the values of `d2xdsdt` and `d2ydsdt` to -1 and -0.5 at all four corners.

    Then plot the resulting figure. You will see that some boundary lines are straight and some are not. One of these curved lines is AD.

(2) Focus on the line AD. This is a "*t*-line." Take a piece of paper and use it as a straight edge to see what the line AD should look like. You should see that the slope of the line at A is about right, but the slope at D is not. Since the line AD is supposed to be straight, what should `dxdt` be at D? Put your estimate of `dxdtD` into your script and run it again. If your estimate is correct, the line AD is a straight line.

(3) Similarly, fix the other curved boundaries.

(4) Look at the line BC. It is straight, but the mesh spacing is not uniform. Since this is a "*t* line," and since it is vertical, values of `dydtB` and `dydtC` may need to be adjusted.

(5) Similarly, adjust the other boundaries if their mesh spacing is not uniform.

Exercise 7.8. Consider the following sketch:

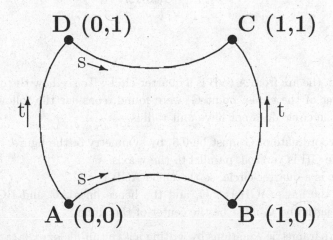

Regard each of the four curves as quarter-circles, so the slopes of each boundary curve at each corner are $\frac{dy}{dx} = \pm 1$. (Do not confuse this with $\frac{dx}{ds}$ or $\frac{dy}{ds}$.) You can see that the four corner points are at the corners of the same unit square you used in Exercise 5.

(1) Addressing the bottom curve first, the curve AB is a quarter-circle (an arc with angle $\pi/2$). It is hard to guess $\frac{dx}{ds}$ and $\frac{dy}{ds}$ of this arc at its ends, so in this part of the exercise you will plot the curve parametrically and compute the derivatives. Write a simple set of parametric equations, analogous to (7.1), for x and y in terms of the variable $s \in [0, 1]$.

You may find the following figure helpful in generating the parametric equations.

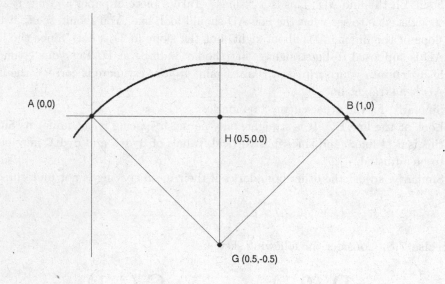

As specified, the arc from A to B is a quarter-circle. To see how the coordinates of the center of the circle, point G, were found, consider the following steps. Note that the circle does not have unit radius.

- The x-coordinate of G must be 0.5, by symmetry of the figure.
- The line GH is vertical, parallel to the y-axis.
- The arc is a quarter-circle, so the angle AGB is $\pi/2$.
- Hence, the angle AGH is $\pi/4$, and the hence lines AG and BG have the same length and G must be the center of the circle.

(2) Check your parametric equations by writing a script m-file exer8a.m to plot the quarter-circular arc as a function of the parameter $s \in [0, 1]$. (The parameter s is *not* the arc length.) You should see an arc AB very similar to the one shown above. From your parametric equations, what are $\frac{dx}{ds}$ and $\frac{dy}{ds}$ evaluated at the points A and B?

(3) Write a script m-file `exer8b.m` based on `exer5.m` but only for the AB (bottom) portion of the outer boundary. Use 25 points for the s values.

- Use the values of x, dx/ds and dy/ds at A and B that you found above.
- Plot the arc at 25 points, with marks showing the intermediate points. You can use a command such as

```
plot(x,y,'r*-')
%plot color red, * at each data point with lines between points
```

- Use `exer8a.m` to plot the quarter-circle on the same plot. You should see 25 points that appear to be uniformly distributed along a red arc connecting the points A and B. The red arc should be close to, but not on top of, the quarter-circle from `exer8a.m`.

(4) Copy the file `exer5.m` to another file `exer8.m` and use it to generate an 25×25 mesh on the curved-sided quadrilateral. Your mesh should clearly show two-fold symmetry (left-right and top-bottom), and the mesh intervals along the perimeter should be roughly uniform. Please include a plot of your mesh with the files you send to me. **Hint**: You already know the parameters for the bottom boundary (AB). Insert these into the file and plot to be sure you have them right. Ignore the other boundaries and the interior lines. Once you have AB correct, put the correct parameters into the BC line, plot to be sure, and continue.

7.5 Matching patches

So far, you have not seen a reason for using Hermite cubic interpolation over any other method. There is a good reason for this choice, namely, that two patches can be placed adjacent to each other and, if the derivatives at the endpoints of the common side are given the same values, the mesh will smoothly transition from one to the other. In the following exercise, there are two patches that touch along one edge. Consider the following sketch.

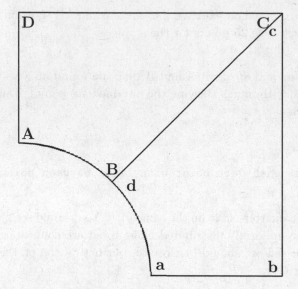

Point	Coordinates
A	$(0,1)$
B	$(\sqrt{2}/2, \sqrt{2}/2)$
C	$(2,2)$
D	$(0,2)$
a	$(1,0)$
b	$(2,0)$
c	$(2,2)$
d	$(\sqrt{2}/2, \sqrt{2}/2)$

Exercise 7.9. In this exercise, you will generate a mesh for the two-patch region in the sketch by generating meshes for the two patches separately and seeing that they naturally match up. The inner boundary in the sketch is circular, but we will be using a Hermite cubic approximation, which is ok but not perfect. If more accuracy were needed for this circular boundary, more patches could be used.

(1) Copy the following commands for patch ABCD to a file `exer9a.m`, or copy from the file `exer9a.txt`.

```
% Generate a mesh for the patch ABCD
sqrt2on2=sqrt(2)/2;

% values for point A
xA        = 0;              yA        = 1;
dxdsA     = pi/4;           dydsA     = 0;
dxdtA     = 0;              dydtA     = 1;
d2xdsdtA = 0;               d2ydsdtA = 0;

% values for point B
xB        = sqrt2on2;       yB        = sqrt2on2;
dxdsB     = sqrt2on2*pi/4;  dydsB     = -sqrt2on2*pi/4;
dxdtB     = 2-sqrt2on2;     dydtB     = 2-sqrt2on2;
d2xdsdtB = 0;               d2ydsdtB = 0;

% values for point C
xC        = 2;              yC        = 2;
dxdsC     = 2;              dydsC     = 0;
```

```
dxdtC    = 2-sqrt2on2;     dydtC    = 2-sqrt2on2;
d2xdsdtC = 0;              d2ydsdtC = 0;


% values for point D
xD       = 0;              yD       = 2;
dxdsD    = 2;              dydsD    = 0;
dxdtD    = 0;              dydtD    = 1;
d2xdsdtD = 0;             d2ydsdtD = 0;
```

(2) Using your earlier exercises as a model, add appropriate commands to complete `exer9a.m` so that it generates and plots a 20×30 mesh on patch ABCD.

(3) Make a copy of `exer9a.m` and call it `exer9b.m`. Add a second set of commands to `exer9b.m` so that it generates a 20×30 mesh on patch abcd. Make sure that the number of points along edge BC is the same as the number of points along edge dc, so that the mesh lines are continuous. Your mesh should be symmetric about the line BC (dc).

(4) Hermite cubics will generate smooth mesh lines if the outline of the region is smooth. The outer boundary DCcb has a corner, and the interior mesh lines clearly echo that corner. Make a copy of `exer9b.m`, call it `exer9c.m`, and "round the corner off" by moving the points C and c from (2,2) to (1.8,1.8), and making the boundary lines at point C (and c) have slope -1. It doesn't matter too much what parametric slopes `dxdsC` and `dydsC` (and `dxdsc` and `dydsc`) you use, so long as the ratio is $dy/dx = (dy/ds)/(dx/ds) = -1$, but parametric slopes of ± 0.5 generate a nice picture.

7.6   Cubic spline interpolation

One disadvantage of the Hermite interpolation scheme is that you need to know the derivatives of your function. But it's very possible that you don't have any formula for your data, just the values at the data points. A second problem is the Hermite interpolant is smooth, but not as smooth as it could be. The discontinuities in the second derivative at the data points might be noticeable.

Cubic splines are piecewise cubic interpolants that are very smooth. They are continuous, with continuous first *and* second derivatives. Only the third derivative is allowed to jump at the join points (called knots).

Suppose you have broken an interval $[a, b]$ into subintervals

$$a = x_1 < x_2 < \ldots < x_n = b$$

then $s(x)$ is a "spline function" if there is a value $m \geq 1$ with

(1) $s(x)$ is a polynomial of degree $\leq (m - 1)$ on each subinterval $[x_{k-1}, x_k]$; and,

(2) $\frac{d^r s}{dx^r}(x)$ is continuous on $[a, b]$ for $0 \leq r \leq (m - 2)$.

Given data values y_k at each of the "knots" x_k, you are interested in interpolating the data, so that $s(x_k) = y_k$.

The continuity and interpolation conditions are almost enough to specify the cubic. If you add one extra condition at each endpoint, such as the value of the derivative, then the spline is determined. Unlike Hermite interpolation, the derivatives are needed only at the endpoints of the curve, not at the endpoints of each subinterval.

As with the other interpolation methods, you will first compute the parameters of the spline interpolant. Given those parameters, you will compute the first derivatives $s'(x_k)$ at the knots, and use the `eval_pherm` function you have already written to evaluate the spline function.

[Atkinson (1978)], pages 168ff, and [Quarteroni *et al.* (2007)], pages 357ff, describe how to find the second derivatives (M or s'') of the "complete" cubic spline interpolant. (These references use indices starting with $k = 0$, but you will obey the MATLAB convention and start indices with $k = 1$.)

Since the desired spline is a piecewise C^2 cubic polynomial, its second derivative is continuous. Thus the values $M_k = s''(x_k)$ are well-defined. Since $s(x)$ is a piecewise cubic polynomial, its second derivative $s''(x)$ is piecewise linear. Thus,

$$s''(x) = \frac{(x_{k+1} - x)M_k + (x - x_k)M_{k+1}}{h_k} \text{ for } k = 1, \ldots, n-1.$$

For each of the intervals $[x_k, x_{k+1}]$, this expression can be integrated twice to yield

$$s(x) = \frac{(x_{k+1} - x)^3 M_k + (x - x_k)^3 M_{k+1}}{6h_k} + C_k(x_{k+1} - x) + D_k(x - x_k),$$

where C_k and D_k are constants of integration. Since $s(x)$ is continuous and interpolates, $s(x_k) = y_k$, so the values of C_k and D_k can be evaluated, yielding the expression

$$s(x) = \frac{(x_{k+1} - x)^3 M_k + (x - x_k)^3 M_{k+1}}{6h_k} + \frac{(x_{k+1} - x)y_k + (x - x_k)y_{k+1}}{h_k}$$
$$- \frac{h_k}{6}\left((x_{k+1} - x)M_k + (x - x_k)M_{k+1}\right). \tag{7.5}$$

So far, the M_k remain unknown and continuity of $s(x)$ and of $s''(x)$ have been used. Continuity of $s'(x)$ yields equations for M_k. Differentiating (7.5) yields, for the interval $[x_k, x_{k+1}]$

$$s'(x) = \frac{-(x_{k+1} - x)^2 M_k + (x - x_k)^2 M_{k+1}}{2h_k} + \frac{y_{k+1} - y_k}{h_k} - \frac{(M_{k+1} - M_k)h_k}{6}, \tag{7.6}$$

where

$$h_k = x_{k+1} - x_k \text{ for } k = 1, 2, \ldots, (n-1). \tag{7.7}$$

Applying (7.6) to evaluate $s'(x_k)$ yields

$$s'(x_k) = -M_k \frac{h_k}{3} - M_{k+1} \frac{h_k}{6} + \frac{y_{k+1} - y_k}{h_k}. \tag{7.8}$$

For $k = 2, 3, \ldots, n - 1$, the knot x_k is a member of *two* intervals, $[x_k, x_{k+1}]$, as above, and also $[x_{k-1}, x_k]$. Replacing k with $(k-1)$ in (7.6), to make it refer to the interval $[x_{k-1}, x_k]$, and applying it to the knot x_k yields

$$s'(x_k) = M_k \frac{h_{k-1}}{3} + M_{k-1} \frac{h_{k-1}}{6} + \frac{y_k - y_{k-1}}{h_{k-1}}. \tag{7.9}$$

Putting these together gives $(n - 2)$ equations for the n values M_k. We still need two more equations. One way would be to require that values for $s'(x_1) = y_1'$ and $s'(x_n) = y_n'$ be specified as part of the problem. Using these two relations and equating (7.8) with (7.9) yields the following system of equations.

$$M_1 \frac{h_1}{3} + M_2 \frac{h_1}{6} = \frac{y_2 - y_1}{h_1} - y_0'$$

$$\frac{M_1}{6} + \frac{h_1 + h_2}{3} M_2 + M_3 \frac{h_2}{6} = \frac{y_3 - y_2}{h_2} - \frac{y_2 - y_1}{h_1}$$

$$\ldots = \ldots \tag{7.10}$$

$$M_{n-2} \frac{h_{n-1}}{6} + M_{n-1} \frac{h_{n-1} + h_{n-1}}{3} + M_n \frac{h_{n-1}}{6} = \frac{y_n - y_{n-1}}{h_{n-1}} - \frac{y_{n-1} - y_n}{h_n}$$

$$M_{n-1} \frac{h_{n-1}}{6} + M_n \frac{h_{n-1}}{3} = y_n' - \frac{y_n - y_{n-1}}{h_{n-1}}.$$

This system is equivalent to the matrix equation

$$AM = D,$$

with the matrix

$$A = \begin{pmatrix} \frac{h_1}{3} & \frac{h_1}{6} & 0 & 0 & \cdots & \\ \frac{h_1}{6} & \frac{h_1+h_2}{3} & \frac{h_2}{6} & 0 & & \\ 0 & \frac{h_2}{6} & \frac{h_2+h_3}{3} & \frac{h_3}{6} & 0 & \\ & \ddots & \ddots & \ddots & \ddots & \ddots \\ & 0 & \frac{h_{n-3}}{6} & \frac{h_{n-3}+h_{n-2}}{3} & \frac{h_{n-2}}{6} & 0 \\ & & 0 & \frac{h_{n-2}}{6} & \frac{h_{n-2}+h_{n-1}}{3} & \frac{h_{n-1}}{6} \\ & \cdots & & 0 & \frac{h_{n-1}}{6} & \frac{h_{n-1}}{3} \end{pmatrix} \tag{7.11}$$

and the vectors

$$D = \begin{pmatrix} \frac{y_2 - y_1}{h_1} - y_1' \\ \frac{y_3 - y_2}{h_2} - \frac{y_2 - y_1}{h_1} \\ \vdots \\ \frac{y_n - y_{n-1}}{h_{n-1}} - \frac{y_{n-1} - y_{n-2}}{h_{n-2}} \\ y_n' - \frac{y_n - y_{n-1}}{h_{n-1}} \end{pmatrix} \tag{7.12}$$

and

$$
M = \begin{pmatrix} M_1 \\ M_2 \\ \vdots \\ M_{n-1} \\ M_n \end{pmatrix}.
$$

Since each diagonal element of A is no smaller than the sum of the off-diagonal elements with strict inequality in some rows, the Gershgorin disk theorem shows that A is nonsingular so this equation can be solved for M.

Exercise 7.10. In this exercise you will write a function to implement the matrix A, the right side D, and solve the resulting matrix system. Then, it will construct the values of the derivatives of the complete cubic spline at the knots. This implementation uses loops rather than vector (componentwise) operations because it is easier to see what is happening with loops. If you wish, you can replace the loops with vector (componentwise) operations. You may find it convenient to copy some code from `exer10.txt`.

(1) Write a file named `ccspline.m` in the following way:

(a) Begin writing a file named `ccspline.m` to compute the values M_k and the values $s'(x_k)$. Its signature and outline should be

```
function sprime=ccspline(xdata,ydata,y1p,ynp)
  % sprime=ccspline(xdata,ydata,y1p,ynp)
  %    ... more comments ...

  % your name and the date

  << code >>

end
```

(b) Choose the proper value of **n** and write a loop to define the vector of interval lengths h_k as given in (7.7). It should be similar to the following:

```
for k=1:n-1
  h(k)= ???
end
```

(c) Write a loop to define the column vector D_k in (7.12). It will use the quantities y'_1 =y1p and y'_n =ynp and be similar to the following:

```
D(1,1)= ???
for k=2:n-1
  D(k,1)= ???
end
D(n,1)= ???
```

(d) Write a loop to define the matrix A in (7.11). Your loop should be similar to the following:

```
A=zeros(n,n);
A(1,1)= ???
A(1,2)= ???
for k=2:n-1
  A(k,k-1)=???
  A(k,k)  =???
  A(k,k+1)=???
end
A(n,n-1)=???
A(n,n)  =???
```

(e) Solve the matrix system $AM = D$ for the vector M using the backslash operator.

(f) Write a loop to define the derivatives of the spline function given in (7.8). This vector should be a row vector if ydata is a row vector and a column vector if ydata is a column vector. Recall that, for a complete cubic spline, $s_1' =$y1p and $s_n' =$ynp. Your loop should look something like the following:

```
sprime=zeros(size(ydata));
sprime(1)=???
for k=2:n-1
  sprime(k)=???
end
sprime(n)=???
```

(2) Test your function using the polynomial $y(x) = x^3$ on the interval $[0, 3]$ using unequal subintervals. Since you are using a cubic polynomial, the resulting spline function is x^3 back again, and you know the derivatives.

```
xdata=[0 1 2  4];
ydata=[0 1 8 64];
y1p=0;
ynp=48;
```

If your results for sprime are correct, go on the next exercise. If not, debug your work using the following strategy.

(a) Double-check your code for sprime against (7.8).

(b) Print the matrix A and check that it is symmetric. If it is not, fix your code and test it again now.

(c) The Gershgorin theorem holds because the sum of the off-diagonal terms in each row is no larger than the diagonal term, and strictly smaller in some rows. Double-check that this fact holds for your matrix.

(d) If A is symmetric, then try your code on the interval $[0, 2]$ using the following data

```
xdata=[0 1 2];
ydata=[0 1 8];
y1p=0;
ynp=12;
```

If your results are correct, then you have an mistake in your subscripts `h(k)`. (This debugging case was chosen because `h(k)=1` for all `k` but not in general.) Fix your code and test it again now.

(e) If the results of the previous test are not correct, consider the case

```
xdata=[0 1 3];
ydata=[0 1 27];
y1p=0;
ynp=27;
```

and compute the 3×3 matrix A and the vector D *by hand* from (7.11) and (7.12). Compare with the MATLAB values, and fix the mistake.

(f) If your A and D are correct, compute `sprime` *by hand* from (7.8). Compare with the MATLAB values, and fix the mistake.

Now that you are confident of your `ccspline` function is correct, you can use it to interpolate a function. Recall that one of the most common applications of spline interpolation is to interpolate tabular data so that computing derivatives is a major difficulty.

Exercise 7.11.

(1) Copy your `test_pherm_interpolate.m` to `test_ccspline_interpolate.m` and modify it so that the derivative values are computed using `ccspline`. You will still need derivatives at the endpoints. You can continue using `eval_pherm` to evaluate your spline approximation because you know both function values (`ydata`) and derivative values (`y1p`, `ynp`, and `sprime`).

(2) Using equally spaced data points from 0 to 5 for `xdata` and the `runge.m` function to estimate the maximum interpolation error by computing the maximum relative difference (infinity norm) between the exact and interpolated values at 4001 evenly spaced points. Fill in the following table, where the column "ratio" is the error on the preceeding line divided by the error on the current line.

Runge function, Complete Cubic Spline		
ndata	error	ratio
5		—
11		
21		
41		
81		
161		
321		
641		

(3) Estimate the rate of convergence by examining the ratio column and estimating the nearest power of two.

7.7 Splines without derivatives

If you have only function data (perhaps tabulated) without any way to compute derivatives, you need another set of equations to completely specify M_k for $k = 1, 2, \ldots, n$. One natural way of doing it is to assume that $M_1 = M_n = 0$, whereby the resulting approximation is assumed linear at the endpoints. Such splines are called "natural cubic splines."

An alternative approach is to require the $s'''(x_k)$ is continuous at the two points $k = 2$ and $k = n - 1$, as well as $s''(x_k)$, $s'(x_k)$, and $s(x_k)$ itself. Since $s(x)$ is piecewise cubic, if those four conditions hold, then $s(x)$ is a *single cubic* on the intervals $[x_{k-1}, x_{k+1}]$, $k = 2$ and $k = n - 1$, not two cubics meeting at x_k. Hence, x_k is not properly a knot, and this is called the "not-a-knot" condition.

In the following two exercises, you will examine each of these alternatives.

Exercise 7.12.
(1) Write an m-file `ncspline.m` to construct a natural cubic spline by following the steps:

    (a) Copy your `ccspline.m` to a file `ncspline.m` and modify the signature and outline to be

```
function sprime=ncspline(xdata,ydata)
  % sprime=ncspline(xdata,ydata)
  %    ... more comments ...

  % your name and the date

  << code >>

end
```

(b) Modify the first and last lines of the matrix A and the first and last components of the vector D so that they embody the equations

$$M_1 = 0, \text{ and}$$
$$M_n = 0.$$

Hint: The matrix system $AM = D$ above is given explicitly by (7.10). Look at the first line of this system. How would you change the matrix A so that the first line of the system in (7.10) becomes $M_1 = 0$? Similarly for the last line.

(c) Use (7.8) to get $s'(x_1)$ and (7.9) to get $s'(x_n)$.

(2) The following case represents a linear equation

```
xdata=[0 1 2 3];
ydata=[0 2 4 6];
```

Use it to test that **ncspline** works correctly for one case.

(3) Temporarily print the solution vector M, pick some nonlinear function to approximate and verify that M(1)=M(n)=0. (You cannot exactly reproduce such a function using natural cubic splines, but you can check the endpoints.) Remove the temporary printing.

(4) Test that **ncspline** and **ccspline** agree when given the same data. Choose

```
xdata=[0,1,2];
ydata=[0,-3,-16];
```

and compute s' using **ncspline** and call it **ncsprime**. Next compute s' using **ccspline** and using **ncsprime(1)** and **ncsprime(end)** for the derivative values. Call the results of **ccspline** **ccsprime**. Check that **ncsprime** and **ccsprime** are the same up to roundoff.

(5) Copy your **test_ccspline_interpolate.m** to **test_ncspline_interpolate.m** and modify it so that the derivative values are computed using **ncspline**.

(6) Using equally spaced data points from 0 to 5 for **xdata** and the **runge.m** function to estimate the maximum interpolation error by computing the maximum relative difference (infinity norm) between the exact and interpolated values at 4001 evenly spaced points. Fill in the following table, where the column "ratio" is the error on the preceeding line divided by the error on the current line.

Runge function, Natural Cubic Spline		
ndata	error	ratio
5		—
11		
21		
41		
81		
161		
321		
641		

(7) Estimate the rate of convergence by examining the ratio column and estimating the nearest power of two.

In the following exercise, you will see how to use a spline generated using the not-a-knot condition. The not-a-knot condition is more difficult to program, providing an opportunity to introduce the MATLAB `spline` function, whose default constructive behavior is to use the not-a-knot condition. Complete cubic splines can also be generated using `spline`, and the syntax for that is presented in a remark following the exercise.

Exercise 7.13.

(1) Copy your `test_ncspline_interpolate.m` to `test_spline_interpolate.m` and modify so that the MATLAB `spline` function is used to evaluate the spline function itself using the following syntax:

```
yval=spline(xdata,ydata,xval);
```

where `xdata` are the data points, `ydata` are the data values, and `xval` are the points at which the interpolant is to be evaluated.
(2) Using equally spaced data points from 0 to 5 for `xdata` and the `runge.m` function to estimate the maximum interpolation error by computing the maximum relative difference (infinity norm) between the exact and interpolated values at 4001 evenly spaced points. Fill in the following table, where the column "ratio" is the error on the preceeding line divided by the error on the current line.

Runge function, Not-a-knot Cubic Spline		
ndata	error	ratio
5		—
11		
21		
41		
81		
161		
321		
641		

(3) Estimate the rate of convergence by examining the ratio column and estimating the nearest power of two.

Remark 7.2. It should be clear that the complete cubic spline is smooth and accurate, but requires some derivative values. The natural cubic spline requires no derivative values, but is less accurate than the complete cubic spline. The not-a-knot spline retains the asymptotic accuracy of the complete cubic spline without requiring any derivative information.

Remark 7.3. It is possible to use the MATLAB `spline` function to compute the complete cubic spline as well as the not-a-knot cubic spline. Syntax for the complete cubic spline is

```
% complete cubic spline for row-vector ydata
yval = spline( xdata, [yp1, ydata, ypn], xval);
```

7.8   Monotone interpolation

Suppose you have a closed metal box containing some water and you heat it while you measure its temperature. You will observe that the temperature rises monotonically until it reaches 100° C. The temperature will then remain constant even as you apply heat until all the water boils away, and then it will begin to rise again. Adding heat will *always* cause a nonnegative temperature change.

If you take this temperature data and attempt to interpolate it using splines, you may find non-physical oscillations in temperature near the boiling point. If the temperature behavior near the boiling point is important to you, you need to interpolate it using a method that respects monotonicity, even if it means lower asymptotic accuracy.

Monotone interpolation is a topic of theoretical importance with a seminal paper by Fritsch and Carlson [Fritsch and Carlson (1980)]. MATLAB provides a function called `pchip` that implements Fritsch and Carlson's algorithm. In the following exercise you will compare a `spline` interpolation with a `pchip` interpolation of data inspired by the water experiment described above.

Exercise 7.14. Consider the function given by

```
xdata=[0, 1, 2, 3, 4];
ydata=[0, 1, 2, 2, 2];
```

where `xdata=2` represents the time just after boiling starts.

(1) Plot the data using circles with no lines between them.
(2) Construct the not-a-knot spline interpolant using the MATLAB function `spline` and plot it on the same frame using at least 100 points.
(3) Construct the monotone cubic interpolant using the MATLAB function `pchip` (using the same syntax as `spline` but with the name `pchip`) and plot it on the same frame using at least 100 points. You should observe that the `pchip` interpolant is more physically logical than the `spline` interpolant in this situation.

Exercise 7.15.

(1) Copy your `test_spline_interpolate.m` to `test_pchip_interpolate.m` and modify so that the MATLAB `pchip` function is used to evaluate the interpolant function.
(2) Using equally spaced data points from 0 to 5 for `xdata` and the `runge.m` function to Estimate the maximum interpolation error by computing the maximum relative difference (infinity norm) between the exact and interpolated values at 4001 evenly spaced points. Fill in the following table, where the column "ratio" is the error on the preceeding line divided by the error on the current line.

Runge function, Monotone Cubic Interpolation		
ndata	error	ratio
5		—
11		
21		
41		
81		
161		
321		
641		

(3) Estimate the rate of convergence by examining the ratio column and estimating the nearest power of two. You should observe a lower rate of convergence than for the MATLAB `spline` function.

Chapter 8

Legendre polynomials and L^2 approximation

8.1 Introduction

With *interpolation* you were given a formula or data about a function $f(x)$, and you made a model $p(x)$ that passed through a given set of data points. This chapter considers *approximation*, in which you are still trying to build a model, but specify some condition of "closeness" that the model must satisfy. It is still likely that $p(x)$ will be equal to the function $f(x)$ at some points, but you will not know in advance which points.

As with interpolation, you build this model out of a set of basis functions. The model is then a linear combination with coefficients c that specify how much of each basis function to use when building the model.

In this chapter you will consider four different selections of basis functions in the space $L^2([-1, 1])$. The first is the usual monomials $1, x, x^2, \ldots$. In this case, the coefficients c are exactly the coefficients MATLAB uses to specify a polynomial. The second is the set of Legendre polynomials, which will yield the same approximations but will turn out to have better numerical behavior. The third selection is the trigonometric functions, and the final selection is a set of piecewise constant or piecewise linear functions. Advantages and disadvantages of each for numerical computations will be presented.

Once you have your basis set, you will consider how you can determine the approximating function $p(x)$ as the "best possible" approximate for the given basis functions, and you will look at the behavior of the approximation error. Since you are working in $L^2[-1, 1]$, you will use the L^2 norm to measure error.

It turns out that approximation by monomials results in a matrix similar to the Hilbert matrix whose inversion can be quite inaccurate, even for small sizes. This inaccuracy translates into poor L^2 approximations. Use of orthogonal polynomials such as the Legendre polynomials, results in a diagonal matrix that can be inverted almost without error, but the right side can be difficult and slow to compute to high accuracy. In addition, roundoff accumulation in function reconstruction can be a serious problem. Fourier approximation substantially reduces the roundoff errors, but is slow to compute and evaluate, although fast methods (not discussed in this chapter) can improve speed dramatically. Approximation by piecewise constants is not subject to these sources of error until ridiculously large numbers of pieces are employed, but can be slow to converge.

You will be attempting to approximate several functions in this chapter, all on the interval $[-1, 1]$. These functions include:

- The Runge example function, $f(x) = 1/(1 + x^2)$, `runge.m`.
- The function

$$f(x) = \begin{cases} 0 & -1 \le x < 0 \\ x(1 - x) & 0 \le x \le 1 \end{cases}. \tag{8.1}$$

For the purpose of this chapter, this function will be called "partly quadratic." It was chosen because it is simple, continuous and satisfies $f(-1) = f(1)$, but is not differentiable. A simple MATLAB function m-file to compute this "partly quadratic" function can be found by downloading `partly_quadratic.m` or by copying the following code:

```
function y=partly_quadratic(x)
  % y=partly_quadratic(x)
  % input x (possibly a vector or matrix)
  % output y, where
  % for x<=0, y=0
  % for x>0,  y=x(1-x)

  y=(heaviside(x)-heaviside(x-1)).*x.*(1-x);
end
```

Remark 8.1. The "heaviside" function (sometimes called the "unit step function") is named after Oliver Heaviside and is defined as

$$f(x) = \begin{cases} 0 & x < 0 \\ 0.5 & x = 0 \\ 1 & 0 < x \end{cases} \tag{8.2}$$

MATLAB includes the heaviside function as one of its special functions.

- A third function is a function whose graph is shaped like the teeth of a saw, similar to one used in Chapter 7:

$$f(x) = \begin{cases} (x + 1) & -1 \le x < 0 \\ 0 & x = 0 \\ (x - 1) & 0 < x \le 1 \end{cases}. \tag{8.3}$$

A simple MATLAB function m-file to compute this sawshape function can be found by downloading `sawshape8.m` or by copying the following code:

```
function y=sawshape8(x)
  % y=sawshape8(x)
```

```
% input x (possibly a vector or matrix)
% output y, where
%    y=(x+1) for -1<=x<0
%    y=0     for x=0
%    y=(x-1) for 0<x<=1

   y= heaviside(-x).*(x+1) + heaviside(x).*(x-1);
end
```

Remark 8.2. In Chapter 7, you saw a similar function defined using the MATLAB `find` command, while `heaviside` is used here. There is no essential reason for this change.

This chapter includes only six exercises, but each exercise except the second involves a different approximation method, with its `coeff`, `eval` and `test` functions, and each exercise examines the three functions mentioned above. Exercise 8.1 involves monomial approximations, Exercise 8.3 involves Legendre approximations, Exercise 8.4 involves Fourier approximations, Exercise 8.5 involves piecewise constant approximations, and Exercise 8.6 involves piecewise linear approximations. Exercises 8.2 and 8.3 are a sequence, otherwise each of the exercises is independent of the others, although the first exercise introduces the general format of the MATLAB script and function files. Exercise 8.6 is less structured than the previous exercises and is aimed at students who have already completed several of the previous exercises. Exercises 8.5 and 8.6 use `bracket.m`, among the downloadable files from Chapter 6.

8.2 Integration in MATLAB

This kind of approximation requires evaluation of integrals. you will use the MATLAB numerical integration function `integral`. Since you will be computing fairly small errors, will replace the default error tolerances with smaller ones. The MATLAB command is

```
q=integral(func,-1,1,'AbsTol',1.e-14,'RelTol',1.e-12);
```

where `func` is a function handle to a function written using vector (array) syntax. This command will result in an approximation, q, satisfying

$$\left| q - \int_{-1}^{1} \text{func}(x)dx \right| \leq \max\{10^{-14}, 10^{-12}|q|\}.$$

Remark 8.3. The `integral` function was introduced into MATLAB in 2012. If your version of MATLAB is older than that, you can use the `quadgk` function for this chapter. The calling sequence for that function is

```
q=quadgk(func,-1,1,'AbsTol',1.e-14,'RelTol',1.e-12, ...
         'MaxIntervalCount',15000);
```

Since these integration functions involve extra parameters and extra typing, It is convenient to create an abbreviation for them for use in this chapter. Create an m-file named `ntgr8.m` in the following way:

```
function q=ntgr8(func)
  % q=ntgr8(func) integrates func over $[-1,1]$ with tight tolerances

  q=integral(func,-1,1,'AbsTol',1.e-14,'RelTol',1.e-12);
end
```

with a similar function if you are using `quadgk`. This function will save you some typing for this chapter. The name looks strange in print, but it is easy to remember because it is a pun. The letter `n` can be pronounced "en," the letter `t` can be pronounced "tee" and `gr8` can be pronounced "gr-eight". Put them together and it sounds like "integrate."

8.3   Least squares approximations in $L^2([-1, 1])$

The problem of L^2 approximation can be described in the following way.

Problem: Given a function $f \in L^2$, and a (complete) set of functions $\phi_\ell(x) \in L^2$, $n = 1, 2, \ldots$, then for a given N, find the set of values $\{c_\ell | \ell = 1, 2, \ldots, N\}$ so that

$$f(x) \approx \sum_{\ell=1}^{N} c_\ell \phi_\ell. \tag{8.4}$$

The approximation is best in the sense of having the smallest L^2 error if the functions are pairwise orthogonal. If the functions ϕ_ℓ are at least linearly independent, the coefficients c_ℓ are unique.

[Quarteroni *et al.* (2007)], in Section 10.7, [Atkinson (1978)], in Section 4.3, and [Ralston and Rabinowitz (2001)], in Section 6.2, discuss least-squares approximation in function spaces such as $L^2([-1, 1])$. The idea is to minimize the norm of the difference between the given function and the approximation. Given a function f and a set of approximating functions (such as the monomials $\{x^{k-1} : k = 1, 2, \ldots, n\}$), for each vector of numbers $\mathbf{c} = \{c_\ell\}$ define a functional

$$F(\mathbf{c}) = \int_{-1}^{1} \left(f(x) - \sum_{\ell=1}^{n} c_\ell x^{n-\ell} \right)^2 dx.$$

This continuous functional becomes large when $\|\mathbf{c}\|$ is large and it is bounded below by 0, so it must have a minimum, and because the function is differentiable as a function of $\mathbf{c}$, the minimum must occur when

$$\frac{\partial F}{\partial c_\ell} = 0 \qquad \text{for } \ell = 1, \ldots, n.$$

This expression is evaluated here for the case of quadratic approximations ($n = 3$).

Consider the functional

$$F = \int_{-1}^{1} \left(f(x) - (c_1 x^2 + c_2 x + c_3) \right)^2 dx$$

where f is the function to be approximated on the interval $[-1, 1]$. Taking partial derivatives with respect to c_i yields the equations

$$\frac{\partial F}{\partial c_1} = -2 \int_{-1}^{1} (f(x) - (c_1 x^2 + c_2 x + c_3)) x^2 dx$$

$$\frac{\partial F}{\partial c_2} = -2 \int_{-1}^{1} (f(x) - (c_1 x^2 + c_2 x + c_3)) x\, dx$$

$$\frac{\partial F}{\partial c_3} = -2 \int_{-1}^{1} (f(x) - (c_1 x^2 + c_2 x + c_3))\, dx$$

and setting each of these to zero yields the system of equations

$$c_1 \int_{-1}^{1} x^4 dx + c_2 \int_{-1}^{1} x^3 dx + c_3 \int_{-1}^{1} x^2 dx = \int_{-1}^{1} x^2 f(x) dx$$

$$c_1 \int_{-1}^{1} x^3 dx + c_2 \int_{-1}^{1} x^2 dx + c_3 \int_{-1}^{1} x\, dx = \int_{-1}^{1} x f(x) dx$$

$$c_1 \int_{-1}^{1} x^2 dx + c_2 \int_{-1}^{1} x\, dx + c_3 \int_{-1}^{1} dx = \int_{-1}^{1} f(x) dx$$

or

$$(2/5)c_1 + \quad 0 \quad + (2/3)c_3 = \int_{-1}^{1} x^2 f(x) dx$$

$$0 \quad + (2/3)c_2 + \quad 0 \quad = \int_{-1}^{1} x f(x) dx \qquad (8.5)$$

$$(2/3)c_1 + \quad 0 \quad + (2/1)c_3 = \int_{-1}^{1} f(x) dx.$$

(Since the interval of integration is symmetric about the origin, the integral of an odd monomial is zero.)

For an arbitrary value of n, Equation (8.5) can be written in the following way, where the indexing corresponds with MATLAB indexing (starting with 1 instead of 0) and with the MATLAB convention for coefficients of polynomials (first coefficient is for the highest power of x).

$$\sum_{\ell=1}^{n} \left(\int_{-1}^{1} x^{(2n-k-\ell)} dx \right) c_\ell = \int_{-1}^{1} x^{n-k} f(x) dx \qquad \text{for } k = 1, \ldots, n. \qquad (8.6)$$

The matrix in (8.6)

$$H_{k,\ell} = \int_{-1}^{1} x^{(2n-k-\ell)} dx = (1 - (-1)^{(n-\ell)+(n-k)+1})/((n-\ell) + (n-k) + 1)$$

is closely related to the Hilbert matrix. In fact, if the derivation above were done over the interval $[0,1]$ instead of $[-1,1]$, the matrix that arose would be the Hilbert matrix. The Hilbert matrix is notorious for having a very poor condition number and being difficult to invert without losing accuracy. The following exercise illustrates this difficulty and its implication for approximation.

Examining Equation (8.6), there are two sets of integrals that need to be evaluated in order to compute the coefficients c_k. On the left side, the integrands involve the products $x^{(n-k)}x^{(n-\ell)}$. On the right side, the integrands involve the products $x^{(n-k)}f(x)$. Please note that these expressions are made more complicated by the fact that they are indexed "backwards." This is done to be consistent with MAT-LAB 's numbering scheme for coefficients. Later in the chapter when you switch to Legendre polynomials and are free to number the coefficients as you wish, you will switch to a simpler numbering scheme.

Once the coefficients c_k have been found, the MATLAB `polyval` function can be used to evaluate the resulting polynomials.

Exercise 8.1.

(1) In this exercise, you will be writing a function m-file to compute the coefficient vector of the best $L^2([-1,1])$ approximation to a function $f(x)$ using Equation (8.6) above. You may find it convenient to download the file `coef_mon.txt` when writing your code. This m-file will begin with

```
function c=coef_mon(func,n)
  % c=coef_mon(func,n)
  % func is a function handle
  % ...  more comments  ...

  % your name and the date
```

and be called `coef_mon.m`. It will solve the system (8.6) by constructing the matrix H and right side **b** and solving the resulting system for c_k. In `coef_mon.m`, `func` refers to a function handle.

  (a) Begin `coef_mon.m` with the above code.
  (b) $\mathbf{b}_k = \int_{-1}^{1} x^{n-k} f(x) dx$.

```
for k=1:n
  % force b to be a column vector with second index
  b(k,1)=ntgr8(@(x) func(x).*x.^(n-k));
end
```

  Warning: You should already have created the abbreviation function `ntgr8.m` according to instructions in Section 8.2 above on MATLAB remarks contained in the introduction to this chapter.

  (c) Complete the following code to compute the matrix elements `H(k,ell)` using the formula $H_{k,\ell} = \int_{-1}^{1} x^{(2n-k-\ell)} dx$. Your code will be similar to the above code for `b(k)`.

```
for k=1:n
  for ell=1:n
    H(k,ell)=ntgr8( ??? )
  end
end
```

Remark 8.4.

i. Using the letter "l" as a variable can be confusing because it looks so much like the number 1. Instead, use `ell`.

ii. Time could be saved in the above code by taking advantage of the fact that `H` is a symmetric matrix.

iii. It is not necessary to write code to compute the quantities $H_{k,\ell}$ because you can easily write out the values of the integrals $\int_{-1}^{1} x^{(2n-k-\ell)} dx$. The approach described above, though, is consistent with (and reinforces the idea behind) the calculation of $\mathbf{b}_k$.

(d) Complete the function by solving for the coefficients with

```
c=H\b;
```

Do not be surprised if MATLAB warns you that `H` is poorly conditioned for larger values of `n`.

(2) Verify your code is correct by computing the best 3-term approximation by monomials for the polynomial $f(x) = 3x^2 - 2x + 1$. The result should be the coefficient vector for the polynomial f itself.

(3) Write a script m-file named `test_mon.m` containing code similar to the following (you may find it convenient to download the file `test_mon.txt`)

```
func=@partly_quadratic;
c=coef_mon(func,n);
xval=linspace(-1,1,10000);
yval=polyval(c,xval);
yexact=func(xval);
plot(xval,yval,xval,yexact)

% relative Euclidean norm is approximating
% the relative integral least-squares (L2 norm)
% using an approximate trapezoid rule
relativeError=norm(yexact-yval)/norm(yexact)
```

and use it to evaluate the approximation for `n=1` and `n=5`. Look at the plot and estimate by eye if the area between the exact and approximate curves is divided equally between "above" and "below." Further, the error for `n=5` should be smaller than for `n=1` and the plot should look much better, but still far from perfect.

Remark 8.5. The reasons `test_mon` is written as a script file instead of a function file are because it contains a plot, because later in this chapter timing code will be added to it, and to give another example of how this kind of test file can be written.

(4) Fill in the following table using the Runge example function. When you report errors, please use at least three significant digits.[1] You should find that the error gets smaller for early values of **n** and then deteriorates. At what value of **n** does the smallest relative error occur? (You may get warnings that the matrix H is almost singular.)

Runge test function	
n	relative error
1	
2	
3	
4	
5	
6	
10	
20	
30	
40	

(5) You should notice that the errors in the Runge case for **n=1** and **n=2** are the same, as are the errors for **n=3** and **n=4**, as well as **n=5** and **n=6**. Explain why this should occur.

(6) Fill in the following table for the partly quadratic function.

partly_quadratic function	
n	relative error
1	
2	
3	
4	
5	
6	
10	
20	
30	
40	

(7) Fill in the following table for the sawshape8 function.

---

[1] One easy way to get good precision is to use `format short e`.

| sawshape8 function ||
n	relative error
1	
2	
3	
4	
5	
6	
10	
20	
30	
40	

You should notice that much smaller errors are associated with the smooth Runge example function than with the non-differentiable partly quadratic function and with the discontinuous sawshape8 function.

You should see that the errors do not seem to be decreasing much for the final values of n, and wonder why the method becomes poor as n gets large. It may not be obvious, but the matrices (8.5) and (8.6) are related to the Hilbert matrix and are extremely difficult to invert. The reason inversion is difficult is because the monomials, all start to look the same as n gets larger, that is, they become almost parallel in the L^2 sense. Even if the integration could be performed without error, you would observe roundoff errors in evaluating the resulting high-order polynomial.

One way to make the approximation problem easier might be to pick a better set of functions than monomials. The following section discusses a good alternative choice of polynomials. These polynomials allow much larger values of n.

8.4   Legendre polynomials

The Legendre polynomials form an $L^2([-1,1])$-orthogonal set of polynomials. You will see below why orthogonal polynomials make particularly good choices for approximation. In this section, you are going to write m-files to generate the Legendre polynomials and you are going to confirm that they form an orthogonal set in $L^2([-1,1])$. Throughout this section, you will be representing polynomials as vectors of coefficients, in the usual MATLAB way.

The Legendre polynomials are a basis for the set of all polynomials, just as the usual monomial powers of x are. They are appropriate for use on the interval $[-1,1]$ because they are orthogonal when considered as members of $L^2([-1,1])$. Polynomials that are orthogonal are discussed [Quarteroni *et al.* (2007)] Chapter 10, with Legendre polynomials discussed in Section 10.1.2, [Atkinson (1978)] starting on page 210, and [Ralston and Rabinowitz (2001)] Section 6.4, pages 254ff. The first few Legendre polynomials are:

$$P_0 = 1$$
$$P_1 = x$$
$$P_2 = (3x^2 - 1)/2 \qquad\qquad (8.7)$$
$$P_3 = (5x^3 - 3x)/2$$
$$P_4 = (35x^4 - 30x^2 + 3)/8.$$

The value at x of any Legendre polynomial P_i can be determined using the following recursion:

$$P_0 = 1,$$
$$P_1 = x, \qquad \text{and}$$
$$P_k = ((2k - 1)xP_{k-1} - (k - 1)P_{k-2})/k.$$

The following recursive MATLAB function computes the values of the kth Legendre polynomial. The file `recursive_legendre.m` is available for download.

```
function yval = recursive_legendre ( k , xval )
  % yval = recursive_legendre ( k , xval )
  % yval = values of the k-th Legendre polynomial
  % at values xval

  if k<0
    error('recursive_legendre: k must be nonnegative.');
  elseif k==0    % WARNING: no space between else and if!
    yval = ones(size(xval));
  elseif k==1    % WARNING: no space between else and if!
    yval = xval;
  else
    yval = ((2*k-1)*xval.*recursive_legendre(k-1,xval) ...
            - (k-1)*recursive_legendre(k-2,xval) )/k;
  end
end
```

Unfortunately, this recursive function is too slow to be used in this chapter. The alternative to recursive calculation of Legendre polynomials is one that uses loops. It is a general fact that any recursive algorithm can be implemented using a loop. The code for the loop is typically more complicated than the recursive formulation. In the following exercise, you will write an algorithm using loops for Legendre polynomials.

Exercise 8.2. In this exercise, you will will write code to find the values of the nth Legendre polynomial P_n using a loop. The strategy will be to first compute the values of P_0 and P_1 from their formulæ, then compute the values of P_k for larger subscripts by building up from lower subscripts, stopping at P_n. You should note that if k is larger than 2, you only need to retain the values of P_{k-1} and P_{k-2} in order to compute the values of P_k.

(1) Write a function m-file `legen.m` using the following steps.

    (a) Use the outline

```
function yval = legen ( n , xval)
  % yval = legen ( n , xval)
  % more comments

  % your name and the date
end
```

    and add appropriate comments.

    Remark 8.6. Why switch from index k to index n? The reason is that a loop on k will be used to build up to the value n.

    (b) Use `if` tests to define the cases `n<0`, `n==0` and `n==1`, and use the formulæ for these cases to compute the vector of values `yval`.

    (c) When `n` is larger than 1, compute the vector of values of P_0 and call it `ykm1` (`ykm1` for "y sub k minus 1"), and compute the vector of values of P_1 and call it `yk`.

    (d) Write a loop `for k=2:n` in which you first put the value of `ykm1` into `ykm2` ("y sub k minus 2") and then the value of `yk` into `ykm1`. You do this because you are changing the value of k to be one larger. Then compute P_k, calling it `yk`, using the values `ykm1` and `ykm2`. This line will be similar to the corresponding line in `recursive_legendre`.

    (e) When the loop is complete, k has the value n. Set `yval=yk;`

(2) Test and verify your `legen` function for P_3 with the following code.

```
xval=linspace(0,1,10);
norm( legen(3,xval)-(5*xval.^3-3*xval)/2 )
```

and showing that the difference is of roundoff size.

(3) Your `legen` function and `recursive_legendre` should agree. Test this agreement for n=10 with the following code.

```
xval=linspace(0,1,20);
norm( legen(10,xval) - recursive_legendre(10,xval) )
```

The difference should be of roundoff size.

8.5   Orthogonality and integration

The Legendre polynomials form a basis for the linear space of polynomials. One desirable characteristic of any set of basis vectors is to be orthogonal. For functions, you use the standard L^2 *dot product*, and say that two functions $f(x)$ and $g(x)$ are orthogonal if their dot product

$$(f, g) = \int_{-1}^{1} f(x)g(x)dx$$

is equal to zero. In MATLAB , you could use `integral` or `quadgk` via the abbreviation `ntgr8` to compute this quantity in the following way:

```
q=ntgr8(@(x) f_func(x).*g_func(x) );
```

where `f_func` gives the function f and `gfunc` gives g.

8.6   Legendre polynomial approximation

Legendre polynomial approximation in $L^2([-1, 1])$ follows the same recipe as monomial approximation:

(1) Compute the matrix $H_{m,n} = \int_{-1}^{1} P_{m-1}(x)P_{n-1}(x)dx$. This matrix is diagonal (as opposed to the Hilbert matrix in the monomial case), with diagonal entries $H_{m,m} = 2/(2m - 1)$, so integration is not necessary!
(2) Compute the right side values $b_m = \int_{-1}^{1} f(x)P_{m-1}(x)dx$.
(3) Solve $\mathbf{d} = H^{-1}\mathbf{b}$ using the formula $d_m = \frac{2m-1}{2}b_m$.
(4) The approximation can be evaluated as

$$f(x) \approx f_{\text{legen}}(x) = \sum_{k=1}^{n} d_k P_{k-1}(x). \tag{8.8}$$

The coefficients d_k are not the same as the monomial coefficients c_k computed earlier, and Equation (8.8) must be used rather than `polyval` to evaluate the resulting approximations.

Exercise 8.3.

(1) Write a function m-file named `coef_legen.m` with outline

```
function d=coef_legen(func,n)
  % d=coef_legen(func,n)
  % comments

  % your name and the date

  << your code >>

end
```

to compute the coefficients of the approximation

$$d_k = \frac{2k-1}{2} \int_{-1}^{1} f(x)P_{k-1}(x)dx.$$

You should use the `ntgr8` abbreviation function in the same way as used in Exercise 8.1.

Remark 8.7. The factor $(2k-1)/2$ comes from the inverse of the diagonal matrix $H_{k,k} = 2/(2k-1)$.

(2) Verify that `coef_legen` is correct by computing the best Legendre approximation to the Legendre function P_3, where $n \geq 4$. (Recall that n is the number of terms, not the degree of the polynomial.) The values you get for d are the coefficients in Equation (8.8), not the coefficients of the polynomial.

(3) Write a function m-file called `eval_legen.m` to be used to evaluate Legendre polynomials. It should evaluate Equation (8.8) and have outline

```
function yval=eval_legen(d,xval)
  % yval=eval_legen(d,xval)
  % comments

  % your name and the date

  << your code >>

end
```

(4) Verify that `eval_legen` is correct by choosing d as the coefficients of P_3 (you computed d for this case above) and comparing the results of `eval_legen` and `legen` at the five values [0,1,2,3,4].

(5) Write an m-file called `test_legen.m`, similar to the `test_mon.m` file you wrote above. It should use `eval_legen` and produce the relative error of the approximation. It is instructive if you plot the approximation as well, but it is not required.

(6) Place the MATLAB command `tic;` at the beginning of the script and the MATLAB command `toc;` at the end. This pair of commands will measure and print the elapsed time.

(7) Fill in the following table for the Runge example function.

Runge test function	
n	relative error
1	
2	
3	
4	
5	
6	
10	
20	
30	
40	
50	

(8) Fill in the following table for the partly quadratic function. Include the elapsed time only for the final three lines.

partly_quadratic function		
n	relative error	elapsed time
5		—
10		—
20		—
40		—
80		—
160		
320		
640		

(9) Fill in the following table for the sawshape8 function. Include the elapsed time only for the final three lines.

sawshape8 function		
n	relative error	elapsed time
5		—
10		—
20		—
40		—
80		—
160		
320		
640		

(10) Based on the above data, roughly estimate the value of p where elapsed time is proportional to n^p. Is $1 \leq p \leq 2$? $2 \leq p \leq 3$? $3 \leq p \leq 4$? $4 \leq p \leq 5$? $5 \leq p \leq 6$?

You should find the same errors as for approximation by monomials for small n, and you can accurately compute with larger values of n using Legendre polynomials than using monomials. However, using large values of n can result in computing times that grow more rapidly than expected as n increases because the `integral` function must compensate for roundoff error arising from rapid oscillations. In fact, the time does grow more rapidly than $O(n^p)$, where you just estimated p.

8.7 Fourier series

There is another set of functions that is orthogonormal in the $L^2[-1,1]$ sense. This is the set of trigonometric functions

$$\frac{1}{\sqrt{2}}, \ \cos(\pi x), \ \sin(\pi x), \ \cos(2\pi x), \ \sin(2\pi x), \ \cos(3\pi x), \ \ldots$$

and they can be used for approximating functions. You have seen trigonometric polynomials before in Chapter 5 in the context of interpolation using $e^{ik\pi x}$ for $k = -n, -n+1, \ldots, -1, 0, 1, \ldots, n-1, n$. Using complex exponentials is equivalent to sin and cos but the trigonometric functions are orthogonormal on $[-1, 1]$ while the complex exponentials are not.

The sum of the first $2n + 1$ terms of the Fourier series for a function f is given as

$$f(x) \approx \frac{z}{\sqrt{2}} + \sum_{k=1}^{n} s_k \sin k\pi x + c_k \cos k\pi x. \tag{8.9}$$

As usual, the coefficients can be found by multiplying both sides by $(1/\sqrt{2})$, $\sin(\ell\pi x)$, or $\cos(\ell\pi x)$ and integrating. Orthonormality leads to the expressions

$$z = \int_{-1}^{1} \frac{f(x)}{\sqrt{2}} dx$$

$$s_k = \int_{-1}^{1} f(x) \sin(k\pi x) dx \tag{8.10}$$

$$c_k = \int_{-1}^{1} f(x) \cos(k\pi x) dx$$

(terms involving $k \neq \ell$ are zero).

Exercise 8.4.

(1) Write a function m-file named `coef_fourier.m` that is similar to `coef_legen` and has outline

```
function [z,s,c]=coef_fourier(func,n)
  % [z,s,c]=coef_fourier(func,n)
  % more comments

  % your name and the date

  << your code >>

end
```

to compute the first $2n + 1$ coefficients of the Fourier series using Equation (8.10).

(2) Test your coefficient function by using $f(x) = 1/\sqrt{2}$, $f(x) = \sin(2\pi x)$ and $f(x) = \cos(3\pi x)$, with $n \geq 3$. Of course, you should get $z = 1$, and all others zero in the first case, $s_2 = 1$ and all others zero in the second case, and $c_3 = 1$ with all others zero in the third case.

Warning: The `integral` function requires that its integrand be a function that returns a vector when its argument (x) is a vector. Be sure that you use such a function in the $f(x) = 1/\sqrt{2}$ case! You can check your function by computing $\int_{-1}^{1}(1/\sqrt{2})^2 \, dx$ and checking that it equals 1.

(3) Write a function m-file called `eval_fourier.m` to evaluate Equation (8.9) and have the outline

```
function yval=eval_fourier(z,s,c,xval)
  % yval=eval_fourier(z,s,c,xval)
  % more comments

  % your name and the date

  << your code >>

end
```

(4) Test `eval_fourier.m` using the same three functions you used above: $f(x) = 1/\sqrt{2}$, $f(x) = \sin(2\pi x)$ and $f(x) = \cos(3\pi x)$. In each case, use `eval_fourier.m` with the appropriate choices of coefficients z, s and c, and compare the approximate values at a selection of values against the true values. Describe the values you chose and the results you obtained.

(5) Write an m-file called `test_fourier.m`, similar to the `test_mon.m` and `test_legen.m` file you wrote above. It should use `eval_fourier` and produce the relative error of the approximation. It is instructive if you plot the approximation as well, but it is not required.

(6) Fill in the following table for the Runge example function.

Runge test function		
n	relative error	elapsed time
1		—
2		—
3		—
4		—
5		—
6		—
10		—
50		—
100		
200		
400		
800		

(7) Fill in the following table for the partly quadratic function.

partly_quadratic function		
n	relative error	elapsed time
1		—
2		—
3		—
4		—
5		—
6		—
10		—
50		—
100		
200		
400		
800		

(8) When you used trigonometric polynomial interpolation in Chapter 7, you looked at the error for a sawshape function and saw Gibbs's phenomenon, which kept the error from going to zero. You have seen good performance of Fourier approximation on differentiable and continuous functions above. A discontinuous function exhibits Gibbs's phenomonon, but when convergence is measured using an integral norm it doesn't prevent convergence (although it slows it down). Fill in the following table for the sawshape8 function.

sawshape8 function		
n	relative error	elapsed time
1		—
2		—
3		—
4		—
5		—
6		—
10		—
50		—
100		
200		
400		
800		

(9) Based on the above data, roughly estimate the value of p where elapsed time is proportional to n^p. Is $1 \le p \le 2$? $2 \le p \le 3$? $3 \le p \le 4$?

You should be convinced that these series do not converge very rapidly, with execution times becoming much too large to achieve high accuracy. This increase in execution time is due to the adaptive quadrature used by `integral` requiring progressively more points. It turns out that increasing n beyond about 2000 results in failure of `integral` to achieve the desired accuracy. Thus, the `sawshape8` and `partly_quadratic` functions cannot be integrated to the accuracy that the Runge example function can achieve.

8.8   Piecewise constant approximation

You have learned that approximation is best done using matrices that are easy to invert accurately, like diagonal matrices. This is the reason for using sets of orthogonal basis functions. One would also like to be able to perform the right side integrals easily as well. A large part of the reason that the orders of approximations in the exercises above have been restricted is that the integrals are difficult to perform accurately because they "wiggle" a lot, a major source of inaccuracy in the approximation. Using the `integral` function hides the inaccuracy, but you pay for it because the integrations require substantial time for higher values of n, as you may have noticed.

In this section, you will look at approximation by piecewise constant functions. Approximation by piecewise linears or higher are also useful, but all the important steps are covered in the simpler piecewise constant case. Furthermore, piecewise constants are very easy to extend to higher dimensions.

Suppose that a number N_{pc} is given and that the interval $[-1, 1]$ is divided into N_{pc} equal subintervals and $N_{\text{pc}} + 1$ points x_k, $k = 1, 2, \ldots, N_{\text{pc}} + 1$. For $k = 1, \ldots, N_{\text{pc}}$, a function $u_k(x)$ can be defined as

$$u_k(x) = \begin{cases} 1 & x_k \leq x < x_{k+1} \\ 0 & x < x_k \text{ or } x > x_{k+1} \end{cases}.$$

These functions clearly satisfy

$$\int_{-1}^{1} u_k(x) u_\ell(x) dx = \begin{cases} 2/N_{\text{pc}} & k = \ell \\ 0 & k \neq \ell \end{cases}. \tag{8.11}$$

This orthogonality immediately implies linear independence. In addition, any function in $L^2([-1, 1])$ can be approximated as a sum of them (this is a deep theorem). As it turns out, these theoretical facts are not compromised by numerical difficulties and for reasonable values of n can be used for numerical approximation.

If a vector of coefficients **a** can be found to represent the piecewise constant approximation to a function $f(x)$, then the approximation can be evaluated as

$$f(x) \approx f_{\text{pc}}(x) = \sum_{j=1}^{N_{\text{pc}}} a_j u_j(x) = a_k \tag{8.12}$$

where k is the index satisfying $x_k \leq x < x_{k+1}$.

In the following exercise, you will follow the same basic recipe as before to compute the coefficients **a** and the approximation to $f(x)$.

Exercise 8.5. In this exercise you will be working with these piecewise constant (pc) functions. You may assume that N_{pc} is even so that $x_k < 0$ for $k \leq N_{\text{pc}}/2$, that $x_k > 0$ for $k > N_{\text{pc}}/2 + 1$ and $x_k = 0$ for $k = N_{\text{pc}}/2 + 1$.

(1) Write a function m-file named `coef_pc.m` with outline

```
function a=coef_pc(func,Npc)
  % a=coef_pc(func,Npc)
  % comments

  % your name and the date

  << your code >>

end
```

to compute the coefficients of the approximation as

$$a_k = \frac{N_{\text{pc}}}{2} \int_{-1}^{1} f(x) u_k(x) dx$$
$$= \frac{N_{\text{pc}}}{2} \int_{x_k}^{x_{k+1}} f(x) dx.$$

Use `integral` (or if you are using an older version of MATLAB `quadgk`), not `ntgr8` to compute these integrals, because the interval of integration is not `[-1,1]`. To write this function, you will need to use `linspace` to generate the points x_k. Be careful not to confuse the number of points with the number of intervals!

(2) Test your `coef_pc` on the function that is equal to one for all values of x. This can be done in MATLAB with `y=ones(size(x))`. Use `Npc=10`. Of course, all $a_k = 1$.

(3) Test `coef_pc` with `Npc=10` on the function $f(x) = x$. You should get

$$a_k = \frac{N_{pc}}{2} \int_{x_k}^{x_{k+1}} x \, dx$$

$$= \frac{N_{pc}}{4} (x_{k+1}^2 - x_k^2)$$

$$= \frac{2k}{N_{pc}} - 1 - \frac{1}{N_{pc}}.$$

(4) In Chapter 6, the function `bracket` is used to determine the values of k for which $x_k \le x < x_{k+1}$. You can download it with the functions from that chapter. Use `bracket` to write a function m-file called `eval_pc.m` to evaluate the piecewise constant approximation to f using Equation (8.12) and has the outline

```
function yval=eval_pc(a,xval)
% yval=eval_pc(a,xval)
% comments

% your name and the date

<< your code >>


end
```

As in `coef_pc`, you will need to use `linspace` to generate the points x_k, and `bracket` to find the values of k corresponding to the values of `xval`.

(5) Generate the coefficients a using the function $f(x) = x$ and `Npc=10` that you used above. Test `eval_pc` using

```
xval=[-0.95,-0.65,-0.45,-0.25,-0.05,0.15,0.35,0.55,0.75,0.95]
```

where these points are chosen to be in the interiors of each of the 10 subintervals of $[-1, 1]$. What values did you get? Be sure your answers are correct before continuing.

(6) Write an m-file called `test_pc.m` similar to `test_mon.m` and `test_legen.m` above. You should use vector (componentwise) statements whenever possible or the calculations might take a long time. `test_pc.m` should:

(a) Confirm that Npc is an even number (and call error if not).

(b) Evaluate the coefficients (a) of the approximation using coef_pc.m.

(c) Use eval_pc.m to evaluate the approximation and then compare the approximation against the exact solution, *Because you will be using large values of* Npc, *choose at least 20000 test points.*

(d) Use tic and toc to measure the time taken by computing the coefficients, computing the approximate and exact solutions, and computing the error.

It will be valuable to plot the approximation because it will help you debug your work and it will illustrate the process.

(7) Execute your code for the Runge function with Npc=8. Look carefully and critically at the plot. You should be able to justify to yourself visually that no other piecewise constant function would produce a better approximation.

(8) Fill in the following table for the Runge example function:

Runge test function		
n	relative error	elapsed time
4		—
8		—
16		—
64		—
256		—
1024		
4096		
16384		

(9) Fill in the following table for the partly quadratic function:

partly_quadratic function		
n	relative error	elapsed time
4		—
8		—
16		—
64		—
256		—
1024		
4096		
16384		

(10) Fill in the following table for the sawshape8 function:

sawshape8 function		
n	relative error	elapsed time
4		—
8		—
16		—
64		—
256		—
1024		
4096		
16384		

(11) Based on the above data, roughly estimate the integer p where **relative error** is proportional to $(1/n)^p$.

(12) Based on the above data, roughly estimate the value of p where **elapsed time** is proportional to n^p.

This approximation may take a while to compute, but it does not deteriorate as Npc gets large! In fact, you should observe linear convergence. Further, the time required does not grow very quickly. For any of these three functions, accuracy higher than 10^{-10} can be achieved.

Remark 8.8. As you might imagine, approximation using piecewise linear functions will converge more rapidly than using piecewise constants. There are alternative approaches for using piecewise linears: piecewise linear functions on each interval with jumps at interval endpoints, as the piecewise constant functions have; and, piecewise linear functions that are *continuous* throughout whole interval. The first retains orthogonality and the diagonal form of the coefficient matrix H. The second sacrifices the diagonal form for a banded form that is almost as easy to solve, as you may see in the next exercise. Continuity, however, can be worth the sacrifice, depending on the application. Even higher-order piecewise polynomial approximation is possible, if the application can benefit.

Remark 8.9. In Exercises 8.3, 8.4, and 8.5 you were asked to estimate the growth of the running time with n. You found in Exercises 8.3 and 8.4 that the running time increased more rapdily than linearly ("superlinear") as n increases, but in Exercise 8.5 it increased linearly. The distinction between superlinear and linear is important when choosing an algorithm to use. When running time increases too fast, algorithms can become too time-consuming to be useful. Since it is clear that approximation algorithms must scale at least linearly in n, Exercise 8.5 shows that piecewise constant approximation can be an attractive choice. In the exercise below, you will find a similar result for piecewise linear approximation.

8.9   Piecewise linear approximation

Piecewise linear approximations improve the rate of convergence over piecewise constant approximations, at the cost of increased work. In addition, piecewise linear approximations are commonly used in finite element approximations to differential equations. In this exercise, you will see how the same approach you saw above can be extended to piecewise linear approximations.

For this presentation, you again break the interval into equal subintervals. Denote the number of intervals by N_{pl}, although it is the same as N_{pc} above. Thus, there are $N_{pl} + 1$ points defined over the interval $[-1, 1]$ according to the MATLAB function x=linspace(-1,1,Npl+1). For each $k = 1, 2, \ldots, N_{pl} + 1$, define a set of "hat" functions t_k as

$$t_k(x) = \begin{cases} (x_{k+1} - x)/(x_{k+1} - x_k) & x_k \leq x \leq x_{k+1} \text{ and } k \leq N_{pl} \\ (x - x_{k-1})/(x_k - x_{k-1}) & x_{k-1} \leq x \leq x_k \text{ and } k \geq 2 \\ 0 & \text{otherwise} \end{cases}$$

so that t_k is a continuous piecewise linear function that is 1 at x_k and zero at all the other points x_ℓ for $\ell \neq k$. It is possible to show that these functions are linearly independent when considered as members of the Hilbert space $L^2([-1, 1])$. Further, an equation analogous to (8.11) is

$$\int_{-1}^{1} t_k(x)t_\ell(x)dx = \begin{cases} 4/(3N_{pl}) & 2 \leq k = \ell \leq N_{pl} \\ 2/(3N_{pl}) & k = \ell = 1 \text{ or } k = \ell = N_{pl} + 1 \\ 1/(3N_{pl}) & |k - \ell| = 1 \text{ (}i.e.\text{ } k = \ell \pm 1) \\ 0 & |k - \ell| > 1 \end{cases}. \tag{8.13}$$

There are $N_{pl} + 1$ functions $t_k(x)$!

There is an equation analogous to (8.4), (8.8), (8.9), and (8.12):

$$f(x) \approx f_{pl}(x) = \sum_{j=1}^{N_{pl}+1} a_j t_j(x). \tag{8.14}$$

Exercise 8.6.

(1) Write a function m-file named **hat.m** to implement t_k.
(2) To see why the functions are called "hat" functions, plot the function $t_3(x)$ when $N_{pl} = 4$. If you use MATLAB for the plot, be sure that the points you use to plot t_3 includes the points x_k or your plot will not look right. You might find that forcing space around the plot makes it look more like a hat. You could use axis([-1.1, 1.1, -0.1, 1.1]).
(3) Write a function **coef_plin** to compute the coefficients a_j in (8.14). The system of equations you will need to generate and solve are analogous to (8.6) and can be constructed by replacing j with ℓ in (8.14), multiplying by $t_k(x)$ and integrating. Do not forget to construct a matrix analogous to $H_{k\ell}$. Devise a simple test meant for debugging and test **coef_plin**.

Hint: The functions $t_k(x)$ have limited support, especially for large N_{pl}. Adjust the limits of integration to reflect the support in order to save time (and, it turns out) improve accuracy. You will need to use the `integral` or `quadgk` commands directly instead of through `ntgr8` since `ntgr8` assumes an integration interval of [-1,1].

(4) Write a function `eval_plin`. Devise a test (perhaps one using $N_{\text{pl}} = 2$ or 3) and test `eval_plin`.

(5) Write a function `test_plin`, including timing, and apply it to the three functions you have been using: `runge partly_quadratic` and `sawshape8`. Use values of `Npl=[4,16,64,256,1024]`. In each case, estimate the rate of convergence and rate of increased time as `Npl` varies.

Remark 8.10. You should observe improved convergence rate for the two continuous functions and no better rate for the discontinuous `sawshape8`. The improved rate is achieved without increasing the rate of increase of execution time. Had you run the `sawshape8` case to more intervals, you would probably have noticed a more rapid increase in execution time than expected.

Chapter 9

Quadrature

9.1 Introduction

The term "numerical quadrature" refers to the estimation of an area, or, more generally, any integral. (The term "cubature" refers to estimating a volume in three dimensions, but most people just use the term "quadrature.") You might want to integrate some function $f(x)$ or a set of tabulated data. The domain could be a finite or infinite interval, or a rectangle or irregular shape, or a multi-dimensional volume.

The first topic is the "degree of exactness" (sometimes called the "degree of precision") of a quadrature rule, and its relation to the "order of accuracy." Simple tests to measure the degree of exactness and the order of accuracy of a rule are discussed, followed by (simple versions of) the midpoint and trapezoidal rules. Then the Newton–Cotes and Gauß–Legendre families of rules are examined, providing more accurate approximations by either increasing the order of the rule or subdividing the interval. Finally, you will see one way of adapting the details of the integration method to meet a desired error estimate. In the majority of the exercises below, you will be more interested in the error (difference between calculated and known value) than in the calculated value itself.

A word of caution. There are three similar-sounding concepts:

- "Degree of exactness:" the largest value of n so that all polynomials of degree n and below are integrated *exactly*. (Degree of a polynomial is the highest power of x appearing in it.)
- "Order of accuracy:" the value of n so that the error is $O(h^n)$, where h measures the subinterval size.
- "Index:" a number distinguishing one of a collection of rules from another.

These can be related to one another, but are not the same thing.

Remark 9.1. There is a nice way to use anonymous functions in MATLAB to streamline a sequence of calculations computing the integrals of ever higher degree polynomials in order to find the degree of exactness of a quadrature rule. The following statement

```
q=midpointquad(@(x) 5*x.^4,0,1,11);1-q
```

first computes $\int_0^1 5x^4 dx$ using the midpoint rule, and then prints the error (=1-q because the exact answer is 1). You would only have to change 5*x.^4 into 4*x.^3 to check the error in $\int_0^1 4x^3 dx$, and you can make the change with judicious use of the arrow and other keyboard keys.

Remark 9.2. Errors should be reported in scientific notation (like 1.234e-3, not 0.0012). You can force MATLAB to display numbers in this format using the command `format short e` (or `format long e` for 15 decimal places). This is particularly important if you want to visually estimate ratios of errors.

Computing ratios of errors should *always* be done using full precision, not the value printed on the screen. For example, you might use code like

```
err20=midpointquad(@runge,-5,5,20)-2*atan(5);
err40=midpointquad(@runge,-5,5,40)-2*atan(5);
ratio=err20/err40
```

to get a ratio of errors without loss of accuracy due to reading numbers off the computer screen.

In general, it easier to compute ratios as "larger divided by smaller," yielding ratios larger than 1. It is easier to recognize that 15 is nearly 2^4 (=16) than to recognize that 0.0667 is nearly 2^{-4} (=0.0625).

This chapter contains a relatively large number of short exercises. Exercises 9.1 and 9.2 are a pair, addressing the midpoint rule. Exercise 9.4 applies the midpoint rule to an integrand with a singularity at one endpoint. Exercise 9.3 is similar to 9.1 and 9.2 but addressing the trapezoid rule. Exercises 9.5 through 9.7 address closed Newton–Cotes rules, and Exercises 9.8 through 9.10 independently address Gauß–Legendre integration. Exercise 9.11 discusses infinite integration limits through change of variables and does not depend on the integration rule. Exercises 9.12 through 9.16 present a simple adaptive approach (based on Gauß–Legendre but easily changed to Newton–Cotes, although Newton–Cotes cannot be used in Exercise 9.16) to choose the subintervals that meet the accuracy criterion, including several examples.

9.2   The midpoint method

In general, numerical quadrature involves breaking an interval $[a, b]$ into subintervals, estimating or modelling the function on each subinterval and integrating it there, then adding up the partial integrals.

Perhaps the simplest method of numerical integration is the midpoint method, presented in [Quarteroni *et al.* (2007)] on page 381, [Atkinson (1978)] on page 269, and [Ralston and Rabinowitz (2001)] on page 120. This method is based on interpolation of the integrand $f(x)$ by the constant $f((a + b)/2)$ and multiplying by the width of the interval. The result is a form of Riemann sum that you probably saw in elementary calculus when you first studied integration.

If the interval $[a, b]$ is broken into $N - 1$ subintervals with endpoints $x_1, x_2, \ldots, x_{N-1}, x_N$ (there is one more endpoint than intervals) then the midpoint rule can be written as

$$\textbf{Midpoint rule} = \sum_{k=1}^{N-1} (x_{k+1} - x_k) f(\frac{x_k + x_{k+1}}{2}). \qquad (9.1)$$

In the exercise that follows, you will be writing a MATLAB function to implement the midpoint rule.

Exercise 9.1.

(1) Write a function m-file called `midpointquad.m` with outline (found in the file `midpointquad.txt`)

```
function quad = midpointquad( func, a, b, N)
  % quad = midpointquad( func, a, b, N)
  % comments

  % your name and the date

  << your code >>

end
```

where `func` indicates the name of a function, `a` and `b` are the lower and upper limits of integration, and `N` is the *number of points*, not the number of intervals. The code for your m-file might look like the following:

```
xpts = linspace( ??? ) ;
h = ??? ; % length of subintervals
xmidpts = 0.5 * ( xpts(1:N-1) + xpts(2:N) );
fmidpts = ???
quad = h * sum ( fmidpts );
```

(2) Test your `midpointquad` routine by computing $\int_0^1 2x\,dx = 1$. Even if you use only one interval (*i.e.,* N=2) you should get the exact answer because the midpoint rule integrates linear functions exactly.

(3) Use your `midpoint` routine to estimate the integral of the familiar Runge example function, $f(x) = 1/(1 + x^2)$, over the interval $[-5, 5]$. The exact answer is 2*atan(5). Fill in the following table, using scientific notation for the error values so you can see the pattern.

Midpoint rule			
N	h	Midpoint result	Error
11	1.0		
101	0.1		
1001	0.01		
10001	0.001		

(4) Estimate the order of accuracy (an integer power of **h**) by examining the behavior of the error when **h** is divided by 10. (In previous chapters, you have estimated such orders by repeatedly doubling the number of subintervals. Here, you multiply by ten. The idea is the same.)

9.3   Exactness

If a quadrature rule can compute exactly the integral of any polynomial up to some specific degree, you will call this its *degree of exactness*. Thus a rule that can correctly integrate any cubic, but not quartics, has exactness 3. See, for example, [Quarteroni *et al.* (2007)], page 429 or [Atkinson (1978)], page 266.

To determine the degree of exactness of a rule, you might look at the approximations of the integrals

$$\int_0^1 1 dx = [x]_0^1 \quad = 1$$
$$\int_0^1 2x dx = [x^2]_0^1 \quad = 1$$
$$\int_0^1 3x^2 dx = [x^3]_0^1 \quad = 1$$
$$\vdots \qquad \qquad \vdots$$
$$\int_0^1 (k+1)x^k dx = [x^{k+1}]_0^1 = 1.$$

Exercise 9.2.

(1) To study the degree of exactness of the midpoint method, use a single interval (*i.e.*, N = 2), and estimate the integrals of the test functions over [0,1]. The exact answer is 1 each time.

func	Midpoint result	Error
1		
$2x$		
$3x^2$		
$4x^3$		

(2) What is the degree of exactness of the midpoint rule?
(3) Recall that you computed the order of accuracy of the midpoint rule in Exercise 1. For some methods, but not all, the degree of exactness is one less than the order of accuracy. Is that the case for the midpoint rule?

9.4 The trapezoid method

The *trapezoid rule* breaks [a,b] into subintervals, approximates the integral on each subinterval as the product of its width times the average function value, and then adds up all the subinterval results, much like the midpoint rule. The difference is in how the function is approximated. The trapezoid rule can be written as

$$\textbf{Trapezoid rule} = \sum_{k=1}^{N-1} (x_{k+1} - x_k) \frac{f(x_k) + f(x_{k+1})}{2}. \tag{9.2}$$

If you compare the midpoint rule (9.1) and the trapezoid rule (9.2), you will see that the midpoint rule takes f at the midpoint of the subinterval and the trapezoid takes the average of f at the endpoints. If each of the subintervals happens to have length h, then the trapezoid rule becomes

$$\frac{h}{2} f(x_1) + \frac{h}{2} f(x_N) + h \sum_{k=2}^{N-1} f(x_k). \tag{9.3}$$

To apply the trapezoid rule, you need to generate N points and evaluate the function at each of them. Then, apply either (9.2) or (9.3) as appropriate.

Exercise 9.3.

(1) Use your `midpointquad.m` m-file as a model and write a function m-file called `trapezoidquad.m` to evaluate the trapezoid rule. The outline of your m-file should be

```
function quad = trapezoidquad( func, a, b, N )
  % quad = trapezoidquad( func, a, b, N )
  % more comments

  % your name and the date

  << your code >>

end
```

You may use either form of the trapezoid rule.

(2) To test your routine and to study the exactness of the trapezoid rule, use a single interval (N = 2), and estimate the integrals of the same test functions used for the midpoint rule over [0,1]. The exact answer should be 1 each time.

func	Trapezoid result	Error
1		
$2x$		
$3x^2$		
$4x^3$		

(3) What is the degree of exactness of the trapezoid rule?
(4) Use the trapezoid method to estimate the integral of the Runge function over $[-5, 5]$, using the given values of N, and record the error using scientific notation.

Runge example function			
N	h	Trapezoid result	Error
11	1.0		
101	0.1		
1001	0.01		
10001	0.001		

(5) Estimate the rate at which the error decreases as h decreases. (Find the power of h that best fits the error behavior.) This is the order of accuracy of the rule.
(6) For some methods, but not all, the degree of exactness is one less than the order of accuracy. Is that the case for the trapezoid rule?

9.5   Singular integrals

The midpoint and trapezoid rules seem to have the same exactness and about the same accuracy. There is a difference between them, though. Some integrals have perfectly well-defined values even though the integrand has some sort of mild singularity. There are some sophisticated ways to perform these integrals, but there is a simple way that can be adequate for the case that the singularity appears at the endpoint of an interval. Something is lost, however.

Consider the integral

$$I = \int_0^1 \log(x)dx = -1, \tag{9.4}$$

where log refers to the natural logarithm. Note that the integrand "is infinite" at the left endpoint, so you could not use the trapezoid rule to evaluate it. The midpoint rule, conveniently, does not need the endpoint values.

Exercise 9.4. Apply the midpoint rule to the integral in (9.4), and fill in the following table.

$\int_0^1 \log(x)\,dx$			
N	h	Midpoint result	Error
11	1.0		
101	0.1		
1001	0.01		
10001	0.001		

Estimate the rate of convergence (power of h) as $h \to 0$. You should see that the singularity causes a loss in the rate of convergence.

9.6   Newton–Cotes rules

Look at the trapezoid rule for a minute. One way of interpreting that rule is to say that if the function f is roughly linear over the subinterval $[x_k, x_{k+1}]$, then the integral of f is the integral of the linear function that agrees with f (*i.e.*, interpolates f) at the endpoints of the interval. What about trying higher-order methods? It turns out that Simpson's rule can be derived by picking triples of points, interpolating the integrand f by a quadratic polynomial, and integrating the quadratic. The trapezoid rule and Simpson's rule are Newton–Cotes rules of index one and index two, respectively. In general, a Newton–Cotes formula uses the idea that if you approximate a function by a polynomial interpolant on uniformly-spaced points in each subinterval, then you can approximate the integral of that function with the integral of the polynomial interpolant. This idea does not always work for derivatives but usually does for integrals. The polynomial interpolant in this case being taken on a uniformly distributed set of points, including the end points. The number of points used in a Newton–Cotes rule is a fundamental parameter, and can be used to characterize the rule. The "index" of a Newton–Cotes rule is commonly defined as one fewer than the number of points it uses, although this common usage is not universal.

You applied the trapezoid rule to an interval by breaking it into subintervals and repeatedly applying a simple formula for the integral on a single subinterval. Similarly, you will be constructing higher-order rules by repeatedly applying Newton–Cotes rules over subintervals. But Newton–Cotes formulæ are not so simple as the trapezoid rule, so you will first write a helper function to apply the rule on a single subinterval.

Over a single interval, all (closed) Newton–Cotes formulæ can be written as

$$\int_a^b f(x)dx \approx Q_N(f) = \sum_{k=1}^N w_{k,N} f(x_k)$$

where f is a function and x_k are N **evenly-spaced** points between a and b. The weights $w_{k,N}$ can be computed from the Lagrange interpolation polynomials $\ell_{k,N}$ as

$$w_{k,N} = (b-a) \int_0^1 \ell_{k,N}(\xi)d\xi.$$

The Lagrange interpolation polynomials arise because you are doing a polynomial interpolation. See, for example, [Quarteroni *et al.* (2007)] page 387, [Atkinson (1978)] page 263, or [Ralston and Rabinowitz (2001)] page 118. The weights do not depend on f, and depend on a and b in a simple manner, so they are often tabulated for the unit interval. In the exercise below, they are provided for download in the form of the function `nc_weight.m`.

Remark 9.3. There are also open Newton–Cotes formulæ that do not require values at endpoints, but there is not time to consider them in this chapter.

Exercise 9.5.

(1) Write a routine called `nc_single.m` with the following outline. (You may find it convenient to download the file `nc_single.txt`.)

```
function quad = nc_single ( func, a, b, N )
  % quad = nc_single ( func, a, b, N )
  % more comments

  % your name and the date

  << your code >>

end
```

There are no subintervals in this case. The coding might look like something like this:

```
xvec = linspace ( a, b, N );
wvec = nc_weight ( N );
fvec = ???
quad = (b-a) * sum(wvec .* fvec);
```

(2) Test your function by showing its exactness is at least 1 for $N = 2$: $\int_0^1 2x\,dx = 1$ exactly.

(3) Fill in the following table by computing the integrals over [0,1] of the indicated integrands using `nc_single`. It is well-known for this case (see, for example [Quarteroni *et al.* (2007)], Theorem 9.2, [Atkinson (1978)], page 266 or [Ralston and Rabinowitz (2001)], page 119) that the degree of exactness is equal to $(N - 1)$ when N is even and the degree of exactness is N when N is odd. Your results should agree, further confirming that your function is correct. (Hint: You can use anonymous functions to simplify your work.)

func	Error ($N = 4$)	Error ($N = 5$)	Error ($N = 6$)
$4x^3$			
$5x^4$			
$6x^5$			
$7x^6$			
Degree			

The objective of numerical quadrature rules is to *accurately* approximate integrals. You have already seen that polynomial interpolation on uniformly spaced points does not always converge, so it should be no surprise that increasing the order of Newton–Cotes integration might not produce accurate quadratures.

Exercise 9.6. Attempt to get accurate estimates of the integral of the Runge function over the interval $[-5, 5]$. Recall that the exact answer is `2*atan(5)`. Fill in the following table

N	nc_single result	Error
3		
7		
11		
15		

The results of Exercise 9.6 should have convinced you that you raising N in a Newton–Cotes rule is not the way to get increasing accuracy. One alternative to raising N is breaking the interval into subintervals and using a Newton–Cotes rule on each subinterval. This is the idea of a "composite" rule. In the following exercise you will use `nc_single` as a helper function for a composite Newton–Cotes routine. You will also be using the `partly_quadratic` function from Chapter 8.

$$f_{\text{partly quadratic}} = \begin{cases} 0 & -1 \leq x < 0 \\ x(1-x) & 0 \leq x \leq 1 \end{cases}$$

whose MATLAB implementation is

```
function y=partly_quadratic(x)
  % y=partly_quadratic(x)
  % input x (possibly a vector or matrix)
  % output y, where
  % for x<=0, y=0
  % for x>0,  y=x(1-x)

  y=(heaviside(x)-heaviside(x-1)).*x.*(1-x);
end
```

Clearly, $\int_{-1}^{1} f_{\text{partly quadratic}}(x)\,dx = \int_{0}^{1} x(1-x)dx = \frac{1}{6}$.

Exercise 9.7.

(1) Write a function m-file called `nc_quad.m` to perform a composite Newton–Cotes integration. Use the following outline.

```
function quad = nc_quad( func, a, b, N, numSubintervals)
  % quad = nc_quad( func, a, b, N, numSubintervals)
  % comments

  % your name and the date

  << your code >>

end
```

This function will perform these steps:

(a) break the interval into `numSubintervals` subintervals;
(b) use `nc_single` to integrate over each subinterval; and,
(c) add them up.

(2) The most elementary test to make when you write this kind of routine is to check that you get the same answer when `numSubintervals=1` as you would have obtained using `nc_single`. Choose at least one line from the table in Exercise 9.6 and make sure you get the same result using `nc_quad`.

(3) Test your routine by computing $\int_{-1}^{1} f_{\text{partly quadratic}}(x)\,dx$ using at least N=3 and `numSubintervals=2`. Explain why your result should have an error of zero or roundoff-sized.

(4) Test your routine by computing $\int_{-1}^{1} f_{\text{partly quadratic}}(x)\,dx$ using at least N=3 and `numSubintervals=3`. Explain why your result should *not* have an error of zero or roundoff-sized.

(5) Test your routine by checking the following value

```
nc_quad(@runge, -5, 5, 4, 10) = 2.74533025
```

(6) Fill in the following table using the Runge function on $[-5, 5]$, the exact integral of which is `2*atan(5)`.

Subintervals	N	nc_quad error	ratio
10	2		
20	2		
40	2		
80	2		
160	2		
320	2		—
10	3		
20	3		
40	3		
80	3		
160	3		
320	3		—
10	4		
20	4		
40	4		
80	4		
160	4		
320	4		—

(7) For each index, estimate the order of convergence by taking the sequence of ratios of the error for num subintervals divided by the error for (2*num) subintervals and guessing p, the power of two that best approximates the limit of the sequence.

In the previous exercise, the table served to illustrate the behavior of the integration routine. Suppose, on the other hand, that you had an integration routine and you wanted to be sure it had no errors. It is not good enough to just see that you can get "good" answers. In addition, it must converge at the correct rate. Tables such as the previous one are one of the most powerful debugging and verification tools a researcher has.

9.7 Gauß quadrature

Like Newton–Cotes quadrature, Gauß–Legendre quadrature interpolates the integrand by a polynomial and integrates the polynomial. Instead of uniformly spaced points, Gauß–Legendre uses optimally-spaced points. Furthermore, Gauß–Legendre converges as degree gets large, unlike Newton–Cotes, as you saw above. In real applications, one does not use higher and higher degrees of quadrature; instead, one uses more and more subintervals, each with some fixed degree of quadrature.

The disadvantage of Gauß–Legendre quadrature is that there is no easy way to compute the node points and weights. [Quarteroni *et al.* (2007)], in Section 10.2, present a program `zplege.m` that does the job. Tables of values are generally available. See, for example, Table 5.10 in [Atkinson (1978)], page 276 or [Ralston and Rabinowitz (2001)], Table 4.1, page 100. One very careful way to compute the node points and weights is given by [Rutishauser (1962)], and is available as a Fortran program from Netlib (toms/125). You will be using a MATLAB function named `gl_weight.m` that is included among the files for this chapter to serve as a table of node points and weights.

Normally, Gauß–Legendre quadrature is characterized by the number of integration points. For example, people speak of "three-point" Gauß.

The following two exercises involve writing m-files analogous to `nc_single.m` and `nc_quad.m`.

Exercise 9.8.

(1) Write a routine called `gl_single.m` with the outline

```
function quad = gl_single ( func, a, b, N )
  % quad = gl_single ( func, a, b, N )
  % comments

  % your name and the date

  << your code >>

end
```

As with `nc_single` there are no subintervals in this case. Your coding might look like something like this:

```
[xvec, wvec] = gl_weight ( a, b, N );
fvec = ???
quad = sum( wvec .* fvec );
```

(2) Test your function by showing its exactness is at least 1 for N=1 and one interval: $\int_0^1 2x\,dx = 1$ exactly. If the exactness is not at least 1, fix your code now.
(3) Fill in the following table by computing the integrals over [0,1] of the indicated integrands using `gl_single`. The degree of exactness of the method is $2N - 1$, (see, for example, [Quarteroni *et al.* (2007)], Corollary 10.2, [Atkinson (1978)], page 272, or [Ralston and Rabinowitz (2001)], page 98–99) and your results should agree, further confirming that your function is correct. (Hint: You can use anonymous functions to simplify your work.)

func	Error ($N = 2$)	Error ($N = 3$)
$3x^2$		
$4x^3$		
$5x^4$		
$6x^5$		
$7x^6$		
Degree		

(4) Get accuracy estimates of the integral of the Runge function over the interval $[-5, 5]$. Recall that the exact answer is `2*atan(5)`. Fill in the following table

N	`gl_single` result	Error
3		
7		
11		
15		

You might be surprised at how much better Gauß–Legendre integration is than Newton–Cotes, using a single interval. There is a similar advantage for composite integration, but it is hard to see for small N. When Gauß–Legendre integration is used in a computer program, it is generally in the form of a composite formulation because it is difficult to compute the weights and integration points accurately for high order Gauß–Legendre integration. The efficiency of Gauß–Legendre integration is compounded in multiple dimensions, and essentially all computer programs that use the finite element method use composite Gauß–Legendre integration rules to compute the coefficient matrices.

Exercise 9.9.

(1) Write a function m-file called `gl_quad.m` to perform a composite Gauß–Legendre integration. Use the following outline.

```
function quad = gl_quad( f, a, b, N, numSubintervals)
  % quad = gl_quad( f, a, b, N, numSubintervals)
  % comments

  % your name and the date

  << your code >>

end
```

This function will perform three steps:

(a) break the interval into `numSubintervals` subintervals;

(b) use `gl_single` to integrate over each subinterval; and,

(c) add them up.

(2) The most elementary test to make when you write this kind of routine is to check that you get the same answer when `numSubintervals=1` as you would have obtained using `gl_single`. Choose at least one line from the table in the previous exercise (9.8) and make sure you get the same result using `gl_quad`.

(3) Test your routine by computing $\int_{-1}^{1} f_{\text{partly quadratic}}(x)\, dx$ using at least N=3 and `numSubintervals=2`. Explain why your result should have an error of zero or roundoff-sized.

(4) Test your routine by computing $\int_{-1}^{1} f_{\text{partly quadratic}}(x)\, dx$ using at least N=3 and `numSubintervals=3`. Explain why your result should *not* have an error of zero or roundoff-sized.

(5) Test your routine by checking the following value

```
gl_quad(@runge, -5, 5, 4, 10) = 2.7468113
```

(6) Fill in the following two tables using the Runge function on $[-5, 5]$, the exact integral of which is `2*atan(5)`.

Subintervals	N	gl_quad error	ratio
10	1		
20	1		
40	1		
80	1		
160	1		
320	1		—
10	2		
20	2		
40	2		
80	2		
160	2		
320	2		—

Subintervals	N	gl_quad error	ratio
45	3		
90	3		—
46	3		
92	3		—
47	3		
94	3		—
48	3		
96	3		—
49	3		
98	3		—

(7) For N=1 and 2, estimate the order of convergence by taking the sequence of ratios of the error for num subintervals divided by the error for (2*num) subintervals and guessing the power of two that best approximates the limit of the sequence.

(8) Estimate the order of accuracy for N=3 by taking the five ratios of errors $r_k = e_k/e_{2k}$ for $k = 45, \ldots, 49$, take their geometric mean $r = (\prod_{k=45}^{49} r_k)^{1/5}$, and guess the power of two that best approximates r.

Remark 9.4. The reason that the N=3 case is different from the others is that the errors become near roundoff and averaging is necessary to smooth out the resulting values. Geometric averaging is appropriate because the theoretical error curve is a straight line on a log–log plot and geometrical averaging is the same as arithmetic averaging of logs.

The proofs you have seen about convergence of Gauß quadrature rely on bounds on higher derivatives of the function. When bounds are not available, higher-order convergence might not be observed.

Exercise 9.10. Consider the integral from Exercise 4

$$I = \int_0^1 \log(x)dx = -1.$$

(1) Use gl_quad to fill in the following table.

$\int_0^1 \log(x)\,dx$			
Subintervals	N	gl_quad error	Error ratio
10	1		
20	1		
40	1		
80	1		—

What is the order of accuracy of the method using N=1?

(2) Use `gl_quad` to fill in the following table.

$\int_0^1 \log(x)\,dx$			
Subintervals	N	gl_quad error	Error ratio
10	2		
20	2		
40	2		
80	2		—

What is the order of accuracy of the method using N=2?

(3) Use `gl_quad` to fill in the following table.

$\int_0^1 \log(x)\,dx$			
Subintervals	N	gl_quad error	Error ratio
10	3		
20	3		
40	3		
80	3		—

What is the order of accuracy of the method using N=3?

It is instructive to see how to compute integrals over infinite intervals. Basically, the best way to do that is to make a change of variables to make the interval finite. There are other ways, such as multiplying the integrand by a weighting function and then using an integration method based on weighted integrals, but you will see how a change of variables works in thee following exercise.

Exercise 9.11. Consider the integral

$$\int_0^\infty \frac{1}{1+x^2}\,dx = \frac{\pi}{2}.$$

Making a change of variables $u = 1/(1+x)$ or $x = (1-u)/u$ yields the integral

$$\int_0^1 \frac{1}{u^2 + (1-u)^2}\,du.$$

Use knowledge of the exact answer and orders of accuracy of various methods to predict the number of points or subintervals needed to evaluate this integral to an accuracy of $\pm 1.e - 8$. Explain why you chose the method you used and how you determined the number of intervals necessary to achieve the specified accuracy.

Remark 9.5. There is no easy way to tell in advance which method to use to achieve a particular accuracy. You can pick a method by trial-and-error, switching methods when you cannot achieve the desired error in the time you are willing to wait. Once you have some trial values, you can use theoretical rates of convergence to help you decide the number of subintervals necessary to reach your desired accuracy.

It is a common strategy in cases where many similar integrals need to be evaluated to choose one case with a known integral, decide which method is most efficient and how many subintervals are needed, and then use that method for all other cases.

In the following section, you will see how accuracy might be improved during the integration process itself, so that a target accuracy can be achieved.

9.8 Adaptive quadrature

Your next task will consider "adaptive quadrature." Adaptive quadrature employs *non-uniform* division of the interval of integration into subintervals of non-equal length. It uses smaller subintervals where the integrand is changing rapidly and larger subintervals where it is flatter. The advantage of this approach is that it minimizes the work necessary to compute a given integral. Adaptive quadrature is discussed, for example, by [Quarteroni *et al.* (2007)], page 402, by [Atkinson (1978)], page 300f, or by [Ralston and Rabinowitz (2001)], page 126f.

In this section, you will investigate a recursive algorithm for adaptively computing quadratures.

Numerical integration is used often with integrands that are very complicated and take a long time to compute. Although this section will use simple integrands for illustrative purposes, you should think of each evaluation of the integrand ("function call") to take a long time. Thus, the objective is to reach a given accuracy with a minimum number of function calls. A good strategy for achieving a specified accuracy efficiently is to attempt to uniformly distribute the error over each subinterval. If no interval is particularly bad, there is no obvious place to improve the estimate and if no interval is particularly good, no work has been wasted.

Recall that, if an integration method has degree of exactness p, then the local error on an integration interval of length h satisfies an expression involving a constant C and a point located somewhere in the interval. Assuming the derivative of the function is roughly constant in an interval, then C can be estimated by dividing the interval into two subintervals of length $h/2$ each, estimating the error in the interval as the sum of the errors on the two subintervals, and then equating the two expressions. Denote by Q_h the integral over the interval of length h, then

$$Q_h = Q + Ch^{p+2}f^{(p+1)}(\xi) + O(h^{p+3})$$

and $Q_{h/2}^L$ and $Q_{h/2}^R$ the two estimates of the integral on the left and right subintervals are

$$Q_{h/2}^L + Q_{h/2}^R = Q + C(h/2)^{p+2}(f^{(p+1)}(\xi^L) + f^{(p+1)}(\xi^R)) + O(h^{p+3})$$

where C is a constant, and ξ, ξ^L, and ξ^R are appropriately chosen. Assuming that $f^{(p+1)}$ is roughly constant, $f^{(p+1)}(\xi) = f^{(p+1)}(\xi^L) = f^{(p+1)}(\xi^R) = f^{(p+1)}$, and assuming that the higher-order terms can be neglected yields

$$Q_h = Q + Ch^{p+2}f^{(p+1)} \tag{9.5}$$

$$Q_{h/2}^L + Q_{h/2}^R = Q + C(h/2)^{p+2}(2f^{(p+1)}). \tag{9.6}$$

Eliminating $Ch^{p+2}f^{p+1}$ from the system (9.5)–(9.6) and defining the error as $|Q_{h/2}^L + Q_{h/2}^R - Q|$ yields the expression

$$\text{error estimate} = \frac{|Q_{h/2}^L + Q_{h/2}^R - Q_h|}{2^{p+1} - 1}. \tag{9.7}$$

Supposing that the error estimate is small enough, should the value (9.5) or (9.6) be used for Q? In principle, either one will do, but it is clear that the error term in (9.6) is smaller (by a factor of $1/2^{p+1}$) than the error term in (9.5), so Q should be estimated from the two half-interval integrals.

The basic structure of one simple adaptive algorithm depends on using the error estimate (9.7) over each integration subinterval. If the error is acceptably small over each interval, the process stops, and, if not, continues recursively. In the following exercise, you will write a *recursive* function to implement this procedure.

Exercise 9.12.

(1) Create a function m-file named `adaptquad.m` with the following code, (available for download as `adaptquad.txt`) and fill in the places marked "???".

```
function [Q,errEst,x,recursions]= ...
          adaptquad(func,x0,x1,tol,recursions)
  % [Q,errEst,x,recursions]=
  %         adaptquad(func,x0,x1,tol,recursions)
  % adaptive quadrature
  %     input parameters
  % func       = function to integrate
  % x0         = left end point
  % x1         = right end point
  % tol        = desired accuracy
  % recursions = number of allowable recursions left
  %
  %     output parameters
  % Q          = estimate of the value of the integral
  % errEst     = estimate of error in Q
  % x          = all intermediate integration points
  % recursions = minimum number of recursions remaining
  %              after convergence
```

```
% your name and the date

% Add a mid-point and re-estimate integral
xmid=(x0+x1)/2;

% Qleft and Qright are integrals over two halves
N=3;
Qboth=gl_single(func,x0,x1,N);
Qleft=gl_single(func,x0,xmid,N);
Qright=gl_single( ??? );

% p=degree of exactness of Gauss-Legendre
p=2*N-1;
errEst= ??? ;

if errEst<tol | recursions<=0  %vertical bar means "or"
  % either ran out of recursions or converged
  Q= ??? ;
  x=[x0 xmid x1];
else
  % not converged -- do it again
  [Qleft,estLeft,xleft,recursLeft]=adaptquad(func, ...
             x0,xmid,tol/2,recursions-1);
  [Qright,estRight,xright,recursRight]=adaptquad(func, ...
             ??? );
  % recursive work is all done, return answers
  % don't want xmid to appear twice in x
  x=[xleft xright(2:length(xright))];
  Q= ??? ;
  errEst= ??? ;
  recursions=min(recursLeft,recursRight);
  end
end
```

Remark 9.6. The input and output parameter `recursions` is not theoretically necessary, but is used to guard against infinite recursion. The output vector `x` is not necessary, either, but will be used to show the effect of the adaptation.

(2) Test `adaptquad` by applying it to the polynomial $f_5(x) = 6x^5$ on the interval $[0, 1]$, using `tol=1.e-5` and `recursions=5`. Because Gauß–Legendre integration of index 3 is exact for this polynomial, the integral should equal 1 (*i.e.*, error should be zero or roundoff), and `recursions=5`, and the values of `x` should be three equally-spaced points in the interval.

(3) Test `adaptquad` by applying it to the polynomial $f_6(x) = 7x^6$ on the interval $[0, 1]$, using `tol=1.e-5` and `recursions=5`. Because Gauß–Legendre integration of index 3 is *not* exact for this polynomial, the integral should be close to 1. It turns out that a single set of refinements is performed, so `recursions=4`, and the values of x should be five equally-spaced points in the interval. The estimated and true errors should agree to at least 3 significant digits. This excellent agreement is because there are no "higher-order terms" in the expressions (9.5) and (9.6) and so (9.7) is almost exact. What are both the estimated and true errors?

(4) Test `adaptquad` by applying it to the Runge function on the interval $[-5, 5]$. Use `recursions=50`. Recall that the exact answer is `2*atan(5)`. Fill in the following table

adaptquad for the Runge function		
tol	error estimate	exact error
1.e-3		
1.e-6		
1.e-9		

You should find that the estimated and exact errors are close in size, and smaller than `tol`. For the two more accurate cases, the estimated error is slightly larger than the exact error. As you can see, the estimated error is not so good for the case that `tol=1.e-3`.

Exercise 9.13. Consider the following situation.

- A quadrature is being attempted with the call

  `[Q,estErr,x,recursions]=adaptquad(@funct,0,1,tol,50);`

- The estimated error is larger than tol at first, so new calls with `tol/2` are made for the intervals $I_{\text{left}} = [0, 0.5]$ and $I_{\text{right}} = [0.5, 1]$.
- Assume that the call for the interval I_{left} satisfies the convergence criterion.
- Assume that the call for the interval I_{right} does not satisfy the convergence criterion, thus requiring two more calls to adaptquad. Assume that each of these calls satisfies the convergence criterion.

What are the final values of x and `recursions` after the `adaptquad` function has completed its work? Explain your reasoning.

The next few exercises will help you look a little more closely at the results of this recursive adaptive algorithm. Some of the points that will be made are listed below.

- You will see the advantage of adaptive algorithms. They save work over fixed algorithms such as gl_quad.
- You will see the pattern of the integration points take. They can be distributed in a very nonuniform fashion.
- You will see what happens when you try "harder" integrands.
- You will see some of the weaknesses of the algorithm.

In the following exercises, you will examine two functions that are more difficult to integrate. The first is a scaled version of the Runge function, $1/(a^2 + x^2)$, where $a = 10^{-3}$. The value of the integral on $[-1, 1]$ of the scaled Runge function is

$$\int_{-1}^{1} \frac{1}{a^2 + x^2} dx = \frac{2}{a} \tan^{-1} \frac{1}{a}.$$

The scaled Runge function has a peak value of $1/a^2 = 10^6$, so is much more strongly peaked than the unscaled Runge function.

The second function is $\sqrt{|x - 0.5|}$, and the value of its integral over the interval $[-1, 1]$ is

$$\int_{-1}^{1} \sqrt{\left| x - \frac{1}{2} \right|} dx = \frac{\sqrt{2}}{6} + \frac{\sqrt{6}}{2}.$$

This function has a singularity in its derivatives at $x = 0.5$, thus invalidating the proof that the error estimator is reliable.

A third function that is difficult to integrate is $x^{-0.99}$. This function has an integrable singularity at $x = 0$ that is close to being nonintegrable.

Exercise 9.14.

(1) Write a function m-file named srunge.m for the scaled Runge function $f(x) = 1/(a^2 + x^2)$ with $a = 10^{-3}$.
(2) Evaluate the integral of srunge on the interval $[-1, 1]$ using the following call.

```
[Q,estErr,x,recursions]=adaptquad(@srunge,-1,1,1.e-10,50);
```

What are the estimated and true errors? Is recursions larger than zero?
(3) Use gl_quad with index=3 on the scaled Runge function. Use trial-and-error to find the number of subintervals required to achieve a true error from gl_quad that is roughly comparable to the true error from adaptquad. How does this compare with length(x)-1, the number of subintervals that adaptquad used?
(4) Plot the sizes of the subintervals that adaptquad used with the following command.

```
xave=(x(2:end)+x(1:end-1))/2;
dx= x(2:end)-x(1:end-1);
semilogy(xave,dx,'*')
```

A semilog plot is appropriate here because of the wide range of interval sizes.

(5) What are the lengths of the largest and smallest intervals? Explain (one sentence) where you might expect to find the smallest intervals for an arbitrary function.

Exercise 9.15.

(1) Approximate the integral of the function $f(x) = \sqrt{|x - 0.5|}$ over the interval $[-1, 1]$ to a tolerance of `1.e-10`. The exact value of this integral is $\sqrt{2}/6 + \sqrt{6}/2$. What are the estimated and true errors?
(2) What is the returned value of **recursions**? It should be positive, indicating that the subinterval convergence criterion was always reached successfully.
(3) Plot the subinterval sizes using the following command

```
xave=(x(2:end)+x(1:end-1))/2;
dx= x(2:end)-x(1:end-1);
semilogy(xave,dx,'*')
```

where x is replaced by the variable name you used. A semilog plot is appropriate here because of the wide range of interval sizes.

In the following exercise you will apply **adaptquad** to a function that is almost not integrable. You will see the benefit of the **recursions** variable, whose reduction to 0 indicates that convergence was not achieved.

Exercise 9.16.

(1) Use **adaptquad** to approximate the integral

$$\int_0^1 x^{-0.99} dx = 100$$

to a tolerance of `1.e-10`. Use **recursions=50**. *Notice that the integration interval is [0,1].* What is the computed value of the integral? What are the estimated error and the true error? What is the value returned for **recursions**?
(2) Do the same approximation starting with **recursions=60**. What are the computed value of the integral, the estimated error, the true error, and the returned value of **recursions**? You will notice essentially no improvement. (You may have to use the command set(0,'RecursionLimit',200) in order that MATLAB will allow more recursions.)

Remark 9.7. If the variable **recursions** were not used, this recursive function would "never" terminate because the convergence test would never be satisfied. (Actually, it would fail because there is a practical limit on the recursion depth.) The reason is that in the presence of the singularity, halving the interval does result in a reduction of error, but the reduction is half or less (*c.f.* Exercise 9.10), so it never passes the convergence test.

Chapter 10

Topics in quadrature and roundoff

10.1 Introduction

The first topic discussed in this multi-topic chapter is Monte-Carlo quadrature. This is a very powerful method that works for functions that are not necessarily smooth and also in very high dimensional spaces. Its disadvantage is that its convergence rate is slow. Monte-Carlo quadrature is a probalistic method. This topic includes Exercises 10.1 and 10.2.

The second topic discussed in this chapter is a simple approach in a two-dimensional adaptive integration routine using square mesh elements and an elementary technique for determining which mesh elements need to be refined in order to meet the error requirements. This topic includes the sequence of Exercises 10.3 through 10.9.

The third topic is a demonstration of how roundoff error arises in a matrix calculation. This topic includes Exercises 10.10 and 10.11.

10.2 Integration by Monte Carlo methods

There is another approach to approximating integrals, one that can be used even when the integrand is not smooth or piecewise smooth, when the integral is taken over a region $\Omega \in \mathbb{R}^n$ that is not easily characterized or when its dimension, n is high. This versatility comes at a cost: the method is probabilistic and also slowly convergent. It is called the "Monte Carlo" method, after the famous casino in Monaco.

The basic idea behind the Monte Carlo method is that the formula for computing the average value of a function over a region

$$\langle f \rangle = \frac{1}{|\Omega|} \int_\Omega f$$

can be used to compute the integral on the right side if the average on the left side is known, and the average can be approximated by taking the average over a randomly-selected set of points

$$\langle f \rangle = \frac{1}{N} \left(\sum_{k=1}^{N} f(\mathbf{x_k}) \right) \tag{10.1}$$

where $\{\mathbf{x}_i \in \Omega\}_{k=1}^{N}$ are randomly selected points.

It is not easy to generate truly random numbers on a computer, so "pseudo-random" numbers are used instead. These are numbers generated from a deterministic formula but which satisfy enough statistical conditions that (10.2) and (10.3) are true. Generating pseudo-random numbers is a complex topic that is well beyond the scope of this workbook, but it is important to know that the pseudo-random numbers are generated depending on a "seed" that determines the particular sequence of numbers to be generated. When MATLAB starts up, a default seed is chosen and the sequence of numbers is fully determined from then on. That means that repeated Monte-Carlo calculations will produce identical results when computed from freshly-started MATLAB programs but only statistically similar results when no restart is performed between them. It is possible to save, change or modify the seed, and to choose the particular algorithm used to generate the numbers with the MATLAB **rng** function.

The Monte Carlo method is discussed briefly by [Quarteroni *et al.* (2007)], pages 416–417, and in many other places. On the web, you can find a very clear description, with considerable detail by [Fitzpatrick (2006)].

An excellent article, [Weisstein (2020b)], essentially states, for $\Omega \in \mathbb{R}^n$ and $f : \mathbb{R}^n \to \mathbb{R}$

$$\int_{\Omega} f(\mathbf{x}) dx_1 \ldots dx_n = Q + \epsilon$$

$$Q \approx |\Omega| \langle f \rangle \tag{10.2}$$

$$\epsilon \approx |\Omega| \sqrt{\frac{\langle f^2 \rangle - \langle f \rangle^2}{N}} \tag{10.3}$$

where, again, the angle brackets denote an average taken over randomly-chosen points $\{\mathbf{x}_k \in \Omega\}_{k=1}^{N}$,

$$\langle f \rangle = \frac{1}{N} \sum_{k=1}^{N} f(\mathbf{x}_k), \text{ and}$$
$$\langle f^2 \rangle = \frac{1}{N} \sum_{k=1}^{N} (f(\mathbf{x}_k))^2 \tag{10.4}$$

and $|\Omega|$ denotes the volume of Ω.

As an example, consider the problem of computing the area of the unit ball in $\mathbb{R}^2$. In this case, f is simply the characteristic function of the ball

$$\phi(x_1, x_2) = \begin{cases} 1 & x_1^2 + x_2^2 \le 1 \\ 0 & \text{otherwise} \end{cases}$$

and the area of the ball can be computed as $\int_{-1}^{1} \int_{-1}^{1} \phi(x_1, x_2) \, dx_1 \, dx_2$. In this case, Ω is the unit square $[-1, 1] \times [-1, 1]$.

The following code, using x and y to denote x_1 and x_2 will estimate the area of the ball. This code is available in the file `montecarlo.txt`.

Remark 10.1. Computing the area of a figure is a simpler problem than computing the integral of a function, but the essential steps are all here.

```
CHUNK=10000;  % chosen for efficiency
NUM_CHUNKS=100;

VOLUME=4;  % outer square is 2 X 2

totalPoints=0;
insidePoints=0;
for k=1:NUM_CHUNKS
  x=(2*rand(CHUNK,1)-1);
  y=(2*rand(CHUNK,1)-1);
  phi=( (x.^2+y.^2) <= 1 ); % 1 for "true" and 0 for "false"
                            % just like characteristic function
  insidePoints=insidePoints+sum(phi);
  totalPoints=totalPoints+CHUNK;
end
average=insidePoints/totalPoints;
a=VOLUME*average;
disp(strcat('approx area=',num2str(a), ...
            ' with true error=',num2str(pi-a), ...
            ' and estimated error=', ...
      num2str(VOLUME*sqrt((average-average^2)/totalPoints))));
```

Remark 10.2. This code uses a programming trick. In MATLAB the value 0 represents "false" and the value 1 represents "true". As a consequence, the characteristic function of the unit ball can easily be calculated by using a logical expression describing the interior of the ball.

Remark 10.3. Note that the points are divided into chunks of 10,000 points each. Then the function evaluation is done using vectors whose lengths are the chunk size. Working with these long vectors, MATLAB will do the calculations efficiently. You don't, however, want to do all 1,000,000 points at once, because such large vectors can take a while just for the memory to be allocated. That is why the problem is first divided into chunks and then repeated a number of times.

Exercise 10.1.

(1) Copy the code above or from the file `montecarlo.txt` to a script m-file and execute it several times. You should observe that the estimated area changes slightly each time and that the estimated error is usually, but not always, larger

than the actual error. (These are probabilistic quantities, after all.) Explain why both $\langle f \rangle$ and $\langle f^2 \rangle$ are given the same value (**average**) in computing the error.

(2) Computing the volume of the intersection of two unit cylinders with orthogonal axes in $\mathbb{R}^3$ is an exercise often given to calculus students because it is difficult, but not impossible, to compute using elementary calculus. This volume is called a Steinmetz Solid and more information can be found on the web [Weisstein (2020c)]. This volume is known to be 16/3. Use Monte Carlo integration to estimate this volume along with both the true and estimated errors. Use enough trials to achieve an estimated accuracy of ± 0.001 or smaller. How many trials did you use?

In the following exercise you will compute integrals, not just areas. It is possible to integrate a function by embedding its graph into a rectangle, or other simple shape, and repeat the above approach to get the integral. In multiple dimensions, this embedding can result in an outer region that is much too large, slowing down the calculation and possibly introducing roundoff errors. A better approach is to use (10.4) and (10.2) to estimate the integral.

Exercise 10.2. In this exercise, you should write three m-files to compute integrals in $\mathbb{R}^1$, $\mathbb{R}^2$, and $\mathbb{R}^3$ based on (10.2) with error estimates based on (10.3). One difference is that the volume of the region of integration, $|\Omega|$, is the true volume and not the volume of a surrounding, simpler region.

(1) Use (10.2) to estimate the following integral

$$\int_0^2 e^x dx$$

along with both the true error and error estimated from (10.3). Use enough trials to achieve an estimated accuracy of ± 0.001 or smaller. How many trials did you use? This function is easily integrated exactly. What is the true error of your estimate? It should be smaller, or at least not much bigger, than your error estimate. **Hint:** The quantity **VOLUME** is 2, the length of the interval $[0, 2]$.

(2) Use (10.2) to estimate the integral

$$\int_\Omega e^{(x^2+y^2)} \, dxdy,$$

where Ω is the unit ball in $\mathbb{R}^2$. Use enough trials to achieve an estimated accuracy using (10.3) of ± 0.001 or smaller. How many trials did you use? This function is easily integrated exactly by changing to polar coordinates, yielding the value $\pi(e - 1)$. What is the true error of your estimate? It should be smaller, or at least not much bigger, than your error estimate. **Hint:** The quantity **VOLUME** is π, the area of the unit ball in the plane.

(3) Compute the integral of e^{x+y+z} over the volume of the intersection of two unit cylinders along with the estimated error. Use enough trials to achieve an estimated accuracy of ± 0.005 or smaller. How many trials did you use?

Hints:

- The quantity VOLUME is $16/3$, the volume of the Steinmetz solid in Exercise 10.1.
- Use the script m-file you wrote for the volume of the Steinmetz solid to choose points at which to do averaging.

Remark 10.4. In the case that the volume of the region is not known in advance, Exercises 10.1 and 10.2 can be combined to compute the integral and the volume of integration at the same time *using the same sets of random coordinates*. Since random number generation itself consumes a large part of the time the integration takes, this amounts to a significant savings.

10.3 Adaptive quadrature

In this section you will construct a MATLAB function to compute the integral of a given mathematical function over a square region in the plane. One way to do such a task would be to regard the square to be the Cartesian product of two one-dimensional lines and integrate using a one-dimensional adaptive quadrature routine such as adaptquad from last chapter. Instead, in this chapter you will be looking at the square as a region in the plane and you will be dividing it up into many (square) subregions, computing the integral of the given function over each subregion, and adding them up to get the integral over the given square.

The basic outline of the method used in this chapter is the following:

(1) Start with a list containing a single "subregion": the square region of integration.

(2) Use a Gaußian integration rule to integrate the function over each subregion in the list and estimate the resulting error of integration. The integral over the whole region is the sum of the integrals over the subregions, and similarly the estimated error is the sum of the estimated errors over the subregions.

(3) If the total estimated error of the integral is small enough, the process is complete. Otherwise, find the subregion with largest error, replace it with four smaller subregions, and return to the previous step.

The way the notion of a "list" is implemented will introduce a data structure (discussed in detail below) that is more versatile than arrays or matrices.

The adaptive quadrature discussed here is built on quadrature and error estimation on a single (square) element. The discussion starts there.

10.3.1   *Two-dimensional Gauß quadrature*

One simple way of deriving a two-dimensional integration formula over a square is to use iterated integration. In this case, the square has lower left coordinate (x, y) and side length h, so the square is $[x, x+h] \times [y, y+h]$. Recall that a one-dimensional Gauß integration rule can be written as

$$\int_x^{x+h} f(x)dx \approx \sum_{n=1}^{N} w_n f(x_n). \tag{10.5}$$

Here, N is the index of the rule. For the case $N = 2$, the points x_n are $x + h/2 \pm h/(2\sqrt{3})$ and the weights are $w_1 = w_2 = h/2$. The degree of precision is 3, and the error is proportional to $h^5 \max |f''''|$. (If you look up the error in a reference somewhere, you will notice that the error is usually given as proportional to h^4, not h^5. The extra power of h appearing in (10.5) comes from the fact that the region of integration itself has length h.) Applying (10.5) twice, once in the x-direction and once in the y-direction gives

$$\int_x^{x+h} \int_y^{y+h} f(x,y)dxdy \approx \sum_{n=1}^{N} \sum_{m=1}^{M} w_n w_m f(x_n, y_m). \tag{10.6}$$

For the case $N = M = 2$, (10.6) becomes

$$\int_x^{x+h} \int_y^{y+h} f(x,y)dxdy \approx \sum_{n=1}^{4} (h^2/4)f(x_n, y_n), \tag{10.7}$$

where the four points $(x_n, y_n) = (x + h/2 \pm h/(2\sqrt{3}), y + h/2 \pm h/(2\sqrt{3}))$. These are four points based on the choices of "+" or "−" signs. Numbering the four choices is up to you. The error is $O(h^6)$ over his $h \times h$ square, and (10.7) is exact for monomials $x^n y^m$ with $n \leq 3$ and $m \leq 3$, and for sums of such monomials. In the following exercise, you will write a MATLAB implementation of this method.

Exercise 10.3.

(1) Write a MATLAB function to compute the integral of a function over a single square element using (10.7) with $(x_n, y_n) = (x + h/2 \pm h/(2\sqrt{3}), y + h/2 \pm h/(2\sqrt{3}))$. Name the function m-file q_elt.m with the following outline.

```
function q=q_elt(f,x,y,h)
  % q=q_elt(f,x,y,h)
  % INPUT
  % f=???
  % x=???
  % y=???
  % h=???
  % OUTPUT
  % q=???
```

```
% your name and the date

<< your code >>
```

end

(2) Test q_elt on the functions 1, $4xy$, $6x^2y$, $9x^2y^2$, and $16x^3y^3$ over the square $[0,1] \times [0,1]$ and show that the result is exact, up to roundoff.

(3) Test q_elt on the function $25x^4y^4$ to see that it is not exact, thus showing the degree of precision is 3.

10.3.2 *Error estimation*

In order to do any sort of adaptive quadrature, you need to be able to estimate the error in one element. Remember, this is only an estimate because without the true value of the quadrature, you cannot get the true error.

Suppose you have a square element with side of length h. If you divide it into four sub-squares with sides of length $h/2$, then you can compute the quadrature twice: once on the single square with side of length h and once by adding up the four quadratures over the four squares with sides of length $h/2$. Consider the following figure.

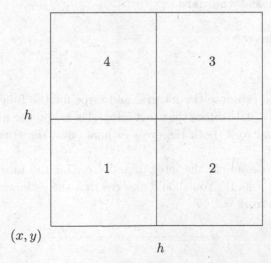

Denote the true integral over this square as q and its approximation over the square with side of length h as q_h. Denote the four approximate integrals over the four squares with sides of length $h/2$ as $q_{h/2}^1$, $q_{h/2}^2$, $q_{h/2}^3$, and $q_{h/2}^4$. Assuming that the fourth derivatives of f are roughly constant over the squares, the following expressions can be written.

$$q_h = q + Ch^6$$
$$q_{h/2}^1 + q_{h/2}^2 + q_{h/2}^3 + q_{h/2}^4 = q + 4C(h/2)^6 = q + (4/64)Ch^6. \qquad (10.8)$$

The second of these is assumed to be more accurate than the first, so use it as the approximate integral,

$$q_{\text{approx}} = q_{h/2}^1 + q_{h/2}^2 + q_{h/2}^3 + q_{h/2}^4. \qquad (10.9)$$

The system of equations (10.8) can be solved for the error in q_{approx} as

$$\text{error in } q_{\text{aprox}} = \frac{4}{64}Ch^6 = \frac{1}{15}\left(q_h - (q_{h/2}^1 + q_{h/2}^2 + q_{h/2}^3 + q_{h/2}^4)\right). \qquad (10.10)$$

In the following exercise you will write a MATLAB function to estimate both the integral and the error over a single element.

Exercise 10.4.

(1) Write an m-file named qerr_elt.m to estimate the both the integral q_{approx} in (10.9) and the error according to (10.10). Use q_elt.m to evaluate the integrals. qerr_elt.m should have the outline

```
function [q,errest]=qerr_elt(f,x,y,h)
  % [q,errest]=qerr_elt(f,x,y,h)
  % more comments

  % your name and the date

  << your code >>

end
```

(2) Use qerr_elt to estimate the integral and error for the function $16x^3y^3$ over the square $[0,1] \times [0,1]$. Since the exact integral is 1, and the method has degree of precision equal to 3, both the error estimate and the true error should be zero or roundoff.

(3) Use qerr_elt to estimate the integral and error for the function $25x^4y^4$ over the square $[0,1] \times [0,1]$. You should observe that the estimated error is within 5% of the true error.

10.3.3 *A versatile data storage method*

Up to now all the MATLAB programs you have used involved fairly simple ways of storing data, involving variables, vectors, and matrices. Another valuable programming tool for storing data is the so-called "structure." In many programming languages, such as Java, C and C++, it is called a "struct," in Pascal it is called a "record," and in Fortran, it is called a "defined type." MATLAB uses the term "structure", although everyone will understand if you call it a "struct."

If you find the description in this section too terse, MATLAB documentation provides considerably more detail.

A structure consists of several named sub-fields, each containing a value, and separated from the variable name with a dot. Rather than going into full detail here, consider just the simple concept of a square region in space, with sides parallel to the coordinate axes. Such a square can be specified with three numerical quantities: the x and y coordinates of the lower left corner point, and the length of a side. These three quantities will be called x, y, and h for the purposes of this chapter. Thus, if a MATLAB (structure) variable named `elt` were to refer to the square $[-1, 1] \times [-1, 1]$, it could be given as

```
elt.x=-1;
elt.y=-1;
elt.h=2;
```

It is important to realize that the value of `elt.x` is unrelated to the value of x that might appear elsewhere in a program. For the purpose of this chapter, two other quantities will be included in this structure: the approximate integral of the function over this element, called `q`, and the estimated error, called `errest`.

In the following exercises, you will be using a subscripted array (a vector) of structures to implement the notion of a "list of elements." Structures can be indexed, and the resulting syntax for the kth entry of the array of structures named `elt` would be `elt(k)`. The sub-fields of `elt(k)` are denoted

```
elt(k).x
elt(k).y
elt(k).h
elt(k).q
elt(k).errest
```

The following exercise is intended to provide an introduction to programming with structures for this application.

Exercise 10.5. In this exercise you will build up a function that estimates the integral of a function and its error over a square by choosing an arbitrary integer n, and dividing the square into n^2 smaller squares, all the same size. The point of this exercise is to introduce you to programming with structures. Subsequent exercises will not use a uniform division of the square.

(1) Begin a function named `q_total.m` with the following code template and correct the lines with **???** in them. This function is incomplete: it ignores `f` and always computes the area of the square, and estimates zero error. You may find it convenient to download the file `q_total.txt`.

```
function [q,errest]=q_total(f,x,y,H,n)
 % [q,errest]=q_total(f,x,y,H,n)
 % more comments
 % n=number of intervals along one side

 % your name and the date

 h=( ??? )/n;
 eltCount=0;
 for k=1:n
   for j=1:n
     eltCount=eltCount+1;
     elt(eltCount).x= ???
     elt(eltCount).y= ???
     elt(eltCount).h= ???
     elt(eltCount).q= elt(eltCount).h^2; % to be corrected later
     elt(eltCount).errest=0;              % to be corrected later
   end
 end
 if numel(elt) ~= n^2
   error('q_total: wrong number of elements!')
 end

 q=0;
 errest=0;
 for k=1:numel(elt);
   q=q+elt(k).q;
   errest=errest+abs(elt(k).errest);
 end
end
```

(2) Test the partially-written function `q_total` by choosing any function `f` (since it is unused so far, it does not matter) and using it to estimate the integral over the square $[0, 1] \times [0, 1]$ using $n = 10$. Since it actually is computing the area of the square, you should get 1.0. If you do not, you have either computed the value of `h` incorrectly or you have somehow generated the wrong number of elements. The length of the vector `elt` should be precisely n^2.

(3) As a second test, apply it to the square $[-1, 1] \times [-1, 1]$ using $n = 13$. Again, you should get the area of the square.

(4) Now that you have some confidence that the code has the correct indexing, use the function `qerr_elt` to estimate the values of `q` and `errest` based on the elemental values of `x`, `y`, and `h` and place them into `elt(elt_count).q` and `elt(elt_count).errest`.

(5) Estimate the integral and error of the function $9x^2y^2$ over the square $[0, 1] \times [0, 1]$ for the value $n = 1$. If you do not get 1.0 with error estimate 0 or roundoff, you have computed `elt(elt_count).x` or `elt(elt_count).y` or `elt(elt_count).h` incorrectly or used `qerr_elt` incorrectly. **Note:** `numel(elt)` is precisely 1 in this case.

(6) Estimate the integral and error of the function $9x^2y^2$ over the square $[-1, 1] \times [-1, 1]$ for the value $n = 1$. If you do not get 4.0, you have computed `elt(elt_count).x` or `elt(elt_count).y` or `elt(elt_count).h` incorrectly. If your estimated error is not zero or roundoff, you have likely used `qerr_elt` incorrectly. Again, `numel(elt)` is precisely 1 in this case.

(7) Estimate the integral and error of the function $16x^3y^3$ over the square $[0, 1] \times [0, 1]$ for the value $n = 2$. If you do not get 1, review your changes to `q_total.m` carefully.

(8) Fill in the following table for the integral of the function $25x^4y^4$ over the square $[0, 1] \times [0, 1]$.

n	integral	estimated error	true error
2			
4			
8			
16			

(9) Are your results consistent with the global order of accuracy of $O(h^4)$?

In order to further test the integration and error estimation, a more complicated function is needed. One function that is neither too easy nor too hard to integrate is the following function.

$$f(x, y) = \frac{1}{\sqrt{1 + 100x^2 + 100(y - 0.5)^2}} +$$
$$\frac{1}{\sqrt{1 + 100(x + 0.5)^2 + 100(y + 0.5)^2}} + \frac{1}{\sqrt{1 + 100(x - 0.5)^2 + 100(y + 0.5)^2}}.$$

This function has three peaks on the square $[-1, 1] \times [-1, 1]$: two at $(\pm 0.5, -0.5)$ and one at $(0, 0.5)$. Its integral over the square $[-1, 1] \times [-1, 1]$ is 1.755223755917299. MATLAB code to effect this function is:

```
function z=three_peaks(x,y)
  % z=three_peaks(x,y)
  % three peaks at (-.5,-.5), (+.5,-.5), (0,.5)
  % the integral of this function over
  %    [-1,1]X[-1,1] is 1.75522375591726
```

```
% M. Sussman

z=1./sqrt(1+ 100*(x+0.5).^2+ 100*(y+0.5).^2)+ ...
  1./sqrt(1+ 100*(x-0.5).^2+ 100*(y+0.5).^2)+ ...
  1./sqrt(1+ 100*(x    ).^2+ 100*(y-0.5).^2);
end
```

A perspective plot of the function is:

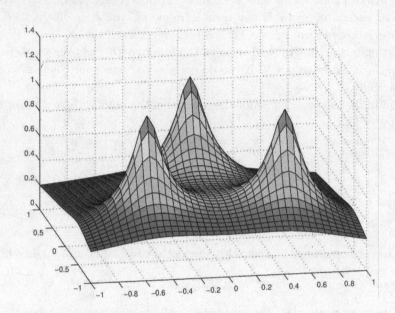

Exercise 10.6.

(1) Use cut-and-paste to copy the above code to a function m-file named `three_peaks.m` or download the file.

(2) Integrate `three_peaks` over the square $[-1, 1] \times [-1, 1]$ using `q_total` and fill in the following table. The true value of the integral is 1.755223755917299. **Warning:** the larger values of n may take some time — be patient.

n	integral	estimated error	true error
10			
20			
40			
80			
160			

(3) Are the true error values consistent with the convergence rate of $O(h^4)$?

(4) Notice that the estimated errors are much larger than the true errors, especially for larger values of n. This is because the elemental errors are

sometimes positive and sometimes negative and should cancel each other, but absolute values are taken in the code for q_total. Make a copy of q_total.m called q_total_noabs.m and remove the absolute value from the summation of the elemental error estimates. Using q_total_noabs, compute the integral of three_peaks over $[-1, 1] \times [-1, 1]$ using $n = 80$. You should observe that the true and estimated errors agree within 0.1%. Nonetheless, use of absolute value is adequate for smaller values of n and is more conservative in all cases.

10.3.4 *An adaptive strategy*

The objective of this quadrature is to present an adaptive strategy for quadrature. You have seen all the pieces and now it is time to put them together. In this strategy, a vector of structures similar to the one used in q_total will be used, *but the way it is used is very different*. The strategy in q_total results in a large number of uniformly-sized squares filling out the unit square. The adaptive strategy below will result in a much smaller number of squares of differing sizes. Small squares will be used only where they are needed to achieve accuracy.

The adaptive strategy used here is the following.

(i) Start with a vector of structures named elt similar to the one used in q_total above. This vector will have only one subscript:

```
elt(1).x = ???
elt(1).y = ???
elt(1).h = ???
elt(1).q = ???
elt(1).errest = ???
```

and the values represent the *entire* given square region over which the integral is to be taken, with elt(1).q and elt(1).errest computed using qerr_elt.

(ii) Add up all the elemental values of q and absolute values of errest to get the total q and errest. If this total errest is smaller than the tolerance, stop and return the values of q and errest.

(iii) If the total estimated error of the integral is too large, find the value of k for which abs(elt(k).errest) is largest and divide it into four subregions.

(iv) Replace elt(k) with values from the upper right of the four smaller square subelements. You can use code similar to the following.

```
x=elt(k).x;
y=elt(k).y;
h=elt(k).h;
```

```
% new values for this element
elt(k).x=x+h/2;
elt(k).y=y+h/2;
elt(k).h=h/2;
[elt(k).q, elt(k).errest]=qerr_elt( ??? )
```

(v) Add three more elements to the vector of elements using code similar to the following for each one.

```
k=numel(elt)+1;
elt(k).x= ???
elt(k).y= ???
elt(k).h=h/2;
[elt(k).q, elt(k).errest]=qerr_elt( ??? )
```

(vi) Go back to the second step above.

Exercise 10.7.

(1) Copy code provided below in Subsection 10.3.5 or download the plotting function `plotelt.m` to display the elements. Elements colored green have small estimated error, elements colored amber have mid-sized error estimates and elements colored red have the largest error estimates. Red elements are candidates for the next mesh refinement.

(2) Write a MATLAB function m-file named `q_adaptive.m` that implements the preceeding algorithm. Your function could use the following outline. You may find it convenient to download the file `q_adaptive.txt` containing the above code and this outline.

```
function [q,errest,elt]=q_adaptive(f,x,y,H,tolerance)
  % [q,errest,elt]=q_adaptive(f,x,y,H,tolerance)
  % more comments

  % your name and the date
  MAX_PASSES=500;

  % initialize elt
  elt(1).x=???
  << more code >>

  for passes=1:MAX_PASSES
    % compute q by adding up elemental values
    % and compute errest by adding up absolute elemental values
    % use a loop for this because the "sum" function doesn't
    % work for structures.
```

```
    << more code >>

    % if error meets tolerance, return
    << more code >>

    % use a loop to find the element with largest abs(errest)
    << more code >>

    % replace that element with a quarter-sized element
    << more code >>

    % add three more quarter-sized elements
    << more code >>

  end
  error('q_adaptive convergence failure.');
end
```

(3) Test q_adaptive by computing the integral of the function $16x^3y^3$ over the square $[0, 1] \times [0, 1]$ to a tolerance of 1.e-3. The result should be exactly correct because the degree of precision is 3, and numel(elt) should be 1.

(4) Test q_adaptive by computing the integral of the function $9x^2y^2$ over the square $[-1, 1] \times [-1, 1]$ to a tolerance of 1.e-3. The result should be exactly 4 because the degree of precision is 3, and numel(elt) should be 1.

(5) Test q_adaptive by computing the integral of the function $25x^4y^4$ over the square $[0, 1] \times [0, 1]$ to a tolerance of 1.e-3. You should see that q is close to 1, that the estimated error is smaller than the tolerance, and that numel(elt) is precisely 4 because only a single refinement pass was required. Use plotelt to plot these four elements.

Debugging hint: *If you do not get the correct number of elements,* you can debug by temporarily setting MAX_PASSES=2 in the code and look at qelt. Is numel(qelt) equal to 4? If not, look at the coordinates of each of the elements in elt. There should be no duplicates or omissions. When you have corrected your error, do not forget to reset MAX_PASSES=500.

(6) Test q_adaptive by computing the integral of the function $25x^4y^4$ over the square $[0, 1] \times [0, 1]$ to a tolerance of 2.e-4. You should see that the integral is close to 1, that the estimated and true errors are close, and that numel(elt) is precisely 7 because two refinement passes were required, with the unit square broken into four subsquares and the upper right subsquare itself broken into four. Use plotelt to plot these 7 elements.

(7) Test q_adaptive by computing the integral of the function $25x^4y^4$ over the square $[0,1] \times [0,1]$ to a tolerance of 1.e-8. You should again see that the integral is very close to 1 and that the estimated and true errors are close.

In the following exercise you will see how the adaptive strategy worked.

Exercise 10.8.

(1) Use q_adaptive to estimate the integral of the function $25x^4y^4$ over the square $[0,1] \times [0,1]$ to an accuracy of 1.e-6. What are the integral, the estimated error, and the true error? How many elements were used? You should observe that the exact and estimated errors are close in size. Use plotelt to plot the final mesh used.
(2) Again estimate the integral of $25x^4y^4$ over the square $[0,1] \times [0,1]$, but to an accuracy of 9.e-7, smaller than before. You should observe that the two large red blocks near but not touching the origin have been refined.
(3) Again estimate the integral of $25x^4y^4$ over the square $[0,1] \times [0,1]$, but to an accuracy of 5.e-7. You can see that the red elements have been refined, green ones were not refined, and the worst remaining elements are in different places.

Exercise 10.9. Use q_adaptive to estimate the integral and error of the function three_peaks over the square $[-1,1] \times [-1,1]$ to a tolerance of 1.e-5.

You should be able to see from the plot that elements near the peaks themselves have been refined, but also elements in the areas between the peaks.

10.3.5　*Code for the* plotelt.m *utility function*
This code is also available for download.

```
function plotelt(elt)
  % plotelt(elt)
  % plot each of the elements in the structure elt
  %   colored according to errest

  % M. Sussman

  if numel(elt)<1
    error('plotelt: elt is empty')
  end

  % Generate a colormap so plotted elements are colored according to
```

```
% how large the estimated error is in the element.
% A colormap is a list of (red,green,blue) triples

NUM_COLORS=64;
VALUE_YELLOW=20;

% generate a colormap
red=[1 0 0];      % large (bad) values
green=[0 1 0];    % small (good) values
yellow=[1 1 0];   % intermediate values

r=[linspace(green(1),yellow(1),VALUE_YELLOW)'
   linspace(yellow(1),red(1),NUM_COLORS-VALUE_YELLOW)'];
g=[linspace(green(2),yellow(2),VALUE_YELLOW)'
   linspace(yellow(2),red(2),NUM_COLORS-VALUE_YELLOW)'];
b=[linspace(green(3),yellow(3),VALUE_YELLOW)'
   linspace(yellow(3),red(3),NUM_COLORS-VALUE_YELLOW)'];

colormap([r g b]);

% Now examine the estimated error in each element and plot
% using the patch function

% find max error
maxErr=abs(elt(1).errest);
for k=2:length(elt)
  maxErr=max(maxErr,abs(elt(k).errest));
end

color=round(abs(elt(1).errest)/maxErr*NUM_COLORS);
patch([elt(1).x,elt(1).x+elt(1).h,elt(1).x+elt(1).h,elt(1).x], ...
      [elt(1).y,elt(1).y,elt(1).y+elt(1).h,elt(1).y+elt(1).h], ...
      color);
hold on
for k=2:length(elt)
color=round(abs(elt(k).errest)/maxErr*NUM_COLORS);
patch([elt(k).x,elt(k).x+elt(k).h,elt(k).x+elt(k).h,elt(k).x], ...
      [elt(k).y,elt(k).y,elt(k).y+elt(k).h,elt(k).y+elt(k).h], ...
      color);
end
hold off
end
```

10.4   Roundoff errors

You have already seen some effects of roundoff errors and you will see more effects later in this workbook. Right now, though, is a good time to look at how some roundoff errors come about.

In the exercise below you will have occasion to use a special matrix called the Frank matrix. Row k of the $n \times n$ Frank matrix has the formula:

$$\mathbf{A}_{k,j} = \begin{cases} 0 & \text{for } j < k - 2 \\ n + 1 - k & \text{for } j = k - 1 \\ n + 1 - j & \text{for } j \geq k \end{cases}.$$

The Frank matrix for $n = 5$ looks like:

$$\begin{pmatrix} 5 & 4 & 3 & 2 & 1 \\ 4 & 4 & 3 & 2 & 1 \\ 0 & 3 & 3 & 2 & 1 \\ 0 & 0 & 2 & 2 & 1 \\ 0 & 0 & 0 & 1 & 1 \end{pmatrix}.$$

The determinant of the Frank matrix is 1, but is difficult to compute numerically. This matrix has a special form called *Hessenberg* form wherein all elements below the first subdiagonal are zero. MATLAB provides the Frank matrix in its "gallery" of matrices, `gallery('frank',n)`, but you will use an m-file `frank.m` that is available for download and is included below in Subsection 10.4.1. The inverse of the Frank matrix also consists of integer entries and an m-file for it is available for download and is included below in Subsection 10.4.1. You can find information about the Frank matrix in [Frank (1958)] and [Golub and Wilkinson (1976)], Section 13, or on the web at [Golub and Wilkinson (1975)], Section 13.

Exercise 10.10. Look carefully at the Frank matrix and its inverse. For convenience, define A to be the Frank matrix of order 6, and Ainv its inverse, computed using `frank` and `frank_inv`, respectively. Similarly, let B and Binv be the Frank matrix of order 24 and its inverse. *Do not use the* MATLAB `inv` *function for this exercise!* You know that both A*Ainv and B*Binv should equal the identity matrices of order 6 and 24 respectively.

(1) What is the result of A*Ainv?
(2) What is the upper left 5×5 square of C=B*Binv? You should see that C is *not* a portion of the identity matrix. What appears to be a mistake is actually the result of roundoff errors.
(3) To see what is going on, let's just look at the top left entry. Compute A(1,:)*Ainv(:,1) and B(1,:)*Binv(:,1). Both of these answers should equal 1. The first does and the second does not.
(4) To see what goes right, compute the terms:

Term	Value
A(1,6)*Ainv(6,1)	
A(1,5)*Ainv(5,1)	
A(1,4)*Ainv(4,1)	
A(1,3)*Ainv(3,1)	
A(1,2)*Ainv(2,1)	
A(1,1)*Ainv(1,1)	
Sum	

Note that the signs alternate, so that when you add them up, each term tends to cancel part of the preceeding term.

(5) Now, to see what goes wrong, compute the terms:

Term	Value
B(1,24)*Binv(24,1)	
B(1,23)*Binv(23,1)	
B(1,22)*Binv(22,1)	
B(1,21)*Binv(21,1)	
B(1,20)*Binv(20,1)	
B(1,16)*Binv(16,1)	
B(1,11)*Binv(11,1)	
B(1,6)*Binv(6,1)	
B(1,1)*Binv(1,1)	

You can see what happens to the sum. The first few terms are huge compared with the correct value of the sum (1). MATLAB uses 64-bit floating point numbers, so you can only rely on the first thirteen or fourteen significant digits in numbers like B(1,24)*Binv(24,1). Further, they are of opposing signs so that there is extensive cancellation. There simply are not enough bits in the calculation to get anything like the correct answer.

Remark 10.5. It would not have been productive to compute each of the products B(1,k)*Binv(k,1) for each k, so only the five largest were chosen and the rest sampled. The sample terms were chosen with an *even* interval between adjacent terms. An odd interval — say every other term — would have obscured the alternating sign pattern. *When you are sampling errors or residuals for any reason,* never *take every other term!*

You see now how roundoff errors can be generated by adding numbers of opposite sign together. It is also possible to generate roundoff by adding small numbers to large numbers, without sign changes. This mechanism is not so dramatic as errors introduced by subtraction. In the following exercise, you will see how roundoff errors can ge generated by adding small numbers to large numbers and how roundoff can be mitigated by grouping the smaller numbers together.

For this exercise, it will be convenient to use *single precision* (32 bit) numbers rather than the usual double precision (64 bit) numbers. This will show the effects of roundoff when using a much smaller number of terms in the sum used below, resulting in much less time in accumulating the sums.

Single precision numbers have only about eight significant digits, in contrast to double precision numbers, which have about fifteen significant digits.

The MATLAB function `single(x)` returns a *single precision* version of its argument `x`. Once single precision numbers `s1` and `s2` have been generated, they can be added (in single precision) in the usual manner (`s3=s1+s2`), and the variable `s3` will automatically be a single precision variable. Even when another variable is a default (double) precision number, adding it to or multiplying it by a single precision number results in a single precision number.

Warning: For those who have programmed in other languages, adding a single precision number to a double precision number in Fortran or C results in a *double precision* value, not single precision. Take care when mixing precisions in arithmetic statements.

Exercise 10.11. You probably have seen the formula for the sum of a geometric series.

$$\sum_{n=0}^{N} x^n = \frac{1 - x^{N+1}}{1 - x}. \tag{10.11}$$

Suppose that `x=0.9999` and `N=100000` (10^5).

(1) Define a *single precision* variable `x=single(0.9999)`.
(2) Compute the sum of the series using the formula on the right of (10.11). Call this value `S`.
(3) Write a loop to accumulate the sum of the series (10.11) by adding up the terms on the left. Call this value `a`.
(4) Write a loop to accumulate the sum of the series (10.11) by adding up the terms on the left *in reverse order* (`for n=N:-1:0`). Call this value `b`.
(5) Using `format long`, how many digits of `a` agree with those of `S`? How many digits of `b` agree with those of `S`? You should find that `b` is substantially closer to `S` than `a`.
(6) Compute the relative errors `abs((a-S)/S)` and `abs((b-S)/S)`. You should find that the error in `a` is more than 100 times the error in `b`.
(7) To see where the error comes from, compute the sum a_{1000} for $N = 1000$ terms. What is a_{1000}? What is the value of the next term in the series, x^{1001}? When you add x^{1001} to a_{1000}, there are only 8 digits of accuracy available, so about four digits of the term x^{1001} are lost in performing the sum! Keeping this kind of loss up for thousands of terms is the source of inaccuracy.

(8) Look at the reversed sum for **b**. In your own words, explain why roundoff error is so much smaller in this case.

10.4.1 *Code for the Frank matrix*

This code is available for download in the files `frank.m` and `frank_inv.m`.

```
function A = frank ( n )
  % function A = frank ( n )
  %    FRANK sets up the Frank matrix.
  %
  % Parameters:
  %    Input, integer N, the order of the matrix.
  %    Output, real A(N,N), the Frank matrix.

  % Author:   John Burkardt
  % Modified by M. Sussman, used by permission
  % $Id: adaptive2d.tex,v 1.12 2021/03/21 03:13:34 mike Exp $

  A = zeros ( n, n );
  for i = 1 : n
    for j = 1 : n
      if ( j < i-1 )
      elseif ( j == i-1 )
        A(i,j) = n+1-i;
      elseif ( j > i-1 )
        A(i,j) = n+1-j;
      end
    end
  end

end

function A = frank_inv ( n )
  % function A = frank_inv ( n )
  % FRANK_INV returns the inverse of the Frank matrix.
  %
  % Parameters:
  %    Input, integer N, the order of the matrix.
  %    Output, real A(N,N), the inverse Frank matrix.

  % Author: John Burkardt
```

```
% Modified by M. Sussman, used by permission
% $Id: adaptive2d.tex,v 1.12 2021/03/21 03:13:34 mike Exp $

A = zeros ( n, n );

for i = 1 : n
  for j = 1 : n
    if ( i == j-1 )
      A(i,j) = - 1.0;
    elseif ( i == j )
      if ( i == 1 )
        A(i,j) = 1.0;
      else
        A(i,j) = n + 2 - i;
      end
    elseif ( i > j )
      A(i,j) = - ( n + 1 - i ) * A(i-1,j);
    else
      A(i,j) = 0.0;
    end
  end
end

end
```

PART 2
Differential Equations and Linear Algebra

Chapter 11

Explicit ODE methods

11.1 Introduction

This chapter presents solution methods for ordinary differential equations (ODEs). You will be looking at two classes of methods that excel when the equations are smooth and derivatives are not too large.

The chapter begins with an introduction to Euler's (explicit) method for ODEs. Euler's method is the simplest approach to computing a numerical solution of an initial value problem. However, it has about the lowest possible accuracy. If you wish to compute very accurate solutions, or solutions that are accurate over a long interval, then Euler's method requires a large number of small steps. Since most of our problems seem to be computed "instantly," you may not realize what a problem this can become when solving a "real" differential equation.

Applications of ODEs are divided between ones with space as the independent variable and ones with time as the independent variable. In this chapter, t will be the independent variable. In Chapter 13 it will be more appropriate to use x as the independent variable because boundary value problems are regarded as being defined over a spatial region.

A number of methods have been developed in the effort to get solutions that are more accurate, less expensive, or more resistant to instabilities in the problem data. Typically, these methods belong to "families" of increasing order of accuracy, with Euler's method (or some relative) often being the member of the lowest order.

In this chapter, you will look at "explicit" methods, that is, methods defined by an explicit formula for u_{k+1}, the approximate solution at the next time step, in terms of quantities derivable from previous time step data. In the next chapter, you will address "implicit" methods that require the solution of an equation in order to find u_{k+1}. You will see the Runge–Kutta and the Adams–Bashforth families of methods, and you will see some of the problems of implementing the higher-order versions of these methods. You will also see how to compare the accuracy of different methods applied to the same problem, and using the same number of steps.

Runge–Kutta methods are "single-step" methods while Adams–Bashforth methods are "multistep" methods. Multistep methods require information from several preceding steps in order to find u_{k+1} and are a little more difficult to use. Nonetheless, both single and multistep methods have been very successful and there are very reliable MATLAB routines (and libraries for other languages) available to solve ODEs using both types of methods.

Exercise 11.1 introduces the explicit or forward Euler method. This exercise serves as an introduction to the subject and also an introduction to MATLAB for students who might be starting with this chapter. This exercise also introduces a way of estimating the accuracy of the method by constructing a table of errors based on increasing numbers (and decreasing step sizes) of subintervals. Exercise 11.2 introduces the "Euler halfstep" method, a second-order Runge–Kutta method. Exercise 11.3 introduces a third-order Runge–Kutta method. Exercise 11.4 illustrates that the functions work just as well for systems of ODEs as for scalar ODEs, and Exercise 11.5 illustrates the system of first-order ODEs arising from motion of a pendulum, a second-order ODE. Exercise 11.6 shows what happens when the conditional stability limits are exceeded. Exercise 11.7 introduces the second-order Adams–Bashforth method and Exercise 11.8 illustrates how one might go about choosing among the various methods. Exercise 11.8 also allows the student to implement an ODE method based on formulas and previous examples. Finally, Exercise 11.9 shows how to construct stability region plots for forward Euler and for the second-order Adams–Bashforth method.

11.2 MATLAB hint

MATLAB vectors can be either row vectors or column vectors. It is important to realize that a matrix can only be multiplied on the right by a *column* vector or on the left by a *row* vector. The reason for the distinction is that vectors are really special cases of matrices with a "1" for the other dimension. A row vector of length n is really a $1 \times n$ matrix and a column vector of length n is really a $n \times 1$ matrix.

You should recall that row vectors are separated by commas when using square brackets to construct them and column vectors are separated by semicolons. It is sometimes convenient to write a column vector as a column. In the following expressions

```
rv = [ 0, 1, 2, 3];
cv = [ 0; 1; 2; 3; 4];
cv1 = [0
       1
       2
       3
       4];
```

The vector `rv` is a row vector and the vectors `cv` and `cv1` are column vectors.

If you do not tell MATLAB otherwise, MATLAB will generate a row vector when it generates a vector. The output from `linspace` is, for example, a row vector.

Similarly, the following code

```
for j=1:10
  rv(j)=j^2;
end
```

results in the vector `rv` being a row vector. If you wish to generate a column vector using a loop, you can either first fill it in with zeros

```
cv=zeros(10,1);
for j=1:10
  cv(j)=j^2;
end
```

or use two-dimensional matrix notation

```
for j=1:10
  cv(j,1)=j^2;
end
```

11.3 Euler's method

A very simple ordinary differential equation (ODE) is the *explicit scalar first-order initial value problem*:

$$\frac{du}{dt} = f_{\text{ode}}(t, u)$$
$$u(t_0) = u_0.$$

The equation is *explicit* because du/dt can be written explicitly as a function of t and u. It is *scalar* because you assume that $u(t)$ is a scalar quantity, not a vector. It is *first-order* because the highest derivative that appears is the first derivative du/dt. It is an *initial value problem* (IVP) because you are given the value of the solution at some time or location t_0 and are asked to produce a formula for the solution at later times.

An *analytic solution* of an ODE is a formula for $u(t)$ that you can evaluate, differentiate, or analyze in any way you want. Analytic solutions can only be determined for a small class of ODE's. The term "analytic" used here is not the same as an analytic function in complex analysis.

A "numerical solution" of an ODE is simply a table of abscissæ and approximate values (t_k, u_k) that approximate the value of an analytic solution. This table is usually accompanied by some rule for interpolating solution values between the abscissæ. With rare exceptions, a numerical solution is *always wrong* because there is always some difference between the tabulated values and the true solution. The important question is, how wrong is it? One way to pose this question is to determine how close the computed values (t_k, u_k) are to the analytic solution, which you might write as $(t_k, u(t_k))$.

The simplest method for producing a numerical solution of an ODE is known as *Euler's explicit method*, or the *forward Euler method*. Given a solution value (t_k, u_k), you estimate the solution at the next abscissa by:

$$u_{k+1} = u_k + hu'(t_k, u_k).$$

(The step size is denoted h here. Sometimes it is denoted dt.) You can take as many steps as you want with this method, using the result from one step as the starting point for the next step.

Typically, Euler's method will be applied to systems of ODEs rather than a single ODE. This is because any higher-order ODE can be written as a system of first-order ODEs, as illustrated for second-order equations in Exercise 11.5. The following MATLAB function m-file implements Euler's method for a system of ODEs. This function is available for download as the function forward_euler.m

```
function [ t, u ] = forward_euler ( f_ode, tRange, uInitial, numSteps )
  % [ t, u ] = forward_euler ( f_ode, tRange, uInitial, numSteps ) uses
  % Euler's explicit method to solve a system of first-order ODEs
  % du/dt=f_ode(t,u).
  % f = function handle for a function with signature
  %    fValue = f_ode(t,u)
  % where fValue is a column vector
  % tRange = [t1,t2] where the solution is sought on t1<=t<=t2
  % uInitial = column vector of initial values for u at t1
  % numSteps = number of equally-sized steps to take from t1 to t2
  % t = row vector of values of t
  % u = matrix whose k-th column is the approximate solution at t(k).

  if size(uInitial,2) > 1
    error('forward_euler: uInitial must be scalar or a column vector.')
  end

  t(1) = tRange(1);
  h = ( tRange(2) - tRange(1) ) / numSteps;
  u(:,1) = uInitial;
  for k = 1 : numSteps
    t(1,k+1) = t(1,k) + h;
    u(:,k+1) = u(:,k) + h * f_ode( t(k), u(:,k) );
  end
end
```

In the above code, the initial value (uInitial) is a *column* vector, and the function represented by f returns a *column* vector. The values are returned in the *columns* of the matrix u, one column for each value of t. The vector t is a row vector.

In the following exercise, you will use `forward_euler.m` to find the solution of the initial value problem

$$\frac{du}{dt} = -u - 3t \qquad (11.1)$$
$$u(0) = 1.$$

The exact analytic solution of this IVP is $u = -2e^{-t} - 3t + 3$.

Exercise 11.1.

(1) If you have not done so already, download or copy (use cut-and-paste) the above code into a file named `forward_euler.m`.
(2) Copy the following code into a MATLAB m-file called `expm_ode.m`.

```
function fValue = expm_ode ( t, u )
  % fValue = expm_ode ( t, u ) is the right side function for
  % the ODE du/dt=-u+3*t
  % t is the independent variable
  % u is the dependent variable
  % fValue represents du/dt

  fValue = -u-3*t;
end
```

(3) Now you can use Euler's method to march from u=uInit at t=0:

```
uInit = 1.0;
[t, u] = forward_euler( @expm_ode,[ 0.0, 2.0 ], uInit, numSteps);
```

for each of the values of `numSteps` in the table below. Use at least four significant figures when you record your numbers (you may need to use the command `format short e`), and you can use the first line as a check on the correctness of the code. In addition, compute the error as the difference between your approximate solution and the exact solution at t=2, u=-2*exp(-2)-3, and compute the ratios of the error for each value of `nstep` divided by the error for the succeeding value of `nstep`. As the number of steps increases, your errors should become smaller and the ratios should tend to a limit.

Euler's explicit method				
numSteps	Step size (h)	Euler	Error	Ratio
10	0.2	-3.21474836	5.5922e-02	
20	0.1			
40	0.05			
80	0.025			
160	0.0125			
320	0.00625			—

Hint: Recall that MATLAB has a special index, end, that always refers to the final index. Thus, u(end) is a short way to write u(numel(u)) when u is either a row vector or a column vector.

(4) Based on the ratios in the table, estimate the order of accuracy of the method, *i.e.*, estimate the exponent p in the error estimate Ch^p, where h is the step size. p is an integer in this case.

Remark 11.1. There is a simple way to estimate the value of p by successively halving h. If the error were *exactly* Ch^p, then by solving twice, once using h and the second time using $h/2$ and taking the ratio of the errors, you would get

$$\frac{\text{error}(h)}{\text{error}(h/2)} = \frac{Ch^p}{C(h/2)^p} = 2^p.$$

Since the error is only $O(h^p)$, the ratio is only approximately 2^p.

11.4   The Euler halfstep (RK2) method

The "Euler halfstep" or "RK2" method is a variation of Euler's method. It is the second-simplest of a family of methods called "Runge–Kutta" methods. As part of each step of the method, an auxiliary solution, one that you don't really care about, is computed halfway, using Euler's method:

$$t_a = t_k + h/2$$
$$u_a = u_k + 0.5hf_{\text{ode}}(t_k, u_k). \tag{11.2}$$

The derivative function is evaluated at this point, and used to take a full step from the original point:

$$t_{k+1} = t_k + h;$$
$$u_{k+1} = u_k + hf_{\text{ode}}(t_a, u_a). \tag{11.3}$$

Although this method uses Euler's method, it ends up having a higher-order of convergence. Loosely speaking, the initial half-step provides additional information: an estimate of the derivative in the middle of the next step. This estimate is presumably a better estimate of the overall derivative than the value at the left end point. The per-step error is $O(h^3)$ and, since there are $O(1/h)$ steps to reach the end of the range, $O(h^2)$ overall. Keep in mind that you do not regard the auxiliary points as being part of the solution. You throw them away, and make no claim about their accuracy. It is only the whole-step points that you want.

In the following exercise you compare the results of RK2 with Euler.

Exercise 11.2.

(1) Write a MATLAB function m-file named `rk2.m` that implements the Euler half-step (RK2) method sketched above in Equations (11.2) and (11.3). Keep the same calling parameters and results as for `forward_euler.m` above. Keeping these the same will make it easy to compare different methods. The following model for the file is based on the `forward_euler.m` file with the addition of the variables `ta` and `ua` representing the auxiliary variables t_a and u_a in Equation (11.2). Add comments to this outline, including explanations of all the variables in the signature line, and fill in expressions where ??? have been left. It may be convenient to download the file `rk2.txt`.

```
function [ t, u ] = rk2 ( f_ode, tRange, uInitial, numSteps )
  % [ t, u ] = rk2 ( f_ode, tRange, uInitial, numSteps )
  % comments including the signature, meanings of variables,
  % math methods

  % your name and the date

  if size(uInitial,2) > 1
    error( ??? )
  end

  t(1,1) = tRange(1);
  h = ( tRange(2) - tRange(1) ) / numSteps;
  u(:,1) = uInitial;
  for k = 1 : numSteps
    ta = ??? ;
    ua = ??? ;
    t(1,k+1) = t(1,k) + h;
    u(:,k+1) = u(:,k) + h * f_ode( ??? );
  end
end
```

(2) Use this file to compute the numerical solution of the model ODE for the exponential, `expm_ode.m`, from Exercise 11.1, from `t = 0.0` to `t = 2.0`, and with the same initial value as in Exercise 11.1, but using Euler's halfstep method, RK2, with stepsizes below. For each case, record the value of the numerical solution at `t = 2.0`; the error, that is, the difference between the numerical solution and the true solution at the end point t=2 (u=-2*exp(-2)-3); and, the ratios of the error for each value of `numSteps` divided by the error for the succeeding value of `numSteps`. Use at least 4 significant digits when you record values.

numSteps	Step size (h)	RK2	RK2 Error	Ratio
10	0.2	-3.274896063	4.2255e-3	
20	0.1			
40	0.05			
80	0.025			
160	0.0125			
320	0.00625			—

(3) Based on the ratios in the table, estimate the order of accuracy of the method, *i.e.*, estimate the exponent p in the error estimate Ch^p. p is an integer in this case.

(4) Compare errors from Euler's method (Exercise 11.1) and Euler's halfstep method for this problem. You should clearly see that Euler's halfstep method (RK2) converges much faster than Euler's method.

numSteps	Step size (h)	Euler Error	RK2 Error
10	0.2		
20	0.1		
40	0.05		
80	0.025		
160	0.0125		
320	0.00625		

(5) Based on the above table, roughly how many steps does Euler require to achieve the accuracy that RK2 has for `numSteps=10`?

(6) You have already found the error for Euler's method is approximately $C_E h^{p_E}$ and the error for RK2 is approximately $C_{RK2} h^{p_{RK2}}$. Based on one or both of these estimates, roughly how many steps would Euler require to achieve the accuracy that RK2 has for `numSteps=320`? Explain your reasoning.

(7) Check that the accuracy obtained using Euler's method with your estimated number of steps is comparable to the accuracy that RK2 has for `numSteps=320`.

11.5   Runge–Kutta methods

The idea in Euler's halfstep method can be thought of as "sampling the water" between where you are and where you are going. This gives us a much better idea of what f is doing, and where our new value of u ought to be. Euler's method (`forward_euler`) and Euler's halfstep method (`rk2`) are the junior members of a family of ODE solving methods known as "Runge–Kutta" methods.

To develop a higher-order Runge–Kutta method, you sample the derivative function f at even more "auxiliary points" between our last computed solution and the next one. These points are *not* considered part of the solution curve; they are just a computational aid. The formulas tend to get complicated, but consider the next one to get and idea of how they work.

The third-order Runge–Kutta method can be described in the following way. Given t, u and a stepsize h, compute two intermediate points:

$$t_a = t_k + .5h$$
$$u_a = u_k + .5hf_{\text{ode}}(t_k, u_k) \tag{11.4}$$
$$t_b = t_k + h$$
$$u_b = u_k + h(2f_{\text{ode}}(t_a, u_a) - f_{\text{ode}}(t_k, u_k))$$

and then estimate the solution as:

$$t_{k+1} = t_k + h$$
$$u_{k+1} = u_k + h(f_{\text{ode}}(t_k, u_k) + 4.0f_{\text{ode}}(t_a, u_a) + f_{\text{ode}}(t_b, u_b))/6.0. \tag{11.5}$$

The global accuracy of this method is $O(h^3)$, and so you say it has "order" 3. Higher-order Runge–Kutta methods are available, with those of order 4 and 5 the most popular.

Exercise 11.3.

(1) Write a MATLAB function m-file called `rk3.m` with the outline

```
function [ t, u ] = rk3 ( f_ode, tRange, uInitial, numSteps )
  % comments including the signature, meanings of variables,
  % math methods

  % your name and the date

  << more code >>

end
```

that implements the above algorithm. You can use `rk2.m` as a model.

(2) Repeat the numerical experiment in Exercise 11.2 (using `expm_ode`) and fill in the following table. Use the first line of the table to confirm that you have written the code correctly.

numSteps	Step size (h)	RK3	RK3 Error	Ratio
10	0.2	-3.27045877	2.1179e-04	
20	0.1			
40	0.05			
80	0.025			
160	0.0125			
320	0.00625			—

(3) Based on the ratios in the table, estimate the order of accuracy of the method, *i.e.*, estimate the exponent p in the error estimate Ch^p. p is an integer in this case.

(4) Compare errors from the RK2 method (Exercise 11.2) and the RK3 method for this problem. You should clearly see that RK3 converges much faster than RK2.

numSteps	Step size (h)	RK2 Error	RK3 Error
10	0.2		
20	0.1		
40	0.05		
80	0.025		
160	0.0125		
320	0.00625		

(5) Based on the above table, roughly how many steps does RK2 require to achieve the accuracy that RK3 has for `numSteps=10`?

(6) You have already found the error for RK2 is approximately $C_{\mathrm{RK2}}h^{p_{\mathrm{RK2}}}$ and the error for RK3 is approximately $C_{\mathrm{RK3}}h^{p_{\mathrm{RK3}}}$. Based on one or both of these estimates, roughly how many steps would RK2 require to achieve the accuracy that RK3 has for `numSteps=320`? Explain your reasoning.

(7) Check that the accuracy obtained using RK2 with your estimated number of steps is comparable to the accuracy that RK3 has for `numSteps=320`.

Exercise 11.4. You have not tested your code for *systems* of equations yet. In this exercise you will do so by solving the "system"

$$\frac{du_1}{dt} = -u_1 - 3t$$

$$\frac{du_2}{dt} = -u_2 - 3t. \tag{11.6}$$

You can see that this "system" is really (11.1) twice, so you can check u_1 and u_2 against each other and against your earlier work.

(1) It turns out that the file `expm_ode.m` will return a vector if it is given a vector for u. To see this fact in action, the following command:

```
fValue = expm_ode(1.0,[5;6])
```

If `fValue` is not a *column vector* of length 2, you have a mistake somewhere. Fix it before continuing. (Hint: First make sure your vector [5;6] has a semicolon in it and is a column vector.)

(2) Solve the system (11.6) using `rk3` and `expm_ode` on the interval $[0,2]$ starting from the initial value *vector* [5;6], with 40 steps. Call the solution `usystem`. What is `usystem(:,end)`?

(3) Solving the system (11.6) amounts to solving (11.1) twice, once with initial value u(0)=5, and once more with initial value u(0)=6. Solve the scalar IVP (11.1) twice using `rk3` and `expm_ode` on the interval $[0,2]$ using 40 steps, once with initial value 5 and once with initial value 6. Call the solutions `u1` and `u2`. What are `u1(end)` and `u2(end)`? If these values are different from those of `usystem(:,end)`, you have a mistake somewhere. Fix your mistake before continuing.

Debugging hint: Check if `u1(1)` and `usystem(1,1)` or `u2(1)` and `usystem(2,1)` also disagree. Use `format long` so you can see all decimal places. One or both of these probably disagree. Using the debugger, look at `ta`, `ua`, `tb`, `ub`, `t(2)` and `u(:,2)` and compare them with the respective results from `expm_ode`. The earliest difference you see comes from an incorrect line of code.

(4) Use the following code to compare the solutions for all t values.

```
norm(usystem(1,:)-u1)/norm(u1)   % should be roundoff or zero
norm(usystem(2,:)-u2)/norm(u2)   % should be roundoff or zero
```

Exercise 11.5. The equation describing the motion of a pendulum can be described by the single dependent variable θ representing the angle the pendulum makes with the vertical. The coefficients of the equation depend on the length of the pendulum, the mass of the bob, and the gravitational constant. Assuming a coefficient value of 3, the equation is

$$\frac{d^2\theta}{dt^2} + 3\sin\theta = 0$$

and one possible set of initial conditions is

$$\theta(t_0) = 1$$
$$\frac{d\theta}{dt}(t_0) = 0.$$

Defining a vector variable u with components $u_1 = \theta$ and $u_2 = d\theta/dt$, this second-order equation can be written as the system

$$\frac{du}{dt} = \begin{pmatrix} u_2 \\ -3\sin u_1 \end{pmatrix}$$

$$u(0) = \begin{pmatrix} 1 \\ 0 \end{pmatrix}.$$

(1) Write a function m-file named `pendulum_ode.m` with outline

```
function fValue = pendulum_ode(t,u)
  % fValue = pendulum_ode(t,u)
  % comments including meanings of variables,
  % math methods, your name and the date

  % your name and the date

  << more code >>

end
```

Be sure that you return a *column* vector.

(2) Compare the solutions using the first-order Euler method, and the third-order RK3 method, using 1000 steps for each method, over the interval 0 to 25, with initial condition `u=[1;0]`. Fill in the following table, where `n` is the value of the subscript for `t(n)=6.25`, *etc.*, within roundoff.

t	n	Euler		RK3	
0.00	1	1.00	0.00	1.00	0.00
6.25					
12.50					
18.75					
25.00					

Hint: You may want to use the MATLAB `find` function. Use "`help find`" or the help menu for details.

(3) Conservation of energy guarantees that θ should stay between -1 and 1. Generate plots of θ *vs.* t for the forward Euler and RK3 solutions. Can you see anything in the plot of the Euler's method solution that might indicate that it is wrong, aside from conservation of energy?

(4) Solve the ODE using Euler's method again, but use 10,000 steps instead of 1000. Plot both θ from the refined Euler solution and θ from the original RK3 solution on the same plot. Your plots should illustrate the fact that refining the mesh helps Euler, but it is still inaccurate compared with RK3.

Message: Euler's method is not very accurate, even on nice problems. Our solution curve from Euler's method looks smooth, but it's fundamentally flawed! You must not accept a solution just because your formulæ seem to be correctly copied and the solution "looks nice."

Remark 11.2. RK3 does not conserve energy, either, but the error is so small that it is hard to see on a plot. If you are interested, try RK3 for the same situation except for $t \in [0, 250]$ with 1000 points. Special methods are available in cases where physical constraints such as conservation of energy are critical.

11.6 Stability

Explicit methods generally are conditionally stable, and require that the step size be smaller than some critical value in order to converge. It is interesting to see what happens when this stability limit is approached and exceeded. In the following exercise you will drive Euler's and RK3 methods unstable using the `expm_ode` function from before. You should observe very similar behavior of the two methods. This behavior, in fact, is similar to that of most conditionally stable methods.

Exercise 11.6.

(1) Use `forward_euler` to solve the ODE using `expm_ode`, over the interval [0,20], starting from u=20 and using `numSteps=40`. This solution is well within the stable regime. Plot this case. What is the step size?

(2) Now solve the same IVP using `numSteps=30, 20, 15, 12` and 10. The consequence of decreasing the numbers of intervals is to increase the step size. Plot these several solutions on the same plot. You should observe increasing "plus-minus" oscillations in the solutions. What are the step sizes of each of these cases?

(3) Same, but using `numSteps=8`. This solution "explodes" in a "plus-minus" fashion. Plot this case. What is the step size of this case?

(4) Use `rk3` to solve the ODE using `expm_ode`, over the interval [0, 20], starting from u=20 and using `numSteps=20, 10, 9` and 8 steps. Plot these on the same plot. The first of these is stable. What are the step sizes of these four cases?

(5) Same, but with `numSteps=7`. This solution "explodes" in a "plus-minus" fashion. Plot it. What is the step size of this case?

The message you should get from the previous exercise is that you can observe poor solution behavior when you are near the stability boundary for any method, no matter what is the theoretical accuracy is. The "poor behavior" often appears as a "plus-minus" oscillation that can shrink, grow, or remain of constant amplitude (unlikely, but possible). It can be tempting to accept solutions with small "plus-minus" oscillations that die out, but it is dangerous, especially in nonlinear problems, where the oscillations can cause the solution to move to a nearby curve with different initial conditions that has very different qualitative behavior from the desired solution.

11.7   Adams–Bashforth methods

Like Runge–Kutta methods, Adams–Bashforth methods estimate the behavior of the solution curve, but instead of evaluating the derivative function at new points close to the next solution value, they use the derivative at old solution values and interpolation ideas, along with the current solution and derivative, to estimate the new solution. This way they don't compute solutions at auxiliary points and then throw the auxiliary values away. The savings can result in increased efficiency. The disadvantage is that the first time steps must use an alternative method and extra storage is needed for the most recent time step results.

Looked at in this way, the forward Euler method is the first-order Adams–Bashforth method, using *no* old points at all, just the current solution and derivative. The second-order method, which we'll call "AB2," adds the derivative at the previous point into the interpolation mix. You might write the formula this way:

$$u_{k+1} = u_k + h(3f_{\text{ode}}(t_k, u_k) - f_{\text{ode}}(t_{k-1}, u_{k-1}))/2.$$

The AB2 method requires derivative values at two previous points, but you only have one when starting out. If you simply used an Euler step, you would pick up a relatively large error on the first step, which would pollute all subsequent results. In order to get a reasonable starting value, you should use the RK2 method, whose per-step error is order $O(h^3)$, the same as the AB2 method.

The following is a complete version of MATLAB code for the Adams–Bashforth second-order method. It is available for download as `ab2.txt`.

```
function [ t, u ] = ab2 ( f_ode, tRange, uInitial, numSteps )
  % [ t, u ] = ab2 ( f_ode, tRange, uInitial, numSteps ) uses
  % Adams-Bashforth second-order method to solve a system
  % of first-order ODEs du/dt=f_ode(t,u).
  % f = name of an m-file with signature
  %    fValue = f_ode(t,u)
  % to compute the right side of the ODE as a column vector
  %
  % tRange = [t1,t2] where the solution is sought on t1<=t<=t2
  % uInitial = column vector of initial values for u at t1
  % numSteps = number of equally-sized steps to take from t1 to t2
  % t = row vector of values of t
  % u = matrix whose k-th row is the approximate solution at t(k).

  % M. Sussman

  if size(uInitial,2) > 1
    error('ab2: uInitial must be scalar or a column vector.')
  end
```

```
t(1) = tRange(1);
h = ( tRange(2) - tRange(1) ) / numSteps;
u(:,1) = uInitial;

k = 1;
  fValue =  f_ode( t(k), u(:,k) );
  thalf = t(k) + 0.5 * h;
  uhalf = u(:,k) + 0.5 * h * fValue;
  fValuehalf = f_ode( thalf, uhalf );

  t(1,k+1) = t(1,k) + h;
  u(:,k+1) = u(:,k) + h * fValuehalf;

  for k = 2 : numSteps
    fValueold=fValue;
    fValue = f_ode( t(k), u(:,k) );
    t(1,k+1) = t(1,k) + h;
    u(:,k+1) = u(:,k) + h * ( 3 * fValue - fValueold ) / 2;
  end
end
```

Exercise 11.7.

(1) Download `ab2.txt` and change its name to `ab2.m` or copy the code to a file called `ab2.m`.
(2) Take a minute to look over this code and see if you can understand what is happening. Insert the following two comment lines into the code in the correct locations:

```
% The Adams-Bashforth algorithm starts here
% The Runge-Kutta algorithm starts here
```

(3) The temporary variables `fValue` and `fValueold` have been introduced here but were not needed in the Euler, RK2 or RK3 methods. Explain, in a few sentences, their role in AB2.
(4) If `numSteps` is 100, then *exactly* how many times will you call the derivative function `f_ode`?
(5) Use `ab2` to compute the numerical solution of the ODE for the exponential (`expm_ode`) from t = 0.0 to t = 2.0, starting at u=1 with step sizes of 0.2, 0.1, 0.05, 0.025, 0.0125 and 0.00625. Recall that the exact solution at t=2, is u=-2*exp(-2)-3. For each case, record the value of the numerical solution at t = 2.0, and the error, and the ratios of the errors. The first line of the table can be used to verify that your code is operating correctly.

numSteps	Step size (h)	AB2 ($t = 2$)	AB2 Error	Ratio
10	0.2	-3.28013993	9.4694e-03	
20	0.1			
40	0.05			
80	0.025			
160	0.0125			
320	0.00625			—

(6) Based on the ratios in the table, estimate the order of accuracy of the method, *i.e.*, estimate the exponent p in the error estimate Ch^p. p is an integer in this case.

Adams–Bashforth methods try to squeeze information out of old solution points. For problems where the solution is smooth, these methods can be highly accurate and efficient. Think of efficiency in terms of how many times you evaluate the derivative function. To compute numSteps new solution points, you only have to compute roughly numSteps derivative values, no matter what the order of the method (remember that the method saves derivative values at old points). By contrast, a third-order Runge–Kutta method would take roughly 3*numSteps derivative values. In comparison with the Runge–Kutta method, however, the old solution points are significantly further away from the new solution point, so the data is less reliable and a little "out of date." So Adams–Bashforth methods are often unable to handle a solution curve which changes its behavior over a short interval or has a discontinuity in its derivative. It's important to be aware of this tradeoff between efficiency and reliability!

11.8   Comparison of several methods

The Heun method is given by the following expressions. This method is presented, for example, in [Quarteroni *et al.* (2007)], p. 483.

$$f_k = f_{ode}(t_k, u_k),$$

$$u_{k+1} = u_k + \frac{h}{2}(f_k + f_{ode}(t_{k+1}, u_k + hf_k)). \qquad (11.7)$$

It should be obvious from this expression that it requires two evaluations of the function f in order to complete a step, *i.e.*, to compute u_{k+1} from t_k, t_{k+1}, and u_k. If you review the expressions defining RK2 and AB2, you will see that RK2 also takes two evaluations, but AB2 takes fewer.

It is customary in evaluating ODE methods to characterize them in terms of their accuracy depending on step size. It is also customary to characterize them in terms of their execution time, assuming that it takes much longer to evaluate the function than to perform any arithmetic operations.

Exercise 11.8. In this exercise, you will be comparing the three methods RK2, AB2, and Heun's. You will be using the ODE

$$\frac{du}{dt} = -u + \sin t \tag{11.8}$$
$$u(0) = 0.$$

The exact solution to this ODE is

$$u_{\text{exact}} = \frac{1}{2}(e^{-t} + \sin t - \cos t). \tag{11.9}$$

You will be solving it on the interval [0,3].

(1) Write a function m-file named **haun.m**, similar to **rk2.m** or **ab2.m** to carry out Heun's method according to (11.7) above.
(2) In order to debug and verify that your **haun.m** is correct, solve the ODE (11.8) on the interval [0,5], and fill in the first column in the following table. In it, **numSteps** refers to the number of steps taken and **Error** refers to the error $|u_{\text{numSteps}+1} - u_{\text{exact}}(5)|$ where u_{exact} comes from 11.9.

	Heun		AB2		RK2	
numSteps	Error	Calls	Error	Calls	Error	Calls
10						
20						
40						
80						
160						
320						

You should confirm that Heun's method is $O(h^2)$. If not, you should carefully examine your code to find the error.

(3) Fill in the rest of the preceeding table using your **ab2.m** and **rk2** functions. The column heading "**Calls**" refers to the total number of function calls needed to carry out the solution.
(4) Based on this experiment, what are the orders of accuracy of each of these three methods?
(5) Comparing Heun's method with AB2, which appears to have the greater accuracy for a given step size? Assuming that each function call takes a relatively long time, which of these two methods would be expected to yield the greater accuracy in the lesser time?
(6) When RK2 is included, which of the three would be the best choice based on accuracy for a given step size? Based on greatest accuracy in the least time?

11.9   Stability region plots

You have seen above that some choices of step size h result in unstable solutions (blow up) and some don't. It is important to be able to predict what choices of h will result in unstable solutions. One way to accomplish this task is to plot the region of stability in the complex plane. An excellent source of further information about stability regions can be found in [LeVeque (2007)], Chapter 7.

When applied to the test ODE $du/dt = \lambda u$, all of the common ODE methods result in linear recurrance relations of the form

$$u_{k+n} + a_{n-1} u_{k+n-1} + \ldots + a_1 u_{k+1} + a_0 u_k = 0 \qquad (11.10)$$

where n is some small integer, $k = 0, 1, \ldots$, and the a_j are constants depending on $h\lambda$. For example, the explicit Euler method results in the recurrance relation

$$u_{k+1} - (1 + h\lambda)) u_k = 0.$$

It is a well-known fact that linear recursions of the form (11.10) have unique solutions in the form

$$u_k = \sum_{j=1}^{n} c_j \zeta_j^k$$

where ζ_j for $j = 1, \ldots, n$ are the distinct roots of the polynomial equation

$$\zeta^n + a_{n-1} \zeta^{n-1} + \ldots + a_1 \zeta^1 + a_0 = 0. \qquad (11.11)$$

If any of the roots is not simple, this expression must be modified slightly. The coefficients c_j are determined by initial conditions. It must be emphasised that the roots ζ are, in general, complex. For the explicit Euler method, for example, $n = 1$ and $a_0 = -(1 + h\lambda)$.

The relationship between (11.10) and (11.11) can be seen by assuming that, in the limit of large k, $u_{k+1}/u_k = \zeta$.

Clearly, the sequence $\{u_k\}$ is stable (bounded) if and only if $|u_{k+1}/u_k| \leq 1$. Hence the ODE method associated with the polynomial (11.11) is stable if and only if all the roots satisfy $|\zeta| \leq 1$.

The equation (11.11) can be solved (numerically if necessary) for $\mu = h\lambda$ in terms of ζ. Then, the so-called "stability region," the set $\{\mu : |\zeta| \leq 1\}$ in the complex plane, can be plotted by plotting the curve of those values of μ with $|\zeta| = 1$. This idea can be turned into the following algorithmic steps.

(i) Plug $f_{\text{ode}} = \lambda u$ into the formula you are investigating to yield (11.11).
(ii) Write (11.11) in terms of $\mu = h\lambda$ and $\zeta = u_{k+1}/u_k$.
(iii) For ζ such that $|\zeta| = 1$, solve for μ. This might involve several branches, *e.g.*, when a square root has been taken.
(iv) For $0 \leq \theta \leq 2\pi$, set $\zeta = e^{i\theta}$ and draw one or more $\mu(\theta)$ curves. Since $|e^{i\theta}| = 1$, these curves bound the stability region.
(v) To identify the *interior* of the stability region, set $\zeta = 0.95 e^{i\theta}$ and plot the resulting $\mu(\theta)$ curve(s).

In the following exercise, you will implement this algorithm.

Exercise 11.9. In the first part of this exercise, you will generate the stability region of the explicit Euler method. In the second part you will generate the stability region of the Adams–Bashforth AB2 method. The third part asks you to interpret the results.

(1) Write a script file to generate a stability region plot, following these steps:

    (a) Stability plots often look like combinations or distortions of circles, so a good beginning is plotting a circle on the complex plane. For real values of θ, the exponential $\zeta = e^{i\theta}$, where $i = \sqrt{-1}$ is the imaginary unit, always satisfies $|\zeta| = 1$. Using 1000 values of θ between 0 and 2π, construct 1000 points on the unit circle in the complex plane. Write a MATLAB script m-file to plot these points and confirm they lie on the unit circle. (Use `axis equal` to get the correct aspect ratio.)

    (b) As remarked above, the explicit Euler method applied to the ODE $du/dt = \lambda u$ yields $u_{k+1} - (1 + h\lambda))u_k = 0$. Since $(h\lambda)$ occurs together, denote the product as $\mu = h\lambda$. Also denote by ζ the ratio $\zeta = u_{k+1}/u_k$. Writing these expressions using pencil and paper, solve this expression for μ in terms of ζ. Replace the plot of the unit circle in your MATLAB m-file with a plot of the curve $\mu(\zeta)$ for 1000 points on the unit circle $|\zeta| = 1$.

    (c) Add the curve $\mu(0.95\zeta)$ to your plot to indicate which part of the complex plane represents *stability* instead of instability. You can use `hold on` to plot two curves on the same plot, and you can change the color of the additional curve using the designator `'c'` for cyan or `'r'` for red, *etc.*

    (d) Finally, add lines representing the t-axis and u-axis.

(2) Make a copy of the MATLAB m-file you just wrote and modify it to display *both* the stability region for the explicit Euler method *and* the stability region for the Adams–Bashforth AB2 method given above as

$$u_{k+1} = u_k + h(3f_{\text{ode}}(t_k, u_k) - f_{\text{ode}}(t_{k-1}, u_{k-1}))/2.$$

(3) Give examples of odes and values of h that:

    (a) Would be stable using both `ab2` and explicit Euler;

    (b) Would be stable using `ab2` but not for explicit Euler; and,

    (c) Would be stable using explicit Euler but not `ab2`.

Explain your reasoning and relate it to the plot in the previous part of this exercise.

Comparing the stability regions for explicit Euler and AB2, you can see that oscillatory solutions (whose λ values lie close to the imaginary axis) can be stably simulated using much larger time steps when using AB2 than when using explicit Euler. In contrast, solutions that do not oscillate at all (λ real) can be stably simulated with explicit Euler using larger time steps than with AB2.

Chapter 12

Implicit ODE methods

12.1 Introduction

The explicit methods discussed in the previous chapter are well suited to handling a large class of ODE's. These methods perform poorly, however, for a class of "stiff" problems that occur all too frequently in applications. In this chapter, you will see *implicit methods* that are suitable for such problems. The implementation of an implicit method has an additional complication over the explicit method: a (possibly nonlinear) equation needs to be solved.

The term "stiff" as applied to ODE's does not have a precise definition. Loosely, it means that there is a very wide range between the most rapid and least rapid (with t) changes in solution components. A reasonably good rule of thumb is that if Runge–Kutta or Adams–Bashforth or other similar methods require much smaller steps than you would expect, based on your expectations for the solution accuracy, then the system is probably stiff. The term itself arose in the context of solution for the motion of a stiff spring when it is being driven at a modest frequency of oscillation. The natural modes of the spring have relatively high frequencies, and the ODE solver must be accurate enough to resolve these high frequencies, despite the low driving frequency.

Exercise 12.1 uses a direction field plot to illustrate stiffness. Exercise 12.2 illustrates stiffness with a numerical solution. Exercise 12.3 introduces the backward or implicit Euler method in the context of a linear equation and Exercise 12.4 shows that backward Euler and forward Euler have the same order of accuracy, assuming the step size is small enough for stability of forward Euler. Exercise 12.5 introduces Newton's method (Chapter 4) as a solver for backward Euler in the nonlinear case. The discussion here is abbreviated but is independent of Chapter 4 and employs an implementation of Newton's method tailored for backward Euler. Exercise 12.6 employs the van der Pol equation as an example of a nonlinear ODE that can be stiff or not depending on a parameter. Exercise 12.7 introduces the trapezoid method, following the outline of backward Euler. Exercise 12.8 applies both backward Euler and trapezoid to the van der Pol equation. Exercise 12.9 shows the effect of initial conditions away from the slowly varying solution. Exercise 12.10

introduces a second-order backward difference method and applies it to the van der Pol equation. Finally, Exercise 12.11 introduces the MATLAB ODE solvers, applying them to the van der Pol equation. This final exercise cannot, at the current time, be completed using `octave`.

12.2　Stiff systems

You have been warned that there are problems that defeat the explicit Runge–Kutta and Adams–Bashforth methods. In fact, for such problems, the higher-order methods perform even more poorly than the low-order methods. These problems are called "stiff" ODE's. You will only look at some very simple examples.

Consider the differential system

$$u' = \lambda(-u + \sin t)$$
$$u(0) = 0 \tag{12.1}$$

whose solution is

$$u(t) = Ce^{-\lambda t} + \frac{\lambda^2}{1 + \lambda^2} \sin t - \frac{\lambda}{1 + \lambda^2} \cos t$$

for C a constant. For the initial condition $u = 0$ at $t = 0$, the constant C can easily be seen to be

$$C = \frac{\lambda}{1 + \lambda^2}.$$

The ODE becomes stiff when λ gets large: at least $\lambda = 10$, but in practice the equivalent of λ might be a million or more. One key to understanding stiffness of this system is to make the following observations.

- For large λ and except very near $t = 0$, the solution behaves as if it were approximately $u(t) = \sin t$, which has a derivative of modest size.
- Small deviations from the curve $u(t) = \sin t$ (because of initial conditions or numerical errors) cause the solution to have large derivatives that depend on λ.

In other words, the interesting solution has modest derivative and should be easy to approximate, but nearby solutions have large (depending on λ) derivatives and are hard to approximate. Again, this is characteristic of stiff systems of ODEs.

You will be using this stiff differential equation in the exercises below, and in the next section you will see a graphical display of stiffness for this equation.

12.3　Direction field plots

One way of looking at a differential equation is to plot its "direction field." At any point in the (t, u) plane, you can plot a little arrow **p** equal to the slope of the solution at (t, u). This is effectively the direction that the differential equation is "telling us to go," sort of the "wind direction." How can you find **p**? If the differential equation is $u' = f_{\text{ode}}(t, u)$, and **p** represents u', then $\mathbf{p} = (dt, du)$, or $\mathbf{p} = h(1, f_{\text{ode}}(t, u))$ for a sensibly-chosen value h. MATLAB has some built-in functions to generate this kind of plot.

Exercise 12.1. In this exercise, you will see a graphical illustration of why a differential equation is "stiff."

(1) Copy the following lines into a file called `stiff4_ode.m`, or download it.

```
function fValue = stiff4_ode ( t, u )
  % fValue = stiff4_ode ( t, u )
  % computes the right side of the ODE
  %    du/dt=f_ode(t,u)=lambda*(-u+sin(t)) for lambda = 4
  % t is independent variable
  % u is dependent variable
  % output, fValue is the value of f_ode(t,u).

  LAMBDA=4;
  fValue = LAMBDA * ( -u + sin(t) );
end
```

(2) Also copy (or download) the following lines into a second file called `stiff4_solution.m`.

```
function u = stiff4_solution ( t )
  % u = stiff4_solution ( t )
  % computes the solution of the ODE
  %    du/dt=f_ode(t,u)=lambda*(-u+sin(t)) for lambda = 4
  % and initial condition u=0 at t=0
  % t is the independent variable
  % u is the solution value

  LAMBDA=4;
  u = (LAMBDA^2/(1+LAMBDA^2))*sin(t) + ...
      (LAMBDA /(1+LAMBDA^2))*(exp(-LAMBDA*t)-cos(t));
end
```

(3) Create new versions of these files, two using LAMBDA=55 and named `stiff55_ode.m` and `stiff55_solution.m`, and two using LAMBDA=10000 and named `stiff10000_ode.m` and `stiff10000_solution.m`.

(4) The solution to our stiff ODE is roughly $\sin t$, so you are interested in values of t between 0 and 2π, and values of u between -1 and 1. The MATLAB `meshgrid` command is designed for that (it is kind of a two-dimensional `linspace`). To evaluate the direction vector $\mathbf{p} = (p_t, p_u) = (dt)(1, f_{\text{ode}}(t, u))$, p_t will be all 1's (use the MATLAB `ones` function), and p_u comes from our right hand side

function. Finally, use the MATLAB command `quiver` (a "quiver" is a container filled with arrows) to display the vector plot. Use the following commands (you may find it convenient to download the file `exer1.txt`) to plot the direction field for the $[0, 2\pi] \times [-1, 1]$ range of (`t,u`) values:

```
h = 0.1;  % mesh size
scale = 2.0; % make vectors longer
[t,u] = meshgrid ( 0:h:2*pi, -1:h:1 );
pt = ones ( size ( t ) );
pu = stiff4_ode ( t, u );
quiver ( t, u, pt, pu, scale )
axis equal  %this command makes equal t and u scaling
```

(5) Finally, to see how the direction field relates to the approximate solution, plot the function $\sin t$ on the same frame.

```
hold on
t1=(0:h:2*pi);
u1=stiff4_solution(t1);
plot(t1,u1,'r')  % solution will come out red.
hold off
```

Look at your direction field full-screen. You can see the solution in red and all the arrows point toward it. Basically, no matter where you start in the rectangle, you head *very rapidly* (long arrows) toward $\sin t$, and then follow that curve as it varies *slowly* (short arrows). Some numerical methods overshoot the solution curve in their enthusiasm to reach it, as you will see in the following exercise, and avoiding this overshoot characterizes methods for stiff systems.

Exercise 12.2. In this exercise, you will be seeing a numerical illustration of why an ODE is "stiff."

(1) Using `LAMBDA=10000` to represent a stiff equation, how many points would be needed in the interval $[0, 2\pi]$ to make a pleasing plot of `stiff10000_solution(t)`?

```
t=linspace(0,2*pi,10);  % try 10, 20, 40, etc.
plot(t,stiff10000_solution(t))
```

Try it, but you will probably agree that it takes about 40 evenly-spaced points to make a reasonable curve.

(2) Now, choose either of `forward_euler.m`, or `rk3.m` from Chapter 11 and attempt to solve `stiff10000_ode.m` over the interval `[0,2*pi]`, starting from the initial condition `u=0` at `t=0` and using 40 steps. Your solutions will blow up (are unstable). (If you are plotting the solutions, look at the scale!) Multiply the

number of steps by 10 repeatedly until you get something reasonable, *i.e.*, try 40, 400, 4000, *etc.* steps. How many steps does it take to get a reasonable plot? It takes many more steps to get a reasonable solution than you would expect based on solution accuracy and this behavior is characteristic of stiff systems.

12.4   The backward Euler method

The backward Euler method is an important variation of Euler's method. take a hard look at the algorithm:

$$t_{k+1} = t_k + h$$
$$u_{k+1} = u_k + h f_{\text{ode}}(t_{k+1}, u_{k+1}). \tag{12.2}$$

You might think there is no difference between this method and Euler's method. But this is *not* a "recipe," the way some formulas are. Since u_{k+1} appears both on the left side and the right side of (12.2), it is an equation that must be solved for u_{k+1}, *i.e.*, the equation defining u_{k+1} is implicit. It turns out that implicit methods are much better suited to stiff ODE's than explicit methods.

In order to use backward Euler to solve a stiff ode equation, some method of solution of the implicit equation is needed. Before addressing this issue in general, consider the special case:

$$t_{k+1} = t_k + h$$
$$u_{k+1} = u_k + h\lambda(-u_{k+1} + \sin t_{k+1}). \tag{12.3}$$

This equation can be solved for u_{k+1} easily enough! The solution is

$$t_{k+1} = t_k + h$$
$$u_{k+1} = (u_k + h\lambda \sin t_{k+1})/(1 + h\lambda).$$

In the following exercise, you will write a version of the backward Euler method that implements this solution, and only this solution. Later, you will look at more general cases.

Exercise 12.3.

(1) Write a function m-file called `back_euler_lam.m` with outline

```
function [t,u]=back_euler_lam(lambda,tRange,uInitial,numSteps)
  % [t,u]=back_euler_lam(lambda,tRange,uInitial,numSteps)
  % comments

  % your name and the date.

  << your code >>

end
```

that implements the above algorithm. You may find it helpful to start out from a copy of your `forward_euler.m` file from Chapter 11, but if you do, be sure to realize that this function has `lambda` as its first input parameter instead of the function name `f_ode` that `forward_euler.m` has. This is because `back_euler_lam` solves only the particular linear equation (12.3).

(2) Test `back_euler_lam` by solving the system (12.1) with $\lambda = 10000$ over the interval $[0, 2\pi]$, starting from the initial condition $u = 0$ at t=0 for 40 steps. Your solution should not blow up and its plot should look reasonable.

(3) Plot the solution you just generated using `back_euler_lam` using 40 steps and the solution you generated in Exercise 12.2 using either `forward_euler` or `rk3` with a large number of steps on the same plot (use `hold on`). These solutions should be very close.

(4) For the purpose of checking your work, the first few values of u are u=[0.0; 0.156334939127; 0.308919855800; 0.453898203657; ...]. Did you get these values?

In the next exercise, you will compare backward Euler with forward Euler for accuracy. Of course, backward Euler gives results when the stepsize is large and Euler does not, but you should be curious about the case that there are enough steps to get answers. Because it would require too many steps to run this test with $\lambda = 10000$, use $\lambda = 55$, a modestly stiff value.

Exercise 12.4.

(1) Fill in the table below, using the value `lambda=55`. Compute the error in the solution versus the exact solution as

```
abs( u(end) - stiff55_solution(t(end)) )
```

Compute the ratios of the errors for each value of `numSteps` divided by the error for the succeeding value of `numSteps`. Use `back_euler_lam` to fill in the following table, over the interval `[0,2*pi]` and starting from `uInit=0`.

lambda=55		
numSteps	back_euler_lam error	ratio[1]
40		
80		
160		
320		
640		
1280		
2560		—

(2) Based on the ratios in the table, estimate the order of accuracy of the method, *i.e.*, estimate the exponent p in the error estimate Ch^p. p is an integer in this case.

(3) Repeat this experiment using `forward_euler`, and fill in the following table. Note that the errors using `euler` end up being about the same size as those using `back_euler_lam`, once `euler` starts to behave.

forward_euler error comparison		
numSteps	forward_euler error	ratio
40		
80		
160		
320		
640		
1280		
2560		—

(In filling out this table, please include at least four significant figures in your numbers. You can use `format long` or `format short e` to get the desired precision.)

(4) Compare the order of accuracy using `forward_euler` with that using `back_euler_lam`.

12.5 Newton's method

You should be convinced that implicit methods are worth your while. How, then, can the resulting implicit equation (usually it is nonlinear) be solved? The Newton (or Newton–Raphson) method is a good choice. (You saw Newton's method in Chapters 3 and 4.) You can also find discussions of Newton's method by [Quarteroni *et al.* (2007)] in Chapter 7.1, [Atkinson (1978)] in Section 2.2, page 58ff, or in a

---

[1]The error on the current line divided by the error on the following line.

Wikipedia article [Newton's Method]. Briefly, Newton's method is a way to solve equations by successive iterations. It requires a reasonably good starting point and it requires that the equation to be solved be differentiable. Newton's method can fail, however, and care must be taken so that you do not attempt to use the result of a failed iteration. When Newton's method does fail, it is frequently because the provided Jacobian matrix is not, in fact, the correct Jacobian for the provided function.

Suppose you want to solve the nonlinear (vector) equation

$$\mathbf{F}(\mathbf{U}) = \mathbf{0} \tag{12.4}$$

and you know a good starting place $\mathbf{U}^{(0)\dagger}$. The following iteration is called Newton iteration.

$$\mathbf{U}^{(n+1)} - \mathbf{U}^{(n)} = \Delta \mathbf{U}^{(n)} = -(\mathbf{J}^{(n)})^{-1}(\mathbf{F}(\mathbf{U}^{(n)})) \tag{12.5}$$

where $\mathbf{J}^{(n)}$ denotes the partial derivative of $\mathbf{F}$ evaluated at $\mathbf{U}^{(n)}$. If $\mathbf{F}$ is a scalar function of a scalar variable, $\mathbf{J} = \partial \mathbf{F}/\partial \mathbf{U}$. In the case that $\mathbf{U}$ is a vector, the partial derivative must be interpreted as the Jacobian matrix, with components defined as

$$\mathbf{J}_{ij} = \frac{\partial \mathbf{F}_i}{\partial \mathbf{U}_j}.$$

Newton's method (usually) converges quite rapidly, so that only a few iterations are required.

There is an easy way to remember the formula for Newton's method. Write the finite difference formula for the derivative as

$$\frac{\mathbf{F}(\mathbf{U}^{(n+1)}) - \mathbf{F}(\mathbf{U}^{(n)})}{\mathbf{U}^{(n+1)} - \mathbf{U}^{(n)}} = \frac{\partial \mathbf{F}}{\partial \mathbf{U}}(\mathbf{U}^{(n)})$$

or, for vectors and matrices,

$$\mathbf{F}(\mathbf{U}^{(n+1)}) - \mathbf{F}(\mathbf{U}^{(n)}) = \left(\frac{\partial \mathbf{F}}{\partial \mathbf{U}}(\mathbf{U}^{(n)})\right)\left(\mathbf{U}^{(n+1)} - \mathbf{U}^{(n)}\right)$$

and then take $\mathbf{F}(\mathbf{U}^{(n+1)}) = 0$, because that is what you are wishing for, and solve for $\mathbf{U}^{(n+1)}$.

On each step, the backward Euler method requires a solution of the equation

$$u_{k+1} = u_k + h f_{\text{ode}}(t_{k+1}, u_{k+1})$$

so that you can take u_{k+1} to be the solution, $\mathbf{U}$, of the system $\mathbf{F}(\mathbf{U}) = \mathbf{0}$, where

$$\mathbf{F}(\mathbf{U}) = u_k + h f_{\text{ode}}(t_{k+1}, \mathbf{U}) - \mathbf{U}. \tag{12.6}$$

When you have found a satisfactorily approximate solution $\mathbf{U}^{(n)}$ to (12.6), take $u_{k+1} = \mathbf{U}^{(n)}$ and proceed with the backward Euler method. Think of each of the values $\mathbf{U}^{(n)}$ as successive corrections to u_{k+1}.

---

†Superscripts in parentheses are used to denote iteration number to reduce confusion among iteration number, exponentiation and vector components.

For the Euler, Adams–Bashforth and Runge–Kutta methods, only a function for computing the right side of the differential equation is needed. In order to carry out the Newton iteration, however, the function also needs to compute the partial derivative of the right side with respect to u. In Chapter 4, the derivative of the function was returned as a second returned variable, and the same convention will be used here.

To summarize, on each time step, start off the iteration by predicting the solution using an explicit method. Forward Euler is appropriate in this case. Then use Newton's method to generate successive correction steps. In the following exercise, you will be implementing this method. This approach is an example of a predictor-corrector method.

Exercise 12.5.

(1) Change your stiff10000_ode.m so that it has an additional returned variable. Add an explanation of this variable to the comments.

```
function [fValue, fPartial]=stiff10000_ode(t,u)
  % [fValue, fPartial]=stiff10000_ode(t,u)
  % comments

  % your name and the date

  << your code >>

end
```

and computes the partial derivative of stiff10000_ode with respect to u, **fPartial**. To do this, write the derivative of the formula for stiff10000_ode out by hand, then program it. *Do not attempt to use the symbolic toolbox for differentiation inside the function* stiff10000_ode.

(2) Similarly, change stiff55_ode. There is no need to change stiff4_ode because you won't be using it again.

(3) Copy the following function to a file named back_euler.m using cut-and-paste, or download the file back_euler.txt and modify it.

```
function [t,u]=back_euler(f_ode,tRange,uInitial,numSteps)
  % [t,u]=back_euler(f_ode,tRange,uInitial,numSteps) computes
  % the solution to an ODE by the backward Euler method
  %
  % tRange is a two dimensional vector of beginning and
  %   final values for t
  % uInitial is a column vector for the initial value of u
  % numSteps is the number of evenly-spaced steps to divide
  %   up the interval tRange
```

```
% t is a row vector of selected values for the
%    independent variable
% u is a matrix. The k-th column of u is
%    the approximate solution at t(k)

% your name and the date

% force t to be a row vector
t(1,1) = tRange(1);
h = ( tRange(2) - tRange(1) ) / numSteps;
u(:,1) = uInitial;
for k = 1 : numSteps
  t(1,k+1) = t(1,k) + h;

  U = (u(:,k)) + h * f_ode( t(1,k), u(:,k));
  [U,isConverged]= newton4euler(f_ode,t(k+1),u(:,k),U,h);

  if ~ isConverged
    error(['back_euler failed to converge at step ', ...
        . num2str(k)])
  end

  u(:,k+1) = U;
end
end
```

(4) Copy the following function to a file named `newton4euler.m`, or download the file `newton4euler.txt`. (Read this name as "Newton for Euler.")

```
function [U,isConverged]=newton4euler(f_ode,tkp1,uk,U,h)
  % [U,isConverged]=newton4euler(f_ode,tkp1,uk,U,h)
  % special function to evaluate Newton's method for back_euler

  % your name and the date

  TOL = 1.e-6;
  MAXITS = 500;

  isConverged= false;
  for n=1:MAXITS
    [fValue fPartial] = f_ode( tkp1, U);
    F = uk + h * fValue - U;
    J = h * fPartial - eye(numel(U));
```

```
    increment=J\F;
    if n>1
      r1=norm(increment,inf)/oldIncrement;
    else
      r1=2; % forces convergence failure
    end
    oldIncrement=norm(increment,inf);
    U = U - increment;
    if norm(increment,inf) < TOL*norm(U,inf)*(1-r1)
    if norm(increment,inf) < TOL*norm(U,inf)
      isConverged= true;  % turns TRUE when converged
      return
    end
  end
end
```

(5) Comments are important:

(a) Insert the following comments into either `back_euler.m` or `newton4euler.m` (or both) where they belong. Some comments belong in both files.

```
% f_ode is the handle of a function whose signature is ???
% TOL = ??? (explain in words what TOL is used for)
% MAXITS = ??? (explain in words what MAXITS is used for)
% When F is a column vector and J a matrix,
%   the expression J\F means ???
% The MATLAB function "eye" is used for ???
% The following for loop performs the spatial stepping (on t)
% The following statement computes the initial guess for Newton
```

(b) Insert a comment in the code indicating which lines implement Equation (12.5).

(c) Insert a comment in the code indicating which lines implement Equation (12.6). (**Note:** This line is specific to the implicit Euler method, and will have to be changed when the method is changed.)

(6) If the function $\mathbf{F}(\mathbf{U})$ is given by $F_1(U_1, U_2) = 4U_1 + 2(U_2)^2$, and $F_2(U_1, U_2) = (U_1)^3 + 5U_2$, and if $U_1 = -2$ and $U_2 = 1$, write out the values of $\mathbf{J}(\mathbf{U}) = \partial\mathbf{F}/\partial\mathbf{U}$ and $\mathbf{F}(\mathbf{U})$ as a MATLAB matrix and vector, respectively. Be careful to distinguish row and column vectors.

(7) The equation (12.5) includes the quantities $\mathbf{J}^{(n)}$, $\Delta\mathbf{U}$, and $\mathbf{F}(\mathbf{U}^{(n)})$. What are the names of the variables in the code representing these mathematical quantities?

(8) In the case that `numel(U)>1`, is U a row vector or a column vector?

(9) If `f_ode=@stiff10000_ode`, `tkp1=2.0`, `uk=1.0`, `h=0.1`, and the initial guess

U=1, write out by hand the (linear) equation that `newton4euler` solves. To do this, start from (12.6), with $\mathbf{F}(\mathbf{U}) = 0$. Plug in the formula for f_{ode} and solve for $\mathbf{U}$ in terms of everything else. What is the value of the solution, $\mathbf{U}$, of this linear equation? Please include at least eight significant digits for the result. Since this equation is linear, it should only take 2 iterations. Does it?

(10) Verify that `newton4euler` is correct by showing it yields the same value you just computed by hand for the case that `f_ode=@stiff10000_ode`, `tkp1=2.0`, `uk=1.0`, `h=0.1`, and the initial guess U=1. (If you discover that the Newton iteration fails to converge, you probably have the derivative in `stiff10000_ode` wrong.)

(11) In order to check that everything is programmed correctly, solve the ODE using `stiff10000_ode`, on the interval $[0, 2\pi]$, with initial value 0, for 40 steps, just as in Exercise 12.3, but use `back_euler`. Compare your solution with the one using `lambda=10000` in `back_euler_lam.m`. You can compare all 40 values at once by taking the norm of the difference between the two solutions generated by `back_euler_lam` and `back_euler`. The two solutions should agree to ten or more significant digits. (Testing code by comparison with previous code is called "regression" testing.)

12.5.1 *van der Pol's equation*

One simple nonlinear equation with quite complicated behavior is van der Pol's equation. This equation is written

$$z'' + a(z^2 - 1)z' + z = e^{-t}$$

where e^{-t} has been chosen as a forcing term that dies out as t gets large. When $z > 1$, this equation behaves somewhat as an oscillator with negative feedback ("damped" or "stable"), but when z is small then it looks like an oscillator with positive feedback ("negatively damped" or "positive feedback" or "unstable"). When z is small, it grows. When z is large, it dies off. The result is a nonlinear oscillator. Physical systems that behave somewhat as a van der Pol oscillator include electrical circuits with semiconductor devices such as tunnel diodes in them and some biological systems, such as the beating of a heart, or the firing of neurons. The original paper [van der Pol and van der Mark (1928)] is on a model of beating of a heart. More complete information can be found in [Kanamaru (2007)].

As you know, van der Pol's equation can be written as a system of first-order ODEs in the following way.

$$\begin{aligned}
u_1' &= f_1(t, u_1, u_2) = u_2 \\
u_2' &= f_2(t, u_1, u_2) = -a(u_1^2 - 1)u_2 - u_1 + e^{-t}
\end{aligned} \tag{12.7}$$

where $z = u_1$ and $z' = u_2$.

The matrix generalization of the partial derivative in Newton's method is the Jacobian matrix:

$$J = \begin{bmatrix} \frac{\partial f_1}{\partial u_1} & \frac{\partial f_1}{\partial u_2} \\ \frac{\partial f_2}{\partial u_1} & \frac{\partial f_2}{\partial u_2} \end{bmatrix}. \tag{12.8}$$

Exercise 12.6. In this exercise you will be solving van der Pol's equation with $a = 11$ using back_euler.

(1) Write an m-file to compute the right side of the van der Pol system (12.7) and its Jacobian (12.8). The m-file should be named vanderpol_ode.m and should have the outline

```
function [fValue, J]=vanderpol_ode(t,u)
% [fValue, J]=vanderpol_ode(t,u)
% more comments
% your name and the date

if size(u,1) ~=2  | size(u,2) ~=1
   error('vanderpol_ode: u must be a column vector of length 2!')
end

a=11;

fValue = ???

df1du1 = 0;
df1du2 = ???
df2du1 = ???
df2du2 = -a*(u(1)^2-1);

J=[df1du1 df1du2
   df2du1 df2du2];

end
```

You may find it convenient to download the file vanderpol_ode.txt. Be sure that fValue is a *column* and that the value of the parameter $a = 11$.

(2) Solve the van der Pol system twice, once using forward_euler and once using back_euler. Solve over the interval from t=0 to t=2, starting from u=[0;0] and using 40 intervals. Plot both solutions on the same frame. (You can use hold on and hold off.) You can see that forward_euler does not give a good solution (it has some up/down oscillations) while back_euler does give a good solution. (**Note:** If you get a message saying that convergence failed, you probably have an error in your calculation of the derivatives in J. Fix it.)

(3) Solve the system again, this time using 640 intervals. Plot both solutions on a single frame. You should see that both methods should remain stable and the answers are close.

12.6 The trapezoid method

You have looked at the forward Euler and backward Euler methods. These methods are simple and involve only function or Jacobian values at the beginning and end of a step. They are both first-order, though. It turns out that the trapezoid method also involves only values at the beginning and end of a step and is second-order accurate, a substantial improvement. This method is also called "Crank–Nicolson," especially when it is used in the context of partial differential equations. As you will see, this method is appropriate only for mildly stiff systems.

The trapezoid method can be derived from the trapezoid rule for integration. It has a simple form:

$$t_{k+1} = t_k + h$$
$$u_{k+1} = u_k + \frac{h}{2}(f_{\text{ode}}(t_k, u_k) + f_{\text{ode}}(t_{k+1}, u_{k+1})) \qquad (12.9)$$

from which you can see that this is also an implicit formula. The backward Euler and Trapezoid methods are the first two members of the "Adams–Moulton" family of ODE solvers.

In the exercise below, you will write a version of the trapezoid method using Newton's method to solve the per-timestep equation, just as with back_euler. As you will see in later exercises, the trapezoid method is not so appropriate when the equation gets very stiff, and Newton's method is overkill when the system is not stiff. The method can be successfully implemented using an approximate Jacobian or by computing the Jacobian only occasionally, resulting in greater computational efficiency. These alternatives are not discussed further in this chapter.

Exercise 12.7.

(1) Write `trapezoid.m` by following the steps:

    (a) Make a copy of your `back_euler.m` file, naming the copy `trapezoid.m`. Modify it to have the signature:

```
function [t, u]= trapezoid (f_ode, tRange, uInitial, numSteps)
```

    (b) Add appropriate comments below the signature line.

    (c) Replace the line

```
[U,isConverged]= newton4euler(f_ode,t(k+1),u(:,k),U,h);
```

    with the new line

```
[U,isConverged]= newton4trapezoid(f_ode,t(1,k),t(1,k+1), ...
                u(:,k),U,h);
```

(2) Write newton4trapezoid.m following the steps:

(a) Make a copy of your newton4euler.m function m-file, and name the copy newton4trapezoid.m.

(b) Change the signature of newton4trapezoid to

function [U,isConverged]=newton4trapezoid(f_ode,tk,tkp1,uk,U,h)

(c) In order to implement the trapezoid method, you need to write the function $F(U)$ that appears in (12.5) and (12.6) so it is valid for the trapezoid rule. In the code you copied from newton4euler, it is written for the backward Euler method (12.6) and must be changed for use in the trapezoid method. To do this, consider (12.9), repeated here

$$u_{k+1} = u_k + \frac{h}{2}(f_{\text{ode}}(t_k, u_k) + f_{\text{ode}}(t_{k+1}, u_{k+1}))$$

and replace u_{k+1} with U and bring everything to the right. Then write

$$F(U) = 0 = \text{right side.}$$

Modify newton4trapezoid.m to solve this function. Do not forget to modify the Jacobian J $J_{ij} = \frac{\partial F_i}{\partial U_j}$ to reflect the new function F. Remember, if you have the Jacobian wrong, the Newton iteration may fail.

(3) Test your newton4trapezoid code for the case f_ode=@stiff55_ode, tk=1.9, tkp1=2.0, uk=1.0, h=0.1, and the initial guess for U=1. Since this equation is linear, it should only take 2 iterations. Does it?

(4) Write out by hand the (linear) equation that newton4trapezoid solves in the case f_ode=@stiff55_ode, tk=1.9, tkp1=2.0, uk=1.0, and h=0.1. Be sure your solution to this equation agrees with the one from newton4trapezoid to at least 10 decimal places. If not, find the mistake and fix it.

(5) Test the accuracy of the trapezoid method by computing the numerical solution of stiff55_ode.m starting from $u = 0$ over the interval $t \in [0, 2\pi]$, and fill in the following table. Compute "error" as the difference between the computed and exact values at $t = 2\pi$.

stiff55		
numSteps	trapezoid error	ratio
10		
20		
40		
80		
160		
320		—

(6) Are your results consistent with the theoretical $O(h^2)$ convergence rate of the trapezoid rule? If not, you have a bug you need to fix.

In the following exercise, you are going to see the difference in accuracy between the trapezoid method and the backward Euler method for solution of the van der Pol equation.

Exercise 12.8.

(1) Use `trapezoid.m` to solve the van der Pol equation (a=11) on the interval [0,10] starting from the column vector [0,0] using 100, 200, 400, and 800 steps, and plot both components of all four solutions on one plot. You should see that the solution for 400 steps is pretty much the same as the one for 800 steps, and the others are different, especially in the second component (derivative). Assume that the final two approximate solutions represent the correct solution. **Hint:** you can generate this plot easily with the following sequence of commands.

```
hold off
[t,u]=trapezoid(@vanderpol_ode,[0,10],[0;0],100);plot(t,u)
hold on
[t,u]=trapezoid(@vanderpol_ode,[0,10],[0;0],200);plot(t,u)
[t,u]=trapezoid(@vanderpol_ode,[0,10],[0;0],400);plot(t,u)
[t,u]=trapezoid(@vanderpol_ode,[0,10],[0;0],800);plot(t,u)
hold off
```

(2) Use `back_euler.m` to solve the van der Pol equation on the interval [0,10] starting from the value [0;0] using 400, 800, 3200, and 12800 steps, and plot all four solutions on one plot. The larger number of points may take a minute. You should see a progression of increasing accuracy in the second "pulse."

(3) Plot the 12800-step `back_euler` solution and the 800-step `trapezoid` solution on the same plot. You should see they are close.

The trapezoid method is unconditionally stable, but this fact does *not* mean that it is good for very stiff systems. In the following exercise, you will apply the trapezoid method to a very stiff system so you will see that numerical errors arising from the initial rapid transient persist when using the trapezoid rule but not for backwards Euler.

Exercise 12.9.

(1) Solve the `stiff10000_ode` system using backwards Euler and using the trapezoid method on the interval [0,10] starting from uInitial=0.1 using 100

steps. *Note that the initial condition is not zero.* Plot both solutions on the same plot.

(2) Use the trapezoid method to solve the same case using 200, 400, 800, and 1600 steps. Plot each solution on its own plot. You should see that the effect of the initial condition is not easily eliminated.

12.7 Backward difference methods

You saw the Adams–Bashforth (explicit) methods in Chapter 11. The second-order Adams–Bashforth method (ab2) achieves higher-order without using intermediate points (as the Runge–Kutta methods do), but instead uses points earlier in the evolution history, using the points u_k and u_{k-1}, for example, for ab2. Among implicit methods, the "backwards difference methods" also achieve higher-order accuracy by using points earlier in the evolution history. It turns out that backward difference methods are good choices for stiff problems.

The backward difference method of second-order can be written as

$$u_{k+1} = \frac{4}{3}u_k - \frac{1}{3}u_{k-1} + \frac{2}{3}hf_{\text{ode}}(t_{k+1}, u_{k+1}). \tag{12.10}$$

It is an easy exercise using Taylor series to show that the truncation error is $O(h^3)$ per step, or $O(h^2)$ over the interval $[0, T]$.

Exercise 12.10. In this exercise, you will write a function m-file named bdm2.m to implement the second-order backward difference method (12.10).

(1) Write a function m-file with the outline

```
function [ t, u ] = bdm2 ( f_ode, tRange, uInitial, numSteps )
  % [ t, u ] = bdm2 ( f_ode, tRange, uInitial, numSteps )
  % ... more comments ...

  % your name and the date

  << your code >>

end
```

Although it would be best to start off the time stepping with a second-order method, for simplicity you should take one backward Euler step to start off the stepping. You should model the function on back_euler.m, and, with some ingenuity, you can use newton4euler without change. (If you cannot see how, just write a new Newton solver routine.)

(2) Test your function by solving the system (12.1) for $\lambda = 55$ over the interval $[0, 2\pi]$ starting at 0. Check the function for 40, 80, 160, and 320 intervals, and show that convergence is $O(h^2)$.

(3) Using $a = 11$, solve **vanderpol_ode** over the interval $[0, 10]$ starting from $[0; 0]$ using 200 points, plot and compare it with the one from **trapezoid**. They should be close.

(4) Using $a = 55$, solve **vanderpol_ode** over the interval $[0, 10]$ starting from $[1.1; 0]$ using 200 points, plot and compare it with the one from **trapezoid**. You should observe that **bdm2** does not exhibit the plus/minus oscillations that can be seen in the **trapezoid** solution, especially in the second component (the derivative).

12.8   MATLAB **ODE solvers**

MATLAB has a number of built-in ODE solvers. These include:

MATLAB **ODE solvers**	
ode23	non-stiff, low-order
ode113	non-stiff, variable-order
ode15s	stiff, variable-order, includes DAE
ode23s	stiff, low-order
ode23t	trapezoid rule
ode23tb	stiff, low-order
ode45	non-stiff, medium-order (Runge–Kutta)

All of these functions use the very best methods, are highly reliable, use adaptive step size control, and allow close control of errors and other parameters. As a bonus, they can be "told" to precisely locate interesting events such as zero crossings and will even allow user-written functions to be called when certain types of events occur. There is very little reason to write your own ODE solvers unless you are actively researching new methods. In the following exercise you will see how to use these solvers in a very simple case.

The ODE solvers you have written have four parameters, the function name, the solution interval, the initial value, and the number of steps. The MATLAB solvers use a good adaptive stepping algorithm, so there is no need for the fourth parameter.

The MATLAB ODE solvers require that functions such as **vanderpol_ode.m** return column vectors and there is no need for the Jacobian matrix. Thus, the same **vanderpol_ode.m** that you used for **back_euler** will work for the MATLAB solvers. However, the format of the matrix output from the MATLAB ODE solvers is the *transpose* of the one from our solvers! Thus, while **u(:,k)** is the (column vector) result achieved by **back_euler** on step **k**, the result from the MATLAB ODE solvers will be **u(k,:)**!

Remark 12.1. The following exercise involves the MATLAB ODE solver ode15s. At the time of this writing, this solver is not available in octave. The functionality of this stiff solver is available in octave using the lsode function, but its syntax is different from the MATLAB solver.

Exercise 12.11. For this exercise, set a=55 in vanderpol_ode.m to exhibit some stiffness.

(1) You can watch a solution evolve if you call the solver without any output variables. Use ode45 to solve the van der Pol problem with a=55 and solve on the interval [0,70] starting from [0;0]. Use the following MATLAB command:

```
ode45(@vanderpol_ode,[0,70],[0;0])
```

(2) You can see the difference between a stiff solver and a non-stiff solver on a stiff equation such as stiff10000_ode by comparing ode45 with ode15s. (These are the author's personal favorites.) You can see both plots with the commands

```
figure(2)
ode45(@stiff10000,[0,8],1);
title('ode45')
figure(3)
ode15s(@stiff10000,[0,8],1);
title('ode15s')
```

You should be able to see that the step density (represented by the little circles on the curves) is less for ode15s. This density difference is most apparent in the smooth portions of the curve where the solution derivative is small.

(3) If you wish to examine the solution of vanderpol_ode in more detail, or manipulate it, you need the solution values, not a picture. You can get the solution values with commands similar to the following:

```
[t,u]=ode15s(@vanderpol_ode,[0,70],[0;0]);
```

If you wish, you can plot the solution, compute its error, *etc.* For this solution, what is the value of u_1 at x=70 (please give at least six significant digits)? How many steps did it take? (The length of t is one more than the number of steps.)

(4) Suppose you want a very accurate solution. The default tolerance is .001 relative accuracy in MATLAB , but suppose you want a relative accuracy of 1.e-8? There is an extra variable to provide options to the solver. It works in the following manner:

```
myoptions=odeset('RelTol',1.e-8);
[t,u]=ode15s(@vanderpol_ode,[0,70],[0;0],myoptions);
```

Use `help odeset` for more detail about options, and use the command `odeset` alone to see the default options. How many steps did it take this time? What is the value of u_1 at `t=70` (to at least six significant digits)?

Remark 12.2. The functionality of the MATLAB ODE solvers is available in external libraries that can be used in Fortran, C, C++ or Java programs. One of the best is the Sundials package from Lawrence Livermore [Hindmarsh *et al.* (2005)] or its predecessor odepack [Hindmarsh (2006)].

Chapter 13

Boundary value problems and partial differential equations

13.1 Introduction

The initial value problem for ordinary differential equations of the previous chapters is only one of the two major types of problem for ordinary differential equations. The other type is known as the "boundary value problem" (BVP). A simple example of such a problem would describe the shape of a rope hanging between two posts. You know the position of the endpoints, and you have a second-order differential equation describing the shape. If the two conditions were both given at the left endpoint, we'd know what to do right away. But how do you handle this "slight" variation?

This chapter is concerned with two of the most common approaches to solving BVPs as well as a combined IVP-BVP for a partial differential equation. An additional exercise introduces a third approach, the shooting method, to solving BVPs. The discussion in this chapter is limited to relatively simple approaches in a single space dimension and is intended to give the flavor of these approaches, each of which could easily be the subject of a full semester's course. Except for the shooting method, these methods are easily extended to two and three space dimensions.

The approaches included in this chapter are the following:

- The finite difference method (FDM), Exercises 13.1 and 13.2;
- The finite element method (FEM), Exercises 13.3 through 13.7, and with 13.8 involving Neumann boundary conditions;
- The method of lines, Exercise 13.9; and,
- The shooting method, Exercise 13.10.

The discussion of the method of lines involves the finite difference method for the spatial variable and back_euler from Chapter 12 for the temporal integration. The MATLAB ODE functions could be substituted for back_euler. The shooting method employs MATLAB functions ode45 and fzero.

13.2   Boundary value problems

A one-dimensional boundary value problem (BVP), is similar to an initial value problem, except that the data you are given isn't conveniently located at a starting point, but rather some is specified at the left end point and some at the right. The independent variable represents space, rather than time, in this setting, so ODE solution methods from previous chapters will be applied here using x rather than t as the independent variable.

The clothesline BVP will serve as the illustrative example for several of the following exercises. This problem consists of a rope attached to the tops of two wooden poles of possibly different heights. If the rope were weightless, or if it were rigid, it would lie along a straight line; however, the rope has a weight and is elastic, so it sags down slightly from its ideal linear shape. You wish to determine the curve described by the rope. Use the variable x to denote horizontal distance and $u(x)$ to denote height of the rope at the point x.

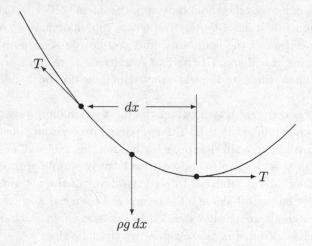

Fig. 13.1   Forces on a clothesline

The equation for the curve described by the rope can be derived[1] by considering the tension in the rope and applying Newton's law ($\mathbf{F} = m\mathbf{a}$). See Figure 13.1. The tension in the rope is T (a constant because the rope is in equilibrium), and consider a tiny piece of the rope of length dx with mass per unit length ρ. The total mass of the differential piece of rope is $\rho\,dx$, so that the force due to gravity is directed downward and is given by $\rho g\,dx$. This piece of rope observes forces on each of its ends. The magnitudes of these forces are equal to the tension, T, and the directions are given by the slope of the curve at the ends of the differential piece. Hooke's law says that the tension is proportional to the amount of strain

---

[1]This discussion is not a proof: it is a description of why you should believe the equation.

in the string, $T = -K\,du/dx$, where K is a constant of proportionality (Young's modulus). Hence, the equation can be written as

$$-K\left.\frac{du}{dx}\right]_{\text{right}} + K\left.\frac{du}{dx}\right]_{\text{left}} = -\rho g\,dx.$$

Dividing both sides by dx and letting $dx \to 0$ yields the equation

$$-\frac{d}{dx}\left(K\frac{du}{dx}\right) = -\rho g.$$

There is no reason that the "constant" of proportionality cannot change from place to place. For the sake of definiteness, assume K varies as $K(x) = 1 + cx$, for constant C, representing a rope (or spring) whose stiffness varies from end to end. The result is the equation

$$(1 + cx)u'' + cu' = \rho g.$$

When $c = 0$, this equation is called the "Poisson equation" and also describes the distribution of heat in a solid bar, among other common physical problems.

For the sake of definiteness, take $\rho g = 0.4$ and $c = 0.05$, the left end of height 1 at $x = 0$, and the right end height of 1.5 at $x = 5$. Thus, the system to be solved is

$$(1 + cx)u'' + cu' = \rho g$$
$$c = 0.05$$
$$\rho g = 0.4 \tag{13.1}$$
$$u(0) = 1$$
$$u(5) = 1.5.$$

The ODE is *linear*. Linearity implies existence and uniqueness of solutions.

13.3   Finite difference method for a BVP

The derivation presented above for the shape of the rope is suggestive of a way to solve for the shape, called the "finite difference method." Assume that you have divided the interval up into N equal intervals of width Δx determined by $N + 1$ points. Denote the spatial points x_n, $n = 0, 1, \ldots, N + 1$. Approximate the value of $u(x_n)$ by u_n. Also approximate the Young's modulus function as $K_n = K(x_n) = (1 + cx_n)$.

Now, consider the nth interval as if it were the differential piece of rope mentioned in the derivation. Using the standard finite difference approximation for a derivative, the slope of the rope at the left of the nth interval could be approximated as $(u_n - u_{n-1})/\Delta x$ and the slope on the right of the nth interval could be approximated as $(u_{n+1} - u_n)/\Delta x$. The difference between these is an approximation of the second derivative

$$u_n'' \approx \frac{u_{n+1} - 2u_n + u_{n-1}}{\Delta x^2}.$$

Along similar lines, approximate the first derivative as

$$u_n' \approx \frac{u_{n+1} - u_{n-1}}{2\Delta x}.$$

Both approximations (of u' and u'') have the same Taylor-series (truncation) error of $O(\Delta x^2)$.

Put all these together into (13.1) to get

$$(1 + cx_n)\frac{u_{n+1} - 2u_n + u_{n-1}}{\Delta x^2} + c\frac{u_{n+1} - u_{n-1}}{2\Delta x} = \rho g$$

and, as above, $c = 0.05$ and $\rho g = 0.4$. You can associate this equation with the solution value at u_n, except for $n = 0$ and $n = N + 1$ (do you see why?). Conveniently, those are the points at which you have boundary conditions specified.

In particular, look at approximating the rope BVP at 6 points. Set up the ODE at points 1, 2, 3, and 4, and associate the boundary conditions with the $n = 0$ and $n = 5$ solution values. Note that $x_n = n\Delta x$. It is convenient to multiply through by Δx^2 to make things look nicer:

$$u_0 = 1$$

$$(1 + c\Delta x - c\Delta x/2)u_0 - 2(1 + c\Delta x)u_1 + (1 + c\Delta x + c\Delta x/2)u_2 = 0.4\Delta x^2$$

$$(1 + 2c\Delta x - c\Delta x/2)u_1 - 2(1 + 2c\Delta x)u_2 + (1 + 2c\Delta x + c\Delta x/2)u_3 = 0.4\Delta x^2$$

$$(1 + 3c\Delta x - c\Delta x/2)u_2 - 2(1 + 3c\Delta x)u_3 + (1 + 3c\Delta x + c\Delta x/2)u_4 = 0.4\Delta x^2 \qquad (13.2)$$

$$(1 + 4c\Delta x - c\Delta x/2)u_3 - 2(1 + 4c\Delta x)u_4 + (1 + 4c\Delta x + c\Delta x/2)u_5 = 0.4\Delta x^2$$

$$u_5 = 1.5.$$

Actually, in Equation (13.2), the quantities u_0 and u_5 are not really variables, being fixed by the boundary conditions. Hence the only variables are u_1, u_2, u_3 and u_4. The system can be rewritten as

$$
\begin{aligned}
-2(1 + c\Delta x)u_1 +(1 + 1.5c\Delta x)u_2 \qquad\qquad +0 \qquad\qquad\qquad +0 &= 0.4\Delta x^2 \\
&\quad -(1 + 0.5c\Delta x)u_0 \\
(1 + 1.5c\Delta x)u_1 -2(1 + 2c\Delta x)u_2 \quad (1 + 2.5c\Delta x)u_3 \qquad\qquad +0 &= 0.4\Delta x^2 \\
0 +(1 + 2.5c\Delta x)u_2 -2(1 + 3c\Delta x)u_3 +(1 + 3.5c\Delta x)u_4 &= 0.4\Delta x^2 \\
0 \qquad\qquad +0 +(1 + 3.5c\Delta x)u_3 -2(1 + 4c\Delta x)u_4 &= 0.4\Delta x^2 \\
&\quad -(1 + 4.5c\Delta x)u_5
\end{aligned}
$$

and this system has been formatted to suggest the matrix equation

$$
\begin{bmatrix}
-2(1 + c\Delta x) & (1 + 1.5c\Delta x) & 0 & 0 \\
(1 + 1.5c\Delta x) & -2(1 + 2c\Delta x) & (1 + 2.5c\Delta x) & 0 \\
0 & (1 + 2.5c\Delta x) & -2(1 + 3c\Delta x) & (1 + 3.5c\Delta x) \\
0 & 0 & (1 + 3.5c\Delta x) & -2(1 + 4c\Delta x)
\end{bmatrix}
\begin{bmatrix} u_1 \\ u_2 \\ u_3 \\ u_4 \end{bmatrix}
$$

$$
= \begin{bmatrix}
0.4\Delta x^2 - (1 + 0.5c\Delta x)u_0 \\
0.4\Delta x^2 \\
0.4\Delta x^2 \\
0.4\Delta x^2 - (1 + 4.5c\Delta x)u_5
\end{bmatrix}. \qquad (13.3)
$$

Discretizing the differential equations has created a set of linear algebraic equations that have the symbolic form $AU = b$. The following MATLAB code can be used to set up and solve (13.3):

```
N = 4;
C = 0.05;
RHOG = 0.4;
% N interior mesh points, N+1 intervals
dx = 5.0 / ( N + 1 );
x = dx * (0:N+1);
A = [ -2*(1+C*dx)      +(1+1.5*C*dx)          0              0;
        +(1+1.5*C*dx)  -2*(1+2*C*dx)    +(1+2.5*C*dx)         0;
            0          +(1+2.5*C*dx)  -2*(1+3*C*dx)   (1+3.5*C*dx);
            0               0         +(1+3.5*C*dx)  -2*(1+4*C*dx) ];
ULeft=1;
URight=1.5;
b = [ RHOG*dx^2-(1+0.5*C*dx)*ULeft
      RHOG*dx^2
      RHOG*dx^2
      RHOG*dx^2-(1+4.5*C*dx)*URight];
U = A \ b;
U = [ULeft; U; URight]
```

Make sure you understand the first and last components in b. You should recall that the backslash notation is shorthand for saying U=inv(A)*b but tells MATLAB to solve the equation A*U=b without actually forming the inverse of A.

Remark 13.1. The vector x is a row vector and the vector U is a column vector! This is the convention that has been followed for the *_ode.m files and will be followed throughout this workbook.

Exercise 13.1. In this exercise, you will be using the above code to solve the rope BVP. You will also be exhaustively checking that the code is correct.

(1) Copy the above code (or download exer1.txt) and paste it into a script m-file named exer1a.m. Execute exer1a to find a solution of Equation (13.3).
(2) Verify that the values of U and b that you found satisfy at least one of the middle four equations in (13.2). To do this, write a script m-file named exer1b.m and plug the values of c, Δx, u_k, $k = 0, \ldots, 5$ into your chosen equation. Show the result is essentially zero.
  Be careful! Lower-case c is upper-case C, Δx is dx and u_k, $k = 0, 1, \ldots 5$ is the column vector U(1:6) in exer1a.
(3) Direct substitution into (13.3) shows that the function $u(x) = 1$ would be a solution if $\rho g = 0$ and $u(5) = 1$. As a second verification step, make a copy

of `exer1a.m` called `exer1c.m` with `rhog=0` and `URight=1` and check that the discrete solution is (exactly or to roundoff) correct.

(4) Direct substitution into (13.3) shows that the function $u(x) = x$ would be a solution if $\rho g = c$, $u(0) = 0$ and $u(5) = 5$. As a third verification step, make a copy of `exer1a.m` called `exer1d.m` with `rhog=C` and and with boundary values `ULeft` and `URight` chosen to match the solution you are testing. Check that the discrete solution is (exactly or to roundoff) correct.

(5) Direct substitution into (13.3) shows that the function $u(x) = x^2$ would be a solution if $\rho g = 2 + 4cx$, $u(0) = 0$ and $u(5) = 25$. As a fourth verification step, make a copy of `exer1a.m` called `exer1e.m` with constant `rhog`, which is used four times, replaced by the four values of the *vector* `2+4*c*x(2:5)'` and with boundary values chosen to agree with $u(x) = x^2$ and check that the discrete solution is (exactly or to roundoff) correct.

(6) Now that you are confident that the code is correct, use `exer1a.m` to solve the unmodified BVP (13.3). Plot `U` versus `x`. It should appear roughly parabolic, like a rope hanging from its ends, and pass through $(0, 1)$ and $(5, 1.5)$ on its ends.

Remark 13.2. The exact solutions $u = 1$, $u = x$, and $u = x^2$ can be used as exact *discrete* solutions and verification tests *only* when the approximation expressions for first and second derivatives are sufficiently accurate. Because the mesh is uniform, the approximate expressions used here for the derivatives yield the same values as using the usual continuous expressions so long as u is a quadratic (or lower) polynomial, so there is no truncation error and solutions are correct to roundoff.

The purpose of the previous exercise is to verify that you have copied the code correctly and to illustrate the powerful verification strategy of checking against known *exact discrete* solutions. You should always use a small, simple problem to verify code by comparison with hand calculations. If possible, you should also compare results with theoretical results and with results achieved using a different method. In the next exercise you will be modifying the above code to handle the case of large `N` and solving a slightly more realistic problem. Of course, you don't want to bother typing in the matrix `A` if `N` is 100 or more, so you will be writing MATLAB code to do it.

Exercise 13.2.

(1) Turn `exer1a.m` into a function m-file.

   (a) Make a copy of the script m-file `exer1a.m` and change it into a function m-file called `rope_bvp.m` with the outline

```
function [x,U] = rope_bvp(N)
  % [x,U] = rope_bvp(N)
  % comments
```

```
% your name and the date

<< your code >>

end
```

(b) Add comments after the signature line and modify the matrix (A) and right side vector (b) generation statements to be valid for arbitrary values of N. Make sure that the vector U is a column vector. (You should also eliminate the line N = 4.) **Hint:** You can use the zeros(N,N) statement to generate an N-by-N matrix of all zeros for A and then fill in the non-zero values. You can use the command ones(N,1) to construct a column vector of length Ṅ containing all ones.

(2) Check your work by running rope_bvp for N=4 and confirming that you get the same values of U as from exer1a.m. One easy way to do this is to first run exer1a, then use the command [x1,U1]=rope_bvp(4) and check that U-U1 is the zero vector.
Debugging hint: If the results are not correct, print the matrix A from rope_bvp and check it against the matrix A from exer1a. Do the same for the vectors b. *Fix any mistakes before continuing.*

(3) You may still have mistakes in the treatment of N. To be sure of your code, make a copy of rope_bvp.m and modify it to get the constant solution $u = 1$ as you did for exer1c.m above. Check your results for N=4. *Fix any mistakes before continuing.* If you cannot find your mistakes by checking your code, re-do Equations (13.2) for $n = 6$ and check the terms against your code, one term at a time.

(4) As a second test, make a copy of rope_bvp.m and modify it to get the linear solution $u = x$ as you did for exer1d.m above. Check your results for N=4. *Fix any mistakes before continuing.* If you cannot find your mistakes by checking your code, re-do Equations (13.2) for $n = 6$ and check the terms against your code, one term at a time.

(5) As a third test, make a copy of rope_bvp.m and modify it to get the quadratic solution $u = x^2$ as you did for exer1d.m above. Check your results for N=4. *Fix any mistakes before continuing.* If you cannot find your mistakes by checking your code, re-do Equations (13.2) for $n = 6$ and check the terms against your code, one term at a time.

(6) Now you are ready to solve the big problem! Use N=119 so that there are 119 unknowns U(1:119). Call the new solution [x2,U2], and plot U2 versus x2. Re-run exer1a to get U and x, and plot U versus x as circles (plot(x,U,'o')) on the same frame (hold on). To help check accuracy, please include the value of U(50) printed using format long.

Remark 13.3. You may wonder why the four-point mesh solution and the 119-point mesh solution seem to agree at the four common points. This behavior is highly unusual. It happens here because the solution of the differential equation is almost quadratic and the difference scheme exactly reproduces quadratic functions. For larger values of c, the solution looks less like a quadratic, and the solution for $N = 4$ agrees less well with the solution for $N = 119$. Try it, if you wish.

13.4   Finite element method

In the previous section, you saw an example of the finite difference method of discretizing a boundary value problem. That method is based on a finite difference expression for the derivatives that appear in the equation itself. The finite difference method results in a list of values that approximate the true solution at the set of mesh points. Approximate values between the mesh points might be generated using interpolation ideas, but the method itself does not depend on any such interpolation. The reason that $N = 119$ was chosen for comparison with $N = 4$ in the previous exercise is because the four x values in the $N = 4$ case appear among the 119 x-values in the $N = 119$ case.

An alternative approach, called the "finite element method" (FEM) is based on approximating the unknown as a sum of simple "shape functions" defined over the mesh intervals. Since the finite element solution is actually a function, it is defined over the same spatial interval as the true solution and much of the machinery of functional analysis is available for proving facts about the method and solutions that arise. As a consequence, the FEM occupies a large part of the mathematics literature. You can find the FEM discussed by [Quarteroni *et al.* (2007)] in Sections 12.4 and 12.5.

In this section, you will see the FEM applied to a particular boundary value problem. The problem is somewhat simpler than the clothesline problem discussed above, but contains the same essential features. Consider the equation

$$u'' + u' + u = f(x) \tag{13.4}$$

defined for x in the interval $[0, 1]$, for $f(x)$ a given function, and with boundary values

$$u(0) = u(1) = 0. \tag{13.5}$$

While the finite difference method attacks (13.4) directly, the FEM starts from the so-called "weak" form of the equation. This form can be constructed from (13.4) by multiplying through by a function $v(x)$, assumed to satisfy the same boundary conditions (13.5), integrating the whole equation and integrating some terms by parts. In this case, the weak form is given by

$$-\int_0^1 u'(x)v'(x)dx + [u'(x)v(x)]_0^1 + \int_0^1 u'(x)v(x)dx + \int_0^1 u(x)v(x)dx = \int_0^1 f(x)v(x)dx.$$

Since $v(x)$ satisfies (13.5), the bracketed term drops out because of boundary conditions and the result is

$$-\int_0^1 u'(x)v'(x)dx + \int_0^1 u'(x)v(x)dx + \int_0^1 u(x)v(x)dx = \int_0^1 f(x)v(x)dx. \quad (13.6)$$

Remark 13.4. In this case, the first term has been integrated by parts in order to eliminate second derivatives, an important step. Some authors might additionally integrate the second term by parts. Doing so would not change the following discussion very much.

To approximate the function u, choose an odd integer N and a set of functions $\phi_n(x)$ for $n = 1, 2, \ldots, N$, defined on the interval $[0,1]$, that form a basis of some reasonable approximating function space. For most finite element constructions, these functions satisfy the following characteristics:

(1) They are continuous and piecewise polynomials.
(2) Each of the functions takes the value 1 at a single mesh node and zero at all other mesh nodes.

In this exercise, the functions will be piecewise quadratic polynomials and the mesh nodes are given by dividing the interval into $N + 1$ subintervals, each of length $h = 1/(N+1)$, so that a sequence of spatial points is given by $x_n = nh$ for $n = 0, 1, \ldots, N, N+1$ ($x_0 = 0$ and $x_{N+1} = 1$). The quadratic Lagrange functions are defined in the following way. (See [Quarteroni *et al.* (2007)] page 562 or [Atkinson (1978)], pages 131–134.)

$$(n \text{ even}) \ \phi_n(x) = \begin{cases} \frac{(x-x_{n-1})(x-x_{n-2})}{(x_n-x_{n-1})(x_n-x_{n-2})} & \text{for } x_{n-2} < x \leq x_n \\ \frac{(x_{n+1}-x)(x_{n+2}-x)}{(x_{n+1}-x_n)(x_{n+2}-x_n)} & \text{for } x_n < x \leq x_{n+2} \ , \\ 0 & \text{otherwise} \end{cases} \quad (13.7)$$

$$(n \text{ odd}) \ \phi_n(x) = \begin{cases} \frac{(x_{n+1}-x)(x-x_{n-1})}{(x_{n+1}-x_n)(x_n-x_{n-1})} & \text{for } x_{n-1} < x \leq x_{n+1} \\ 0 & \text{otherwise} \end{cases} \quad (13.8)$$

This collection of functions is known to form a basis for a function space that includes all constant, linear, and quadratic functions on $[0,1]$ and it has good approximation properties. It is also true that each of the functions $\phi_n(x)$, for $n = 1, \ldots, N$ satisfies the boundary conditions (13.5).

Remark 13.5. In many introductions to the finite element method, a much simpler set of piecewise *linear* shape functions is used, simplifying the programming task. Here, piecewise *quadratic* functions are used along with an integration method that results in exact (up to roundoff) integration. The result is a method that can be tested by reproducing theoretical results *without approximation error*. Theoretical agreement up to roundoff is a powerful debugging tool and confidence builder.

Assume that an approximate solution to (13.6) can be written as

$$u(x) = \sum_{n=1}^{N} u_n \phi_n(x) \tag{13.9}$$

for (as yet unknown) constants u_n. Plugging (13.9) into (13.6) and choosing $v(x) = \phi_m(x)$ yields N equations of the form

$$\sum_{n=1}^{N} \underbrace{\left(-\int_0^1 \phi_n'(x)\phi_m'(x)dx + \int_0^1 \phi_n'(x)\phi_m(x)dx + \int_0^1 \phi_n(x)\phi_m(x)dx \right)}_{a_{mn}} u_n$$

$$= \underbrace{\int_0^1 f(x)\phi_m(x)dx}_{f_m}. \tag{13.10}$$

Regarding the values u_n as the components of a (column) vector $\mathbf{U}$, the values a_{mn} as the components of a matrix $\mathbf{A}$, and the values f_m as the components of a (column) vector $\mathbf{F}$, then (13.10) can be written as the matrix equation

$$\mathbf{AU} = \mathbf{F}. \tag{13.11}$$

Solving the matrix equation (13.11) completes construction of the approximate solution (13.9).

In the following exercises, you will write MATLAB functions to construct the basis functions $\phi_n(x)$ in (13.7) and (13.8), to evaluate the matrix elements a_{mn} and vector components f_n in (13.10), and solve the matrix equation (13.11).

Remark 13.6. When the FEM is programmed, the integrals in (13.11) are typically performed element-by-element. This is particularly important in multidimensional cases. Nonetheless, you will be using a conceptually simpler approach here for the integrations.

Exercise 13.3. In this exercise, you will construct the Lagrange quadratic basis functions. The values x_k used in (13.7) and (13.8) will be evaluated as $x_k = kh$, where $h = 1/(N+1)$. The expression kh is valid even when $k = 0$, although MATLAB does not allow subscripts equal to zero.

In the MATLAB functions below, you should regard the variable x as a *scalar* value, not a vector. Attempting to write vector (componentwise) code only complicates matters here.

(1) Write a MATLAB function m-file for $\phi_n(x)$ by completing the following outline. (You may find it convenient to download phi.txt.)

```
function z=phi(n,h,x)
  % z=phi(n,h,x)
  % Lagrange quadratic basis functions
```

```
% your name and the date

if numel(x) > 1
  error('x is a scalar, not a vector, in phi.m');
end

if mod(n,2)==0   % n is even

  if (n-2)*h < x & x <= n*h
    z= ??? code implementing first part of (13.7) ???
  elseif n*h < x & x <= (n+2)*h
    z= ??? code implementing second part of (13.7) ???
  else
    z=0;
  end

else  % n is odd

  ??? code implementing (13.8) ???

end
end
```

(2) Plot some of your functions using the following code. (You may find it convenient to download the file exer3.txt.)

```
N=7;
h=1/(N+1);
x=linspace(0,1,97);
mesh=linspace(0,1,N+2);
for k=1:numel(x)
  u3(k)=phi(3,h,x(k));
  u4(k)=phi(4,h,x(k));
end
plot(x,u3,'b')
hold on
plot(x,u4,'r')
plot(mesh,zeros(size(mesh)),'*')
hold off
```

You should observe that each ϕ takes the value 1 at a *single* mesh node ($x(k)$, indicated with an asterisk), takes the value zero at all other mesh nodes, is continuous, and is parabolic or zero between any two mesh nodes.

(3) Examine the definitions (13.7) and (13.8) and show that $\phi_n(x)$ is a continuous function by showing that the pieces match up at x_{n-2}, x_n, and x_{n+2} for even n and at x_{n-1} and x_{n+1} for odd n. (You don't need MATLAB to do this.) Similarly, show that

$$\phi_n(x_m) = \begin{cases} 1 & m = n \\ 0 & \text{otherwise} \end{cases}.$$

(4) Write a MATLAB function m-file similar to `phi.m` for the derivative $\phi_n'(x)$. Differentiate (13.7) and (13.8) by hand to find $\phi_n'(x)$ and use your formulæ for the function `phip.m` with the following outline

```
function z=phip(n,h,x)
  % z=phip(n,h,x)
  % derivative of Lagrange quadratic basis functions

  % your name and the date

  << your code >>

end
```

(5) For the case N=7, plot $\phi_3(x)$ and $h\phi_3'(x)$ (multiply by h to get a better scaling) on the same plot. Examine the plot carefully and convince yourself that ϕ_3' appears to be the derivative of ϕ_3.

(6) Similarly, plot ϕ_4 and ϕ_4'. Examine both cases from (13.7).

(7) For the case N=7 and the point $x = 0.4$, use the finite difference expression

$$\frac{\phi_3(x + \Delta x) - \phi_3(x - \Delta x)}{2\Delta x}$$

with $\Delta x = 0.01$ to estimate $\phi_3'(x)$. Does it agree up to roundoff with the result from `phip`? Similarly for ϕ_4.

In the following three exercises, you will write m-files to construct and verify the three pieces of the matrix $\mathbf{A}$ in (13.10). Following that, you will construct the full matrix $\mathbf{A}$ and solve for the finite element solution of the given problem (13.4).

Exercise 13.4. In this exercise, you will generate the first part of the matrix $\mathbf{A}$,

$$a_{mn}^{(1)} = -\int_0^1 \phi_m'(x)\phi_n'(x)dx. \tag{13.12}$$

Some of the mathematical facts about this quantity provide ways to check that your code is correct.

In order to do the integrations, you will be using code provided for you that provides a uniform way to do the required integrations. Download a copy of a special integration function gaussquad.m. This code takes the handles of two functions (such as @phi or @phip) along with their appropriate subscripts and the value of the mesh spacing h and integrates the resulting product over the interval [a,b]. Its signature is

```
function q=gaussquad(f1,k1,f2,k2,h,a,b)
```

This gaussquad function will provide the *exact* value, not merely an approximate value, of the integral for the cases considered in this chapter: piecewise low degree polynomials.

(1) Write a script m-file named exer4.m to compute $a_{mn}^{(1)}$ in (13.12). Choose N=7 and h=1/(N+1), and compute the matrix values $a_{mn}^{(1)}$ in (13.12). Call the resulting matrix A1. Do not overlook the fact that the basis function derivatives appear in (13.12), not the basis functions themselves, and there is that pesky minus sign in front of the integral.

Remark 13.7. The following two remarks concern program efficiency. Real programs intended to solve large problems should be concerned with efficiency, but the first time you write a program you should strive for simplicity and clarity. It is easier to make a correct program run fast than it is to make a fast program run correctly.

Remark 13.8. The support of $\phi_n(x)$ is contained in the interval $[(n-2)h, (n+2)h]$. You could use this fact to shrink your integration limits and improve the efficiency of the integration, but it is not required.

Remark 13.9. Because the support of $\phi_n(x)$ and of $\phi_m(x)$ do not intersect when $|n - m| > 4$, you can take advantage of this fact to avoid computing components $a_{mn}^{(1)}$ that must be zero, but it is not required.

(2) (13.12) indicates that the matrix A1 is symmetric. Check that your computation is symmetric by showing that

```
norm(A1-A1','fro')
```

is zero or roundoff. Add this checking code to exer4.m.

(3) For the case N=7 and h=1/(N+1), add code to exer4.m to compute the function $\psi(x) = \sum_{n=1}^{N} \phi_n(x)$ for the 97 values x=linspace(0,1,97) and plot it. You should observe that it is equal to 1 except near the endpoints of the interval. As a consequence, it has zero derivative, except near the endpoints of the interval.

(4) Since A1*ones(N,1)$= \sum_{n=1}^{N} \int_0^1 \phi_n'(x)\phi_m'(x)dx = \int_0^1 \psi'(x)\phi_m'(x)dx$, and you just saw that ψ' is zero except for $x_0 \le x \le x_2$ and $x_6 \le x \le x_8$. Add code to print the zeros in positions $m = 3, 4, 5$.

(5) Noting that $a_{mn}^{(1)} = \int_0^1 \phi_m(x)\phi_n''(x)dx$, you should be able to see why the following code

```
N=7;
h=1/(N+1);
v=(1:N)'*h; % v=x
A1*v
```

should yield a vector that is zero except in the positions $m = 6, 7$. Add this code to `exer4.m` and print the result.

(6) Recall that the BVP $-u'' = 2$ with $u(0) = u(1) = 0$ has solution $x(1 - x)$. You can solve this BVP using FEM. Write a MATLAB function m-file with the signature

```
function z=rhs4(n,h,x)
  % z=rhs4(n,h,x)

  % your name and the date

  << your code >>

end
```

to compute the constant function equal to (-2) everywhere. (This is an almost trivial exercise. It is needed so that `gaussquad.m` can be used.) Add code to `exer4.m` to use the `gaussquad.m` function to compute the vector components $(f_4)_m = \int_0^1 f_4(x)\phi_m(x)dx$ where f_4 is `rhs4`, and call the resulting vector RHS4. Since the quadratic function $x(1 - x)$ satisfies the boundary conditions, it can be written *exactly* as a combination of the ϕ_n! Hence, the following code should yield the zero vector.

```
N=7;
h=1/(N+1);
xx=(1:N)'*h;       % the variable x has already been used.
v=xx.*(1-xx);
norm(v-A1\RHS4)   % should be zero
```

Add this code to `exer4.m`. What are the values of RHS4?

The tests in `exer4.m` construct and test the matrix A1, so you should now be reasonably sure that A1 is correct. In the following exercise, you will construct and test A3. In the subsequent exercise, you will construct and test A2 so that you have all of the matrix A.

Exercise 13.5.

(1) Start writing an m-file named `exer5.m` similar to `exer4.m` but for the terms

$$a_{mn}^{(3)} = \int_0^1 \phi_m(x)\phi_n(x)dx.$$

Call the resulting matrix `A3`.

Warning: If you copy code from `A1` for `A3`, don't forget that `A1` has a minus sign in it from the integration by parts but that `A3` does not.

(2) Add code to check that `A3` is symmetric.

(3) Noting that the boundary value problem

$$u'' + u = -2 + x(1 - x) \tag{13.13}$$

with $u(0) = u(1) = 0$ has the exact solution $u = x(1 - x)$, and this quadratic function can be expressed exactly as a sum of the ϕ_n. Write a MATLAB function m-file `rhs2.m` with signature

```
function z=rhs5(k,h,x)
  % z=rhs5(k,h,x)

  % your name and the date

  << your code >>

end
```

to compute the function $f_5(x) = -2 + x(1 - x)$. Add code to `exer5.m` to use `rhs5.m` and `gaussquad.m` to compute the (column) vector $\int_0^1 f_5(x)\phi_m(x)dx$, and call the resulting vector `RHS5`. What are these values?

(4) The matrix `A1` was computed by `exer4.m`, and `A2` and `RHS5` are computed by `exer5.m`. Add code to `exer5.m` to solve the matrix equation `(A1+A3)*U=RHS5`. You have just solved (13.13), and your solution should equal $x_n(1 - x_n)$ at each of the nodes $x_n = nh$ *exactly*, with only roundoff errors. Add code to `exer5.m` to check that this is true. If it is not true, there is a mistake somewhere. Fix it before continuing.

Exercise 13.6.

(1) Write another m-file named `exer6.m` to compute the terms

$$a_{mn}^{(2)} = \int_0^1 \phi_m(x)\phi_n'(x)dx. \tag{13.14}$$

Call this matrix `A2`.

(2) Integrating (13.14) by parts and applying the boundary conditions shows that the matrix A2 is skew-symmetric (*i.e.*, A2'=-A2. Add code to exer6.m to confirm this is true.

(3) Adding the terms A1, A2 and A3 together generates the matrix A, and the boundary value problem

$$u'' + u' + u = -2 + (1 - 2x) + x(1 - x) \tag{13.15}$$

with boundary values $u(0) = u(1) = 0$ has exact solution $u = x(1 - x)$. Write a MATLAB function m-file rhs6.m with signature

```
function z=rhs6(k,h,x)
  % z=rhs6(k,h,x)


  % your name and the date


  << your code >>


end
```

to compute the function $f_6(x) = -2 + (1 - 2x) + x(1 - x)$. Add code to exer6.m to compute the (column) vector $\int_0^1 f_6(x)\phi_m(x)dx$, and call the resulting vector RHS6.

(4) Solve the matrix equation A*U=RHS6. You have just solved (13.15), and your solution should equal $x(1 - x)$ *exactly*, with only roundoff errors.

Remark 13.10. Writing test scripts like exer4.m, exer5.m and exer6.m allows you to re-test your work at any time. This stratagem helps you maintain confidence in your code and also allows you to propose and test modifications easily.

Exercise 13.7. The boundary value problem

$$u'' + u' + u = x + 1$$

with boundary values $u(0) = u(1) = 0$ has exact solution $u = x - e^{-0.5(x-1)} \sin \omega x / \sin \omega$, with $\omega = \sqrt{3}/2$.

(1) Write a function m-file exact7.m to evaluate the above exact solution.
(2) Write a function m-file rhs7.m to evaluate the right side function $z = x + 1$.
(3) Write a function m-file solve7.m with outline

```
function [x,U]=solve7(N)
  % [x,U]=solve7(N)
  %    ... more comments ...


  % your name and the date
```

```
    << your code >>

end
```

that performs the following tasks:

- (a) Compute the finite element matrix A.
- (b) Compute the right side vector RHS7.
- (c) Solve the system A*U=RHS7 for U.
- (d) Compute the spatial coordinate vector (1:N)'*h.

(4) Fill in the following table and estimate the rate of convergence of this method. Measure the error as the maximum absolute value of the difference between the calculated and true solutions at the nodes $x_n = nh$, and take the ratio as the error(h) divided by error(h/2).

N	h	error	ratio
7	1.2500e-1		
15	6.2500e-2		
31	3.1250e-2		
61	1.6129e-2		
121	8.1967e-3		—

Remark 13.11. You may find that the cases for larger N take a very long time. If so, consider implementing either or both of Remarks 13.8 and 13.9.

Remark 13.12. The rate of convergence appears higher than expected from theory. This rate is an artifact of the way the error is computed. To do it properly, one needs to compute L^2 integral errors that involve more than just the node points. Convergence at the node points alone is higher-order than expected here, a feature called "superconvergence."

13.5 Neumann boundary conditions for FEM

Suppose, instead of the Dirichlet boundary conditions $u(0) = u(1) = 0$ that were considered above in the finite element solution for (13.4), you have a Neumann boundary condition at the right side.

$$u(0) = 0$$
$$\bar{u}'(1) = 0.$$

The construction of the weak form (13.6) involved an integration by parts that resulted in the term

$$\left[u'(x)v(x)\right]_0^1.$$

Before, with Dirichlet boundary conditions at both 0 and 1, you needed to assume that $v(0) = v(1) = 0$ in order to make this term disappear. Now, with a Neumann boundary condition at 1, you already have $u'(1) = 0$, so *no condition on v(1) is required*! Dropping the assumption that $v(1) = 0$ results in *one more* shape function $\phi_{N+1}(x)$ being included in the expansion of the solution, so the vectors are now of length $N + 1$ and the matrices are now $(N + 1) \times (N + 1)$.

Exercise 13.8.

(1) Copy your `solve7.m` m-file to `solve8.m` and then modify it so that it solves the equation $u'' + u' + u = x + 1$ using the finite element method with Dirichlet-Neumann boundary conditions $u(0) = u'(1) = 0$. Take particular care that your integrations do not extend to values $x > 1$, or that $\phi_n(x) = 0$ for $x > 1$.

Remark 13.13. The solution $u = x(2 - x)$ can be used as a debugging case because it is a quadratic polynomial that satisfies the boundary conditions.

(2) For the case $N = 21$, plot your solution. Does it appear that $u'(1) = 0$?
(3) Since, by assumption, $u(x) = \sum_{n=1}^{N+1} U_n \phi_n(x)$, then it must be true that $u'(x) = \sum_{n=1}^{N+1} U_n \phi_n'(x)$. For the case $N = 21$, what is the value of $u'(1)$?
(4) What is the maximum absolute value of the difference between your computed solution (for $N = 21$) and the exact solution at the nodes

$$u(x_n) = x_n + Ce^{-.5x_n} \sin \omega x_n$$

for $\omega = \sqrt{3}/2$ and $C = -e^{.5}/(\omega \cos \omega - .5 \sin \omega)$?

13.6 The Burgers equation

A partial differential equation (PDE) involves derivatives of a function u which depends on more than one independent variable. One interesting PDE is the one-dimensional Burgers equation, a one-dimensional nonlinear equation whose nonlinear term is similar to the one in the Navier–Stokes equations of fluid flow. In this case, the variable will be called u and it is a function of space (x) and time (t), $u(x, t)$. The variable u is called "velocity" in the Navier–Stokes equations. The Burgers equation on the spatial interval $[0, 1]$ can be written as

$$\frac{\partial u}{\partial t} + u\frac{\partial u}{\partial x} = \nu\frac{\partial^2 u}{\partial x^2} \tag{13.16}$$

where ν is a constant. Boundary conditions can be taken as $u(0,t) = 1$ and $u(1,t) = 0$, both for all time. Two space boundary conditions are necessary because the equation is second-order in space. Since the equation is first-order in time, only one initial condition is needed.

To get an idea of what the solution to the Burgers equation might look like, first imagine that $\nu = 0$ and also that the coefficient $u = v$ is a constant, so that the equation becomes the wave equation

$$\frac{\partial u}{\partial t} + v\frac{\partial u}{\partial x} = 0.$$

If $v > 0$, this equation represents a right-going wave that moves without changing shape.[2] Changing ν to a small, positive number means that the wave propagates as before, but slowly spreads and decays to zero. Finally, since the coefficient v is *not* constant, the wave propagates faster where u is larger and slower where u is smaller. Thus a wave that is larger to the left and smaller to the right will steepen as it propagates to the right.

13.7   The method of lines

Look at (13.16) and pretend, just for a moment, that time is frozen. The equation suddenly starts looking like the boundary value problem (13.1) that you solved before, only with a extra term on the left and with $\partial u/\partial t$ replacing the right side. This observation is the basis for the "method of lines," wherein the spatial discretization is performed separately from the temporal. Consider the function $u(t,x)$, but think of it as a function of x first. The resulting BVP can be written with primes denoting spatial differentiation so it looks more like what you have been doing.

$$\nu u'' - uu' = \frac{\partial u}{\partial t},$$

where you are focussing on the left side for the moment, along with spatial boundary conditions $u(0,t) = 1$ and $u(1,t) = 0$.

You solved a BVP a lot like this one in Exercises 13.1 and 13.2 using finite differences. You broke the interval $[0,1]$ into $N+1$ subintervals and labeled the $(N+2)$ resulting points $x_n, n = 0, 1, 2, ..., (N+1)$. Next, you defined u_n as being the approximate solution at x_n. Keep in the back of your mind, though, that u_n is really still a function of t, $u_n(t)$. Denote the vector of values u_n by U, because the unsubscripted u has already been used to denote the continuous solution. Keep in mind that $U = U(t)$ is a function of time. Use the same finite difference discretization as before for the term u''

$$u'' \approx \frac{u_{n+1} - 2u_n + u_{n-1}}{\Delta x^2} \tag{13.17}$$

and choose a natural discretization for the first-order nonlinear term

$$uu' \approx u_n\frac{u_{n+1} - u_{n-1}}{2\Delta x}. \tag{13.18}$$

---

[2]If f is an arbitrary differentiable function of a single variable, $u(x,t) = f(tv - x)$ is a solution of the wave equation.

(The truncation error of each of these forms is $O(\Delta x^2)$.) The resulting discrete equations become

$$\frac{\partial u}{\partial t} = \nu \frac{u_{n+1} - 2u_n + u_{n-1}}{\Delta x^2} - u_n \frac{u_{n+1} - u_{n-1}}{2\Delta x}. \qquad (13.19)$$

But the ground remains familiar: (13.19) is just a system of IVPs! You know several different methods to solve it. It turns out that the system is moderately stiff, and will become more stiff when N is taken larger and larger, so backward Euler is a good choice to solve this system. Recall that backward Euler requires an m-file to evaluate both the function $\mathbf{F}$ and its partial derivative (Jacobian)

$$\mathbf{J}_{mn} = \frac{\partial \mathbf{F}_m}{\partial \mathbf{U}_n}. \qquad (13.20)$$

In MATLAB notation, the variable U will be a matrix whose entries U_{kn} approximate the values $u_n(t_k)$. In MATLAB notation, for the kth time interval, U(:,k) represents the (column vector of) values at the locations x(:). The initial condition for U should be a column vector whose values are specified at the locations x(:).

Remark 13.14. Exercise 13.9 can be done using one of the MATLAB ODE solvers, such as ode45 or ode15s, instead of back_euler. Doing so would require changing the plotting commands presented below.

Exercise 13.9. In this exercise you will write, debug, and solve an m-file for the solution of the Burgers equation in the form of (13.19) with $\nu = 0.001$ and using N=500.

(1) Write a function m-file called burgers_ode.m to construct the spatial discretization of the Burgers equation and its gradient using the following steps. The function $\mathbf{F}$ refers to the right side of (13.19).

(a) Begin with the following outline (available as burgers_ode.txt).

```
function [F,J]=burgers_ode(t,U)
  % [F,J]=burgers_ode(t,U)
  % compute the right side of the time-dependent ODE arising
  % from a method of lines reduction of Burgers equation
  % and its derivative, J.
  % Boundary conditions are fixed =0 at the
  % endpoints x=0 and x=1.
  % A fixed number of spatial points (=N) is used.
  % The variable t is not used, but is kept as a place holder.
  % U is the vector of the approximate solution
  % output F is the time derivative of U (column vector)
  % output J is the Jacobian matrix of
  %   partial derivatives of F with respect to U
```

```
% your name and the date

% spatial intervals
N=500;
NU=0.001;
dx=1/(N+1);
ULeft=1;  % left boundary value
URight=0; % right boundary value

F=zeros(N,1);  % force F to be a column vector
J=zeros(N,N);  % matrix of partial derivatives

% construct F and J in a loop
for n=1:N
   if n==1  % left boundary
      F(n)  = ??? Function, left endpoint ???
      J(?,?)= ??? Derivative, left endpoint ???
   elseif n<N  % interior of interval
      F(n)  = ??? Function, interior points ???
      J(?,?)= ??? Derivative, interior points ???
   else  % right boundary
      F(n)  = ??? Function, right endpoint???
      J(?,?)= ??? Derivative, right endpoint???
   end
  end
end
```

(b) It is easiest to treat the *interior* points (the case n<N) first. Replace the line

```
F(n)= ??? Function, interior points ???
```

with the discretization for F(n) given in (13.19).

(c) Replace the line

```
J(?,?)= ??? Derivative, interior points ???
```

with the values J(n,n), J(n,n+1), and J(n,n-1) according to the formula that was given above in (13.20) and is repeated here.

$$\mathbf{J}_{nm} = \frac{\partial \mathbf{F}_n}{\partial \mathbf{U}_m}. \tag{13.21}$$

The variable m will take on the three values n, n+1, and n-1. To do, for example, J(n,n+1), write out the expression for F(n) and differentiate it with respect to U(n+1). It is easy to see from the formula that $\mathbf{J}_{nm} = 0$ for $m < n - 1$ or $m > n + 1$.

(d) When n=1, the variable corresponding with n-1 is the left boundary value ULeft. With this in mind, replace the lines

```
F(n)  = ??? Function, left endpoint ???
J(?,?)= ??? Derivative, left endpoint ???
```

with the expressions for F(n), J(n,n) and J(n,n+1).

(e) When n=N, the variable corresponding with n+1 is the right boundary value URight. With this in mind, replace the lines

```
F(n)  = ??? Function, right endpoint???
J(?,?)= ??? Derivative, right endpoint???
```

with the expressions for F(n), J(n,n) and J(n,n-1).

(2) The spatial values represent a uniform mesh

```
N=500;  % must agree with value inside burgers_ode.m
x=linspace(0,1,N+2);
x=x(2:N+1);
```

and you will assume an initial velocity distribution that looks like a shallowly sloped wave.

```
UInit=((1-x).^3)';
```

(3) Test your version of burgers_ode.m by calling it with UInit (the value of t does not matter) and comparing the result with the following values:

results from burgers_ode(0,UInit)				
n	F(n)	J(n-1,n)	J(n,n)	J(n+1,n)
1	2.9761711475		-499.01396011	498.51296011
2	2.9465759259	1.99800798	-499.02590030	497.02789232
250	0.0976958603	219.12262501	-501.24899902	282.12637400
499	2.395210e-05	251.00094622	-502.00194821	251.00100199
500	1.197605e-05	251.00098406	-502.00198406	

(4) Retrieve your copies of back_euler.m and newton4euler.m and use back_euler to solve the Burgers equation starting from UInit. (Alternatively, you could use ode45 or ode15s, but calling sequences and the solution matrix U will differ.) You should use 100 steps from time t=0 to time t=1.

```
[t,U]=back_euler(@burgers_ode,[0,1],UInit,100);
```

The solution should converge at each step. If you get the "failed to converge" message, your derivative (J) is probably wrong. To help with grading, please print the value of U(200,50).

(5) Recall that the columns of U represent velocities at different *places* all at the same time and rows of U represent velocities at different *times* all at the same place. To see a snapshot of the solution at timestep k, use the command

```
plot(x, U(:,k) )
```

where x is the value assigned above. Make at least one plot, for k=50.

(6) You can see a "flicker picture" of the evolution with the following steps

```
plot(x,U(:,1))
axis([0,1,0,1.5])
for k=2:100
  pause(0.1);
  plot(x,U(:,k));
  axis([0,1,0,1.5])
end
```

You should be able to see the "wave" steepen as it moves to the right.

Remark 13.15. The boundary condition is inappropriate for the case that the "wave" actually reaches the right boundary, and the solution will fail if the time interval is long enough for the wave to reach $x = 1$.

13.8   Shooting methods

Shooting methods solve a BVP by reformulating it as an IVP, using initial values as parameters. The BVP is solved by finding the parameters that reproduce the desired boundary values. For this exercise, return to considering the BVP of a hanging rope.

Taking the example of the rope BVP (13.1), how much violence do you have to do to it in order to make it look like an IVP? An IVP requires *two* conditions at the left point and none at the right point. So, temporarily consider the related problem, with a guessed boundary condition at the initial value of $x = 0$:

$$(1 + cx)u'' + cu' = \rho g, \qquad c = 0.05, \qquad \rho g = 0.4$$
$$u(0) = 1 \tag{13.22}$$
$$u'(0) = \alpha.$$

The following exercise attacks this system using a method called "shooting." The strategy behind this method is the following.

- For each value of α, the problem can be solved to get a numerical solution at the right endpoint.
- Since every value of α determines a (numerical) solution $u(5)$, regard the difference between the value obtained and the desired value, as a function $F(\alpha) = u(5) - 1.5$. (Where $u(5)$ should really be something like $u_\alpha(5)$, to emphasize that the solution depends on the parameter.)

- The desired BVP solution has the property that $F(\alpha) = 0$. The MATLAB function `fzero` can be used to find α.

As you know, the ODE (13.1) can be written as a system of first-order differential equations. This is done by identifying $y_1 = u$ and $y_2 = u'$, yielding the system

$$y_1' = y_2$$
$$y_2' = (\rho g - cy_2)/(1 + cx).$$

For this exercise, you will be using built-in MATLAB ODE routines to solve the initial value problem that arises from a guessed initial condition and then you will use the MATLAB function `fzero` to solve for the correct initial condition.

Exercise 13.10.

(1) Write a function m-file named `rope_ode.m` with outline

```
function fValue=rope_ode(x,y)
  % fValue=rope_ode(x,y) computes the
  % rhs of the first-order system

  % your name and the date

  << your code >>


end
```

Note that since you plan to use `ode45` there is no need to add the Jacobian matrix to `rope_ode.m`! This is a great convenience, but in many cases you would have to provide a function for the Jacobian or `ode45` (or `ode15s`, *etc.*) might fail. In that case, setting an option allows the Jacobian computation.

(2) Choose a provisional value $\alpha = 0$ and use the MATLAB function `ode45` to solve the system (13.22) on the interval [0,5]. What is the value of $u(5) - 1.5$? Is it positive or negative? (Be careful: which of the components of y corresponds with u?)

(3) By trial and error, find a second value of α for which the value of $u(5) - 1.5$ is of the opposite sign as for $\alpha = 0$. The correct value of α lies between the two values you just found.

(4) Write an m-file called `rope_shoot.m` that accepts a value of `alpha` and evaluates `F(alpha)`, for the rope BVP. The file should have outline

```
function F = rope_shoot ( alpha )
  % F = rope_shoot ( alpha )
  % comments

  % your name and the date
```

```
<< your code >>

end
```

and this code should do the following:

- Use the input value of `alpha` as the initial condition for $y'(0)$;
- Use `ode45` to compute the solution `[x,y]` of the IVP (13.22) defined by the initial conditions, and the right hand side function `rope_ode`, for $x \in [0,5]$;
- Return in the function value `F` the value of $u(5) - 1.5$.

For a given value of α, the function you just wrote will return $y(5) - 1.5$. When α is just right, it will return 0.

(5) Test that `rope_shoot` returns the same value you obtained above when `alpha=0`.
(6) Use the MATLAB function `fzero` to find the value of α that makes $F(\alpha) = y(5) - 1.5 = 0$. `fzero` requires two parameters, a function handle (@) first and second the vector `[alpha1,alpha2]` of the two values of α that you just found for which F has opposite signs. What is the value of `alpha` you found?
(7) Plot the solution you found. Does the curve have a height of 1 at $x = 0$ and a height of 1.5 at $x = 5$?
(8) Return to your solution for `N=119` in Exercise 13.2. Use a finite difference expression for the derivative to estimate the derivative of `U2` at the left endpoint. How does it compare with the value α you just computed?

Chapter 14

Vectors, matrices, norms and errors

14.1 Introduction

Linear systems and their solutions are expressed in terms of vectors and matrices. In this chapter, several preliminary concepts and MATLAB functions are introduced in anticipation of the following chapters. In order to make statements about the sizes of these objects, and the errors you make in solutions, you need to be able to describe the "sizes" of vectors and matrices, which you do by using *norms*.

In error analysis, it is important to have bounds for a matrix-vector product in terms of bounds on the matrix and on the vector. In order for this to happen, you will need to use matrix and vector norms that are *compatible*.

You will then see the notions of *forward error* and *backward error* in a linear algebra computation.

From the definitions of norms and errors, you can define the *condition number* of a matrix, that will give you an objective way of measuring how "bad" a matrix is, and how many digits of accuracy you can expect when solving a particular linear system.

You will see one useful matrix example: a tridiagonal matrix with well-known eigenvalues and eigenvectors. Finally, you will write our own determinant function using expansion by minors. Both of these exercises anticipate factorization and solution methods in future chapters.

The exercises in this chapter are basic and somewhat less lengthy than in other chapters: one must secure one's bathing suit before diving. Exercise 14.1 concerns vector norms, Exercise 14.2 concerns matrix norms and compatibility between them. Matrix norms, especially the L^2 norm, can be expensive to compute and 14.3 points that out. Exercise 14.4 looks at the spectral radius and Exercise 14.5 illustrates how $A^k x$ can grow even when it has small spectral radius, hinting at why the spectral radius is not a norm. Exercise 14.6 illustrates that the sizes of the true error and residual errors may not be closely related. Exercise 14.7 illustrates why a relative error norm should be used when estimating convergence rates. Exercise 14.8 illustrates the connection between condition numbers and roundoff error. Exercise 14.9 studies a simple tridiagonal matrix that will be used repeatedly as an example for matrix methods, and Exercise 14.10 continues that introduction along with practice in manipulating matrix entries. Exercise 14.11 provides additional practice in manipulating matrix entries in order to compute determinants using the Laplace rule (expansion by minors).

14.2   Vector norms

A *vector norm* assigns a size to a vector, in such a way that scalar multiples do what you expect, and the triangle inequality is satisfied. There are four common vector norms in n dimensions:

- The L^1 vector norm

$$\|x\|_1 = \sum_{i=1}^n |x_i|.$$

- The L^2 (or "Euclidean") vector norm

$$\|x\|_2 = \sqrt{\sum_{i=1}^n |x_i|^2}.$$

- The L^p vector norm

$$\|x\|_p = \left(\sum_{i=1}^n |x_i|^p \right)^{1/p}.$$

- The L^∞ vector norm

$$\|x\|_\infty = \max_{i=1,\dots,n} |x_i|.$$

MATLAB commands to compute the norm of a vector x include:

- $\|x\|_1 = $ norm(x,1);
- $\|x\|_2 = $ norm(x,2)$=$ norm(x);
- $\|x\|_p = $ norm(x,p);
- $\|x\|_\infty = $ norm(x,inf).

(Recall that `inf` is the MATLAB name corresponding to ∞.)

Exercise 14.1. For this and the following exercise, you may find it convenient to download the file `exer1-2.txt`. For each of the following column vectors:

```
x1 = [ 4; 6; 7 ]
x2 = [ 7; 5; 6 ]
x3 = [ 1; 5; 4 ]
```

compute the vector norms, using the appropriate MATLAB commands. Be sure your answers are reasonable.

	L^1	L^2	L^∞
x1			
x2			
x3			

14.3 Matrix norms

A *matrix norm* assigns a size to a matrix, again, in such a way that scalar multiples do what you expect, and the triangle inequality is satisfied. However, what's more important is that you want to be able to mix matrix and vector norms in various computations. So you are going to be very interested in whether a matrix norm is *compatible* with a particular vector norm, that is, when it satisfies the condition:

$$\|Ax\| \leq \|A\| \, \|x\|.$$

There are four common matrix norms and one "almost" norm:

- The L^1 or "max column sum" matrix norm:

$$\|A\|_1 = \max_{j=1,\ldots,n} \sum_{i=1}^{n} |A_{i,j}|.$$

Remark 14.1. This is *not the same* as the L^1 norm of the vector of dimension n^2 whose components are the same as $A_{i,j}$.

- The L^2 matrix norm:

$$\|A\|_2 = \max_{j=1,\ldots,n} \sqrt{\lambda_i}$$

where λ_i is a (necessarily real and non-negative) eigenvalue of $A^H A$ or

$$\|A\|_2 = \max_{j=1,\ldots,n} \mu_i$$

where μ_i is a singular value of A;
- The L^∞ or "max row sum" matrix norm:

$$\|A\|_\infty = \max_{i=1,\ldots,n} \sum_{j=1}^{n} |A_{i,j}|.$$

Remark 14.2. This is *not the same* as the L^∞ norm of the vector of dimension n^2 whose components are the same as $A_{i,j}$.

- There is no L^p *matrix* norm in MATLAB .

- The "Frobenius" matrix norm:

$$\|A\|_{\texttt{fro}} = \sqrt{\sum_{i,j=1,\ldots,n} |A_{i,j}|^2}.$$

Remark 14.3. This *is the same* as the L^2 norm of the vector of dimension n^2 whose components are the same as $A_{i,j}$.

- The spectral radius (not a norm):

$$\rho(A) = \max |\lambda_i|$$

(only defined for a square matrix), where λ_i is a (possibly complex) eigenvalue of A.

MATLAB commands to compute the norm of a matrix A include:

- $\|A\|_1 = $ `norm(A,1)`;
- $\|A\|_2 = $ `norm(A,2)`=`norm(A)`;
- $\|A\|_\infty = $ `norm(A,inf)`;
- $\|A\|_{\texttt{fro}} = $ `norm(A,'fro')`; and
- See below for computation of $\rho(A)$ (the spectral radius of A).

14.4   Compatible matrix norms

A matrix can be identified with a linear operator, and the norm of a linear operator is usually defined in the following way.

$$\|A\| = \max_{x \neq 0} \frac{\|Ax\|}{\|x\|}.$$

(It would be more precise to use sup rather than max here but the surface of a sphere in finite-dimensional space is a compact set, so the supremum is attained, and the maximum is correct.) A matrix norm defined in this way is said to be "vector-bound" to the given vector norm.

In order for a matrix norm to be consistent with the linear operator norm, the following condition must hold:

$$\|Ax\| \leq \|A\| \, \|x\| \tag{14.1}$$

but this expression *is not necessarily true* for an arbitrarily chosen pair of matrix and vector norms. When it is true, then the two are "compatible."

If a matrix norm is vector-bound to a particular vector norm, then the two norms are guaranteed to be compatible. Thus, for any vector norm, there is always at least one matrix norm that you can use. But that vector-bound matrix norm is not always the only choice. In particular, the L^2 matrix norm is difficult (time-consuming) to compute, but there is a simple alternative.

- The L^1, L^2 and L^∞ matrix norms can be shown to be vector-bound to the corresponding vector norms and hence are guaranteed to be compatible with them;
- The Frobenius matrix norm is not vector-bound to the L^2 vector norm, but is compatible with it; the Frobenius norm is much faster to compute than the L^2 matrix norm (see Exercise 14.3).
- The spectral radius is not really a norm and is not vector-bound to any vector norm, but it "almost" is. It is useful because you often want to think about the behavior of a matrix as being determined by its largest eigenvalue, and it often is. But there is no vector norm for which it is always true that

$$\|Ax\| \le \rho(A)\|x\|.$$

A well-known theorem theorem (see, for example, [Layton and Sussman (2020)], Appendix A, and also Exercise 14.4) states that, for any matrix A, and for any chosen value of $\epsilon > 0$, a norm can be found so that $\|Ax\| \le (\rho(A) + \epsilon)\|x\|$. The norm depends both on ϵ and A.

Exercise 14.2. Consider each of the following column vectors (you may find it convenient to download the file `exer1-2.txt`):

```
x1 = [ 4; 6; 7 ];
x2 = [ 7; 5; 6 ];
x3 = [ 1; 5; 4 ];
```

and each of the following matrices

```
A1 = [38    37    80
      53    49    49
      23    85    46];
```

```
A2 = [77    89    78
       6    34    10
      65    36    26];
```

Check that the compatibility condition (14.1) holds for each case in the following table by computing the ratios

$$r_{p,q}(A, x) = \frac{\|Ax\|_q}{\|A\|_p \|x\|_q}$$

and filling in the following tables. The third column should contain the letter "S" if the compatibility condition is satisfied and the letter "U" if it is unsatisfied. The subsequent columns should be filled in with the symbol ">" if $r_{p,q}(A, x) > 1$ or "<=" if $r_{p,q}(A, x) \le 1$.

Suggestion: This exercise is a good candidate for a script m-file.

Matrix norm (p)	Vector norm (q)	S/U	A=A1 x=x1	A=A1 x=x2	A=A1 x=x3	A=A2 x=x1	A=A2 x=x2	A=A2 x=x3
1	1							
1	2							
1	inf							
2	1							
2	2							
2	inf							
inf	1							
inf	2							
inf	inf							
'fro'	1							
'fro'	2							
'fro'	inf							

The Euclidean norm (two norm) for matrices is a natural norm to use, but it has the disadvantage of requiring more computation time than the other norms. However, the two norm is compatible with the Frobenius norm, so when computation time is an issue, the Frobenius norm should be used instead of the two norm.

The two norm of a matrix is computed in MATLAB as the largest singular value of the matrix. (See [Atkinson (1978)], pages 478ff, [Quarteroni *et al.* (2007)], pages 17–18, or a Wikipedia article [Singular Value Decomposition] for a discussion of singular values of a matrix.) You will visit singular values again in Chapter 18 and you will see some methods for computing the singular values of a matrix. You will see in the following exercise that computing the singular value of a large matrix can take a considerable amount of time.

MATLAB provides a "stopwatch" timer that measures elapsed time for a command. The command `tic` starts the stopwatch and the command `toc` prints (or returns) the time elapsed since the `tic` command. When using these commands, you must take care not to type them individually on different lines since then you would be measuring your own typing speed.

Exercise 14.3.

(1) You have encountered the `magic` command earlier. It generates a "magic square" matrix. A magic square of size n contains the integers $1, \ldots, n^2$. Compute the magic square of size 1000 with the command

```
A=magic(1000); % DO NOT LEAVE THE SEMICOLON OUT!
```

(2) Use the command

`tic;x=norm(A);toc`

to discover how long it takes to compute the two norm of A.

Remark 14.4. You may be interested to note that `x = sum(diag(A))` = `trace(A)` = `norm(A,1)` = `norm(A,inf)` because A is a magic square matrix.

(3) How long does it take to compute the Frobenius norm of A?
(4) Which takes longer, the two norm or Frobenius norm?

14.5 On the spectral radius

In the above text there is a statement that the spectral radius is not vector-bound with any norm, but that it "almost" is. In this section you will see by example what this means.

In the first place, let's try to see why it isn't vector-bound to the L^2 norm. Consider the following *false "proof"*. Given a matrix A, for any vector $\mathbf{x}$, break it into a sum of eigenvectors of A as

$$\mathbf{x} = \sum x_i \mathbf{e}_i$$

where $\mathbf{e}_i$ are the eigenvectors of A, normalized to unit length. Then, for λ_i the eigenvalues of A,

$$\|A\mathbf{x}\|_2^2 = \|\sum \lambda_i x_i \mathbf{e}_i\|_2^2$$
$$\leq \max_i |\lambda_i|^2 \sum |x_i|^2$$
$$\leq \rho(A)^2 \|\mathbf{x}\|^2$$

and taking square roots completes the "false proof."

Why is this "proof" false? The error is in the statement that all vectors can be expanded as sums of eigenvectors. If you knew, for example, that A were positive definite and symmetric, then, the eigenvectors of A would form an orthonormal basis and the proof would be correct. Hence the term "almost." On the other hand, this observation provides the counterexample.

Exercise 14.4. Consider the following (nonsymmetric) matrix

```
A=[0.5  1    0    0    0    0    0
   0    0.5  1    0    0    0    0
   0    0    0.5  1    0    0    0
   0    0    0    0.5  1    0    0
   0    0    0    0    0.5  1    0
   0    0    0    0    0    0.5  1
   0    0    0    0    0    0    0.5]
```

This matrix can be generated with the MATLAB command

```
A=0.5*diag( ones(7,1) ) + diag( ones(6,1) ,1);
```

You should recognize this matrix as a single Jordan block. Any matrix can be decomposed into one or more such blocks by a change of basis.

(1) Use the MATLAB routine [V,D]=eig(A) (recall that the notation [V,D]= is the way that MATLAB denotes that the function — eig in this case — returns two quantities) to get the eigenvalues (diagonal entries of D) and eigenvectors (columns of V) of A. How many linearly independent eigenvectors are there?

(2) You just computed the eigenvalues of A. What is the spectral radius of A?

(3) For the vector x=[1;1;1;1;1;1;1] = ones(7,1), what are $\|x\|_2$, $\|Ax\|_2$, and $(\rho(A))(\|x\|_2)$?

(4) For this case, is $\|Ax\|_2 \le \rho(A)\|x\|_2$?

It should be clear that if the spectral radius of a matrix A is smaller than 1.0 then $\lim_{n\to\infty} A^n = 0$.

Exercise 14.5.

(1) For x=[1;1;1;1;1;1;1] = ones(7,1) compute and plot $\|A^k x\|_2$ for $k = 0, 1, \ldots, 40$.

(2) What is the smallest value of $k > 2$ for which $\|A^k x\|_2 \le \|Ax\|_2$?

(3) What is $\max_{0 \le k \le 40} \|A^k x\|_2$?

14.6   Types of errors

A natural assumption to make is that the term "error" refers always to the difference between the computed and exact "answers." You are going to have to discuss several kinds of error, so you refer to this first error as "solution error" or "forward error" or "true error." Suppose you want to solve a linear system of the form $Ax = b$, (with exact solution x) and you computed x_{approx}. You define the solution error as $\|x_{approx} - x\|$.

Usually the solution error cannot be computed, and you would like a substitute whose *behavior* is acceptably close to that of the solution error. One example would be during an iterative solution process where you would like to monitor the progress of the iteration but you do not yet have the solution and cannot compute the solution error. In this case, you are interested in the "residual error" or "backward error," which is defined by $\|Ax_{approx} - b\| = \|b_{approx} - b\|$ where, $b_{approx} = Ax_{approx}$. Another way of looking at the residual error is to see that it's telling you the difference between the right hand side that would be exact for $x$approx versus the right hand side you have.

If you think of the right hand side as being a target, and your solution procedure as determining how you should aim an arrow so that you hit this target, then

- The solution error is telling you how badly you aimed your arrow;
- The residual error is telling you how much you would have to move the target in order for your badly aimed arrow to be a bullseye.

There are problems for which the solution error is huge and the residual error tiny, and all the other possible combinations also occur.

Exercise 14.6. For each of the following cases, compute the Euclidean norm (two norm) of the solution error and residual error and characterize each as "Large" (L) or "Small" (S). (For the purpose of this exercise, take "Large" to mean greater than 1 and "Small" to mean smaller than 0.01.) In each case, you must find the true solution first (Pay attention! In each case you can solve in your head for the true solution, xTrue.), then compare it with the approximate solution xApprox. You may find it convenient to download the file `exer5.txt`.

(1) `A=[1,1;1,(1-1.e-12)], b=[0;0], xApprox=[1;-1]`
(2) `A=[1,1;1,(1-1.e-12)], b=[1;1], xApprox=[1.00001;0]`
(3) `A=[1,1;1,(1-1.e-12)], b=[1;1], xApprox=[100;100]`
(4) `A=[1.e+12,-1.e+12;1,1], b=[0;2], xApprox=[1.001;1]`

Case	Residual	L/S	xTrue	Error	L/S
1					
2					
3					
4					

The norms of the matrix and its inverse exert some limits on the relationship between the forward and backward errors. Assuming you have compatible norms:

$$\|x\text{approx} - x\| = \|A^{-1}A(x\text{approx} - x)\| \leq (\|A^{-1}\|)(\|b\text{approx} - b\|)$$

and

$$\|Ax\text{approx} - b\| = \|Ax\text{approx} - Ax\| \leq (\|A\|)(\|x\text{approx} - x\|).$$

Put another way,

$$(\text{solution error}) \leq \|A^{-1}\|(\text{residual error})$$

$$(\text{residual error}) \leq \|A\|(\text{solution error}).$$

14.6.1 *Relative error*

It's often useful to consider the size of an error relative to the true quantity. This quantity is sometimes multiplied by 100 and expressed as a "percentage." If the true solution is x and you computed $x_{\text{approx}} = x + \Delta x$, the relative solution error is defined as

$$\text{(relative solution error)} = \frac{\|x_{\text{approx}} - x\|}{\|x\|}$$

$$= \frac{\|\Delta x\|}{\|x\|}.$$

Given the computed solution x_{approx}, you know that it satisfies the equation $A x_{\text{approx}} = b_{\text{approx}}$. If you write $b_{\text{approx}} = b + \Delta b$, then you can define the relative residual error as:

$$\text{(relative residual error)} = \frac{\|b_{\text{approx}} - b\|}{\|b\|}$$

$$= \frac{\|\Delta b\|}{\|b\|}.$$

These quantities depend on the vector norm used, they cannot be defined in cases where the divisor is zero, and they are problematic when the divisor is small.

Actually, relative quantities are important beyond being "easier to understand." Consider the boundary value problem (BVP) for the ordinary differential equation

$$u'' = -\frac{\pi^2}{100} \sin(\frac{\pi x}{10})$$
$$u(0) = 0 \qquad\qquad\qquad (14.2)$$
$$u(5) = 1.$$

This problem has the exact solution $u = \sin(\pi x/10)$. In the following exercise you will be computing the solution for various mesh sizes and using vector norms to compute the solution error. You will see why relative errors are most convenient for error computation.

Exercise 14.7.

(1) Download the file bvp.m or copy the following code to bvp.m.

```
function [x,U]=bvp(N)
% [x,U]=bvp(N)
% Solves the system (14.2) using finite differences
% N=number of interior points.
% x is the space variable (row vector)
% U is the solution (column vector)


% M. Sussman
```

```
ULeft=0;
URight=1;
xLeft=0;
xRight=5;

% N interior mesh points, N+1 intervals, N+2 total points
dx = (xRight - xLeft) / ( N + 1 );
x=xLeft+(1:N)*dx; % interior points only, ROW vector
% construct matrix
A=-2*diag( ones(N,1) ) + diag( ones(N-1,1) ,1) + ...
    diag( ones(N-1,1) ,-1);
% construct RHS
b = -pi^2/100*sin(pi*x'/10)*dx^2;
b(1) = b(1) - ULeft;   % adjust for left  boundary value
b(N) = b(N) - URight;  % adjust for right boundary value
% solve for interior values
U = A \ b;
% add in boundary values to U and boundary positions to x
U = [ULeft; U; URight];
x = [xLeft, x, xRight];
end
```

This code is similar to `rope_bvp.m` file from Chapter 13.

(2) As a test, solve the system for `npts=10`, plot the solution and compare it with `sin(pi*x/10)`. You can use the following code to do so.

```
[x,u]=bvp(5);
plot(x,u,x,sin(pi*x/10));
legend('computed solution','exact solution','location','east');
```

The two solutions curves should lie *almost* on top of one another.

(3) Use code such as the following to compute solution errors for various mesh sizes, and fill in the following table. You may find it convenient to download the file `exer7.txt`. Recall that the exact solution is $\sin \pi x/10$.

```
sizes=[10 20 40 80 160 320 640];
for k=1:numel(sizes)
  [x,u]=bvp(sizes(k));
  error(k,1)=norm(u'-sin(pi*x/10));
  relative_error(k,1)=error(k,1)/norm(sin(pi*x/10));
end
disp([error(1:6)./error(2:7) , ...
    relative_error(1:6)./relative_error(2:7)])
```

Euclidean (L^2) vector norm

	error ratio	relative error ratio
10/ 20		
20/ 40		
40/ 80		
80/160		
160/320		
320/640		

(4) Repeat the above for the L^1 vector norm.

(L^1) vector norm

	error ratio	relative error ratio
10/ 20		
20/ 40		
40/ 80		
80/160		
160/320		
320/640		

(5) Repeat the above for the L^∞ vector norm.

(L^∞) vector norm

	error ratio	relative error ratio
10/ 20		
20/ 40		
40/ 80		
80/160		
160/320		
320/640		

(6) The method used to solve this boundary value problem converges as $O(h^2)$. Since the number of mesh points is about $1/h$, then *doubling* the number of mesh points should *quarter* the error. To see which of the above calculations yields this rate correctly, fill in the following table with the words "error," "relative error" or "both."

	Rate= $O(h^2)$?
L^2	
L^1	
L^∞	

As you will see, convergence rates are an important component of this book, and you can see it is almost always best to use relative errors in computing convergence rates of vector quantities. You may have noticed that relative norms have not always been used previously in this book in order to avoid introducing an extra step into the computations.

The reason that the L^1 and L^2 norms give different results is that the dimension of the space, n creeps into the calculation. For example, if a function is identically one, $f(x) = 1$, then its L^2 norm, $\int_0^1 |f(x)|^2 dx$ is one, but if a vector of dimension n has all components equal to one, then its L^2 norm is $\sqrt{n}$. Taking relative norms eliminates the dependence on n.

When you are doing your own research, you may have occasion to compare theoretical and computed convergence rates. When doing so, you may use tables such as those above. If you find that your computed convergence rates differ from the theoretical ones, before looking for some error in your theory, you should first check that you are using the correct norm!

14.7 Condition numbers

Given a square matrix A, the L^2 condition number $k_2(A)$ is defined as:

$$k_2(A) = \|A\|_2 \|A^{-1}\|_2 \tag{14.3}$$

if the inverse of A exists. If the inverse does not exist, then you say that the condition number is infinite. Similar definitions apply for $k_1(A)$ and $k_\infty(A)$.

MATLAB provides three functions for computing condition numbers: cond, condest, and rcond. cond computes the condition number according to Equation (14.3), and can use the one norm, the two norm, the infinity norm or the Frobenius norm. This calculation can be expensive, but it is accurate. condest computes an estimate of the condition number using the one norm. rcond uses a different method to estimate the "reciprocal condition number," defined as

$$\texttt{rcond(A)} = \begin{cases} 1/k_1(A) & \text{if } A \text{ is nonsingular} \\ 0 & \text{if } A \text{ is singular} \end{cases}.$$

So as a matrix "goes singular," rcond(A) goes to zero in a way reminiscent of the determinant.

For now, don't worry about the fact that the condition number is somewhat expensive to compute, since it requires computing the inverse or (possibly) the singular value decomposition (a topic to be studied later). Instead, concentrate on what it's good for. You already know that it's supposed to give you some idea of how singular the matrix is. But it has a central role in error estimation for the linear system problem, too.

Suppose that you are interested in solving the linear system

$$Ax = b$$

but that the right hand side you give to the computer has a small error or "perturbation" in it. You might denote this perturbed right hand side as $b + \Delta b$. You can then assume that your solution will be "slightly" perturbed, so that you are justified in writing the system as

$$A(x + \Delta x) = b + \Delta b.$$

The question is, if Δb is really small, can you expect that Δx is small? Can you actually guarantee such a limit?

If you are content to look at the *relative errors*, and if the norm used to define $k(A)$ is compatible with the vector norm used, it is fairly easy to show that:

$$\frac{\|\Delta x\|}{\|x\|} \leq k(A)\frac{\|\Delta b\|}{\|b\|}.$$

You can see that you would like the condition number to be as small as possible. *(It turns out that the condition number is bounded below. What is the smallest possible value of the condition number?)* In particular, since MATLAB provides about 14 digits of accuracy by default, if a matrix has a condition number of 10^{14}, or `rcond(A)` of 10^{-14}, then an error in the last significant digit of any element of the right hand side has the potential to make all of the digits of the solution wrong!

In the exercises below you will have occasion to use a special matrix called the Frank matrix, the same matrix used in Chapter 10. Row i of the $n \times n$ Frank matrix has the formula:

$$\mathbf{A}_{i,j} = \begin{cases} 0 & \text{for } j < i - 2 \\ n + 1 - i & \text{for } j = i - 1 \\ n + 1 - j & \text{for } j \geq i \end{cases}.$$

The Frank matrix for $n = 5$ looks like:

$$\begin{pmatrix} 5 & 4 & 3 & 2 & 1 \\ 4 & 4 & 3 & 2 & 1 \\ 0 & 3 & 3 & 2 & 1 \\ 0 & 0 & 2 & 2 & 1 \\ 0 & 0 & 0 & 1 & 1 \end{pmatrix}.$$

The determinant of the Frank matrix is 1, but is difficult to compute numerically. This matrix has a special form called *Hessenberg* form wherein all elements below the first subdiagonal are zero. MATLAB provides the Frank matrix in its "gallery" of matrices, `gallery('frank',n)`, but you will use an m-file `frank.m`. You can find information about the Frank matrix in [Frank (1958)] and [Golub and Wilkinson (1976)], Section 13, or on the web at [Golub and Wilkinson (1975)], Section 13.

Exercise 14.8. To see how the condition number can warn you about loss of accuracy, let's try solving the problem $Ax = b$, for `x=ones(n,1)`, and with A being the Frank matrix. Compute

```
b=A*x;
xSolved=A\b;
difference=xSolved-x;
```

Of course, `xSolved` would be the same as `x` if there were no arithmetic rounding errors, but there are rounding errors, so `difference` is not zero. This loss of accuracy is the point of the exercise. For convenience, use the MATLAB function `cond(A)` for the L^2 condition number, and assume that the relative error in b is `eps`, the machine precision. (Recall that MATLAB understands the name `eps` to indicate the smallest number you can add to 1 and get a result different from 1.) You expect that the relative solution error to be bounded by `eps * cond(A)`. Since `cond` uses the Euclidean norm by default, use the Euclidean norm in constructing the table.

Matrix size	`cond(A)`	`eps*cond(A)`	$\|\texttt{difference}\|/\|\texttt{x}\|$
6			
12			
18			
24			

Your second and third columns should be roughly comparable in size (within a factor of 100 or so). Thus, the condition number gives some idea of the effects of rounding error when solving a matrix equation.

14.8 A matrix example

It is a good idea to have several matrix examples at hand when you are thinking about methods. One excellent example is a class of tridiagonal matrices that arise from second-order differential equations. You saw matrices of this class in Section 13.3. An example 5×5 matrix in this class is given as

$$A = \begin{pmatrix} 2 & -1 & 0 & 0 & 0 \\ -1 & 2 & -1 & 0 & 0 \\ 0 & -1 & 2 & -1 & 0 \\ 0 & 0 & -1 & 2 & -1 \\ 0 & 0 & 0 & -1 & 2 \end{pmatrix}. \tag{14.4}$$

More generally, the $N \times N$ matrix A has components a_{kj} given as

$$a_{kj} = \begin{cases} 2 & k = j \\ -1 & k = j \pm 1 \\ 0 & |k - j| \geq 2 \end{cases}. \tag{14.5}$$

This matrix is tridiagonal (non-zero entries can be found only in three diagonals, the main diagonal and first superdiagonal and first subdiagonal). It turns out that the eigenvectors and eigenvalues of this matrix are well-known.

Exercise 14.9.

(1) Consider a set of N vectors, $\{\mathbf{v}^{(j)} \mid j = 1, 2, \ldots, N\}$, each vector of which is N-dimensional and has components $v_k^{(j)}$ given as

$$v_k^{(j)} = \sin\left(j\pi \frac{k}{N+1}\right). \tag{14.6}$$

Recalling that the "sum of angles" formula for sin is given as

$$\sin(\theta + \phi) = \sin(\theta)\cos(\phi) + \cos(\theta)\sin(\phi)$$

show by a direct (symbolic) calculation that $A\mathbf{v}^{(j)} = \lambda_j \mathbf{v}^{(j)}$, and find the value λ_j.

Note: Do this calculation by hand or use the MATLAB symbolic toolbox or use Maple or another computer algebra program.

Hint: Choose one row of the vector equation $A\mathbf{v}^{(j)} = \lambda_j \mathbf{v}^{(j)}$ and insert the expressions (14.5) and (14.6) into the row equation. Simplifying should yield an expression for λ_j.

(2) Complete the following MATLAB function so that it generates the tridiagonal matrix A according to (14.5).

```
function A=tridiagonal(N)
  % A=tridiagonal(N) generates the tridiagonal matrix A

  % your name and the date

  A=zeros(N);

  << code corresponding to equation (14.5) >>

end
```

(3) Check that your code is correct by generating the tridiagonal matrix for N=5 and comparing with (14.4).

(4) Test your hand calculation is correct by using MATLAB to compute with $N = 25$ and for the values j=1 and j=10, $\|A\mathbf{v}^{(j)} - \lambda_j \mathbf{v}^{(j)}\|_2$.

———————————

Exercise 14.10. It may not be obvious from the eigenvalues you found in the previous exercise, but the determinant of the tridiagonal matrix A given in (14.5) is $(N+1)$. (Recall the determinant is the product of the eigenvalues.) In the case $N = 5$, this can easily be seen by computing the determinant by "row reduction."

Starting from the original matrix, add 1/2 times the first row to the second to get

$$\det(A) = \det \begin{pmatrix} 2 & -1 & 0 & 0 & 0 \\ -1 & 2 & -1 & 0 & 0 \\ 0 & -1 & 2 & -1 & 0 \\ 0 & 0 & -1 & 2 & -1 \\ 0 & 0 & 0 & -1 & 2 \end{pmatrix} = \det \begin{pmatrix} 2 & -1 & 0 & 0 & 0 \\ 0 & \frac{3}{2} & -1 & 0 & 0 \\ 0 & -1 & 2 & -1 & 0 \\ 0 & 0 & -1 & 2 & -1 \\ 0 & 0 & 0 & -1 & 2 \end{pmatrix}. \tag{14.7}$$

Next, adding 2/3 times the second row to the third gives

$$\det(A) = \det \begin{pmatrix} 2 & -1 & 0 & 0 & 0 \\ 0 & \frac{3}{2} & -1 & 0 & 0 \\ 0 & 0 & \frac{4}{3} & -1 & 0 \\ 0 & 0 & -1 & 2 & -1 \\ 0 & 0 & 0 & -1 & 2 \end{pmatrix}. \tag{14.8}$$

Continuing this process gives

$$\det(A) = \det \begin{pmatrix} 2 & -1 & 0 & 0 & 0 \\ 0 & \frac{3}{2} & -1 & 0 & 0 \\ 0 & 0 & \frac{4}{3} & -1 & 0 \\ 0 & 0 & 0 & \frac{5}{4} & -1 \\ 0 & 0 & 0 & 0 & \frac{6}{5} \end{pmatrix}. \tag{14.9}$$

Since the determinant of an upper triangular matrix is the product of the diagonals, the determinant of this matrix is 6. In the following steps, you will transform this process into a calculational "proof."

(1) Complete the following MATLAB function so that it performs one Row Reduction Step (rrs) of a matrix A for row j. You may find it convenient to download the file rrs.txt.

```
function A=rrs(A,j)
  % A=rrs(A,j) performs one row reduction step for row j
  % j must be between 2 and N

  % your name and the date
  N=size(A,1);
  if (j<=1 | j>N)  % | means "or"
    j
    error('rrs: value of j must be between 2 and N');
  end

  factor=1/A(j-1,j-1);  % factor to multiply row (j-1) by

    << include code to multiply row (j-1) by factor and >>
    << add it to row (j).  Row (j-1) remains unchanged. >>
  end
```

Check your work by comparing it with (14.7) through (14.9).

(2) For N=25, use your `tridiagonal.m` code to generate the matrix A. Use the MATLAB `det` function to compute its determinant.

(3) Take one row reduction step using `rrs.m` with j=2. Using `det`, find the determinant of this matrix.

(4) Using a MATLAB script m-file, write a loop to continue the row reduction process until it is complete. Using `det`, has the determinant changed?

(5) Use the MATLAB function `diag` to extract the diagonal entries of your reduced matrix. Check that some of them are given by $(j+1)/j$ for row number j. Use the MATLAB `prod` command to compute the determinant as the product of the diagonal entries.

14.9   The determinant

[Quarteroni *et al.* (2007)], page 10, state that the Laplace rule for computing the determinant of an $N \times N$ matrix A with entries a_{kj} is given by

$$\det(A) = \begin{cases} a_{11} & N = 1 \\ \sum_{j=1}^{N} a_{kj}\Delta_{kj} & N > 1 \end{cases} \tag{14.10}$$

where k is any fixed index, and $\Delta_{kj} = (-1)^{k+j}\det(A_{kj})$ and A_{kj} is the $(N-1) \times (N-1)$ matrix whose entries are obtained from A but with all of row k and column j omitted. For example, if

$$A = \begin{pmatrix} 1 & 2 & 3 & 4 \\ 5 & 6 & 7 & 8 \\ 9 & 10 & 11 & 12 \\ 13 & 14 & 15 & 16 \end{pmatrix}$$

then $a_{24} = 8$ and

$$A_{24} = \begin{pmatrix} 1 & 2 & 3 \\ 9 & 10 & 11 \\ 13 & 14 & 15 \end{pmatrix}.$$

The quantity Δ_{kj} is sometimes called a "minor" of A.

Exercise 14.11.

(1) Write a recursive MATLAB function to compute the determinant of an arbitrary matrix A. Use your function to compute the determinant of the matrix A=magic(5) and check that your value of the determinant agrees with the one from the MATLAB `det` function.

(2) This is an extremely inefficient way to compute a determinant. The time taken by this algorithm is $O(N!)$, a very large value. Time the computations of the determinants of `magic(7)`, `magic(8)`, `magic(9)` and `magic(10)`, calling them T_7, T_8, T_9, and T_{10}. Show that $T_8 \approx 8T_7$, $T_9 \approx 9T_8$ and $T_{10} \approx 10T_9$, confirming the $O(N!)$ estimate.

Chapter 15

Solving linear systems

15.1 Introduction

In many numerical applications, notably ordinary and partial differential equations, the single most time-consuming step is the solution of a system of linear equations. You now begin a major topic in Numerical Analysis, *Numerical Linear Algebra*. Assuming familiarity with the concepts of vectors and matrices, matrix multiplication, vector dot products, and the *theory* of the solution of systems of linear equations, the matrix inverse, eigenvalues *etc.*, this chapter focusses on the practical aspects of carrying out the computations required to do linear algebra in a numerical setting.

In Section 15.2, several test matrices that will be used in this and subsequent chapters are introduced. Exercise 15.1 illustrates that matrix inversion and use of the inverse to solve a system requires $O(n^3)$ operations. Exercise 15.2 does the same for only the solution step (using the MATLAB backslash command). Exercise 15.2 also shows that solving a system is about three times faster than computing the inverse and using it to solve the system. Exercise 15.3 illustrates that systems involving sparse matrices can often be solved in less time than $O(n^3)$. Sparse matrices will arise again and be studied in more detail in Chapter 19. Exercise 15.4 shows that some matrices give rise to systems that can be accurately solved and others give rise to systems that require special consideration in order to allow accurate solution. Exercises 15.5 through 15.8 form a sequence presenting factorization of a matrix A into factors P, a permutation matrix, L, a lower triangular matrix, and U, an upper triangular matrix. Exercise 15.12 returns to this topic for the more advanced student, illustrating a detail of the proof of correctness of factorization arising from pivoting and also investigating details of the pivoting strategy presented in Exercise 15.8. Exercises 15.9 and 15.10 show how to use the factorization in constructing a solution. Exercise 15.11 illustrates one benefit of separating the factorization and solution steps in solving a time-dependent ODE.

15.2   Test matrices

This section introduces a few simple test matrices that you can use to study the behavior of the algorithms for solution of a linear system. You can use such problems as a quick test that you haven't made a serious programming error, but more importantly, you also use them to study the accuracy of the algorithms. You want test problems which can be made any size, and for which the exact values of the determinant, inverse, or eigenvalues can be determined. MATLAB itself offers a "rogues' gallery" of special test matrices through the MATLAB function `gallery`, and it offers several special matrices. You will use your own functions for some matrices and the `gallery` functions for others.

Some test matrices you will be using are:

- The second difference matrix.
- The Frank matrix.
- The Hilbert matrix.
- The Pascal matrix.
- The magic square matrix.

15.2.1   *The second difference (tridiagonal) matrix*

The second difference matrix arises in approximating the second derivative on equally spaced data. It is a tridiagonal matrix, and you have seen it in Section 14.8, and you have written an m-file named `tridiagonal.m` in Exercise 14.9. The formula is:

$$\mathbf{A}_{i,j} = \begin{cases} 2 & \text{for } j = i \\ -1 & \text{for } |j - i| = 1 \\ 0 & \text{for } |j - i| > 1 \end{cases}.$$

So that the matrix looks like:

$$\begin{pmatrix} 2 & -1 & & & & \\ -1 & 2 & -1 & & & \\ & -1 & 2 & -1 & & \\ & & & \ddots & & \\ & & & -1 & 2 & -1 \\ & & & & -1 & 2 \end{pmatrix}$$

where blanks represent zero values. As you have seen, the matrix is positive definite, symmetric, and tridiagonal. The determinant is $(n + 1)$ where n denotes the size (A is $n \times n$). MATLAB provides a function to generate this matrix called `gallery('tridiag',n)`, but you will use the m-file `tridiagonal.m` that you wrote for Exercise 14.9.

Do not use the `gallery('tridiag',n)` form in these chapters except when instructed to use it. MATLAB has two forms of storage for matrices, "sparse" and "full." Everything you have done so far has used the "full" form, but `gallery('tridiag',n)` returns the "sparse" form. The "sparse" matrix form requires some special handling.

15.2.2 The Frank matrix

Row i of the $n \times n$ Frank matrix has the formula:

$$\mathbf{A}_{i,j} = \begin{cases} 0 & \text{for } j < i-2 \\ n+1-i & \text{for } j = i-1 \\ n+1-j & \text{for } j \geq i \end{cases}.$$

The Frank matrix for $n = 5$ looks like:

$$\begin{pmatrix} 5 & 4 & 3 & 2 & 1 \\ 4 & 4 & 3 & 2 & 1 \\ 0 & 3 & 3 & 2 & 1 \\ 0 & 0 & 2 & 2 & 1 \\ 0 & 0 & 0 & 1 & 1 \end{pmatrix}.$$

The determinant of the Frank matrix is 1, but is difficult to compute accurately because of roundoff errors. This matrix has a special form called *Hessenberg* form wherein all elements below the first subdiagonal are zero. MATLAB provides the Frank matrix in its "gallery" of matrices, `gallery('frank',n)`, but you will use the code given in Section 10.4.1, available for download with the name `frank.m`.

15.2.3 The Hilbert matrix

The Hilbert matrix is related to the interpolation problem on the interval [0,1]. The matrix is given by the formula $\mathbf{A}_{i,j} = 1/(i+j-1)$. For example, with $n = 5$ the Hilbert matrix is:

$$\begin{pmatrix} \frac{1}{1} & \frac{1}{2} & \frac{1}{3} & \frac{1}{4} & \frac{1}{5} \\ \frac{1}{2} & \frac{1}{3} & \frac{1}{4} & \frac{1}{5} & \frac{1}{6} \\ \frac{1}{3} & \frac{1}{4} & \frac{1}{5} & \frac{1}{6} & \frac{1}{7} \\ \frac{1}{4} & \frac{1}{5} & \frac{1}{6} & \frac{1}{7} & \frac{1}{8} \\ \frac{1}{5} & \frac{1}{6} & \frac{1}{7} & \frac{1}{8} & \frac{1}{9} \end{pmatrix}.$$

The Hilbert matrix arises in interpolation and approximation contexts because it happens that $\mathbf{A}_{i,j} = \int_0^1 (x^{i-1})(x^{j-1})dx$. The Hilbert matrix is at once nice because its inverse has integer elements and also not nice because it is difficult to compute the inverse accurately using the usual formulæ to invert a matrix.

MATLAB has special functions for the Hilbert matrix and its inverse, called `hilb(n)` and `invhilb(n)`, and you will be using these functions in this chapter.

15.2.4   *The Pascal matrix*

The Pascal matrix is generated by a form of Pascal's triangle. It is defined as $P_{i,j} = \binom{i+j-2}{j-1}$ (the binomial coefficient). For example, with $n = 5$, the Pascal matrix is

$$P = \begin{pmatrix} 1 & 1 & 1 & 1 & 1 \\ 1 & 2 & 3 & 4 & 5 \\ 1 & 3 & 6 & 10 & 15 \\ 1 & 4 & 10 & 20 & 35 \\ 1 & 5 & 15 & 35 & 70 \end{pmatrix}.$$

As you can tell, the bold portion of this matrix is a Pascal triangle, written on its side.

A related lower triangular matrix can be written (for $n = 5$) as

$$L = \begin{pmatrix} 1 & 0 & 0 & 0 & 0 \\ 1 & -1 & 0 & 0 & 0 \\ 1 & -2 & 1 & 0 & 0 \\ 1 & -3 & 3 & -1 & 0 \\ 1 & -4 & 6 & -4 & 1 \end{pmatrix}.$$

Except for signs, this matrix is a Pascal triangle written asymmetrically. The matrix

$$L_1 = \begin{pmatrix} 1 & 0 & 0 & 0 & 0 \\ 1 & 1 & 0 & 0 & 0 \\ 1 & 2 & 1 & 0 & 0 \\ 1 & 3 & 3 & 1 & 0 \\ 1 & 4 & 6 & 4 & 1 \end{pmatrix}$$

is truly a Pascal triangle written asymmetrically. MATLAB provides a function `pascal` so that `P=pascal(5)`, `L=pascal(5,1)`, and `L1=abs(L)`.

The Pascal matrix, like the Frank matrix, is very ill-conditioned for large values of n. In addition, the Pascal matrix satisfies the conditions

- $\det(P) = \det(L) = \det(L_1) = 1$.
- $L^2 = I$ (I is the identity matrix).
- $P = LL^T = L_1 L_1^T$.

More information is available in [Pascal matrix].

15.3 The linear system "problem"

The linear system "problem" can be posed in the following way. Find an n-vector $\mathbf{x}$ that satisfies the matrix equation

$$\mathbf{Ax} = \mathbf{b} \tag{15.1}$$

where $\mathbf{A}$ is a $m \times n$ matrix and $\mathbf{b}$ is a m-vector. Both $\mathbf{x}$ and $\mathbf{b}$ are MATLAB column vectors.

You probably already know that there is usually a solution if the matrix is square, (that is, if $m = n$). You will concentrate on this case for now, but you may wonder if there is an intelligent response to this problem for the cases of a square but singular matrix, or a rectangular system, whether overdetermined or underdetermined.

At one time or another, you have probably been introduced to several algorithms for producing a solution to the linear system problem, including Cramer's rule (using determinants), constructing the inverse matrix, Gauß-Jordan elimination and Gauß factorization. You will see that it is usually not a good idea to construct the inverse matrix and you will focus on Gauß factorization for solution of linear systems.

This discussion will focus on several topics:

Efficiency: What algorithms produce a result with less work?
Accuracy: What algorithms produce an answer that is likely to be more accurate?
Difficulty: What makes a problem difficult or impossible to solve?
Special cases: How do you solve problems that are big? symmetric? banded? singular? rectangular?

15.4 The inverse matrix

A classic result from linear algebra is the following:

Theorem 15.1. *The linear system problem (15.1) is uniquely solvable for arbitrary* $\mathbf{b}$ *if and only if the inverse matrix* $\mathbf{A}^{-1}$ *exists. In this case the solution* $\mathbf{x}$ *can be expressed as* $\mathbf{x} = \mathbf{A}^{-1}\mathbf{b}$.

So what's the catch? There are a few:

- Computing the inverse takes a lot of time — more time than is necessary if you only want to know the solution of a particular system.
- In cases when the given matrix actually has very many zero entries, as is the case with the `tridiagonal` matrix, computing the inverse is enormously more difficult and time consuming than merely solving the system. And it is quite often true that the inverse matrix requires far more storage than the matrix itself. The second difference matrix, for example, is an "M-matrix" and it can be proved that all the entries in its inverse are positive numbers. You only need

about $3n$ numbers to store the `tridiagonal` matrix because you can often avoid storing the zero entries, but you need n^2 numbers to store its inverse.

- Sometimes the inverse is so inaccurate that it is not worth the trouble to multiply by the inverse to get the solution.

In the following exercises, you will see that constructing the inverse matrix takes time proportional to n^3 (for an $n \times n$ matrix), as does simply solving the system, but that solving the system is several times faster than constructing the inverse matrix. You will also see that matrices with special formats, such as tridiagonal matrices, can be solved very efficiently. And you will see even simple matrices can be difficult to numerically invert.

You will be measuring elapsed time in order to see how long these calculations take. The problem sizes are designed to take a modest amount of time on newer computers.

WARNING: Most newer computers have more than one processor (core). By default, MATLAB will use all the processors available by assigning one "thread" to each available processor. This procedure will mess up your timings, so you should tell MATLAB to use only a single thread. To do this, you should start up MATLAB with the command line

```
matlab -singleCompThread
```

If you do not know how to start up MATLAB from the command line, use the following command at the beginning of your session.

```
maxNumCompThreads(1);
```

MATLAB may warn you that this command will disappear in the future. If it is not available in your version of MATLAB then the timings in the following exercises may not agree with theory.

As you have seen, MATLAB provides the commands `tic` and `toc` to measure elapsed time. The `tic` command starts the stopwatch, and the `toc` command stops it, either printing the elapsed time or returning its value as in the expression `elapsedTime=toc;`. The times are in seconds.

Remark 15.1. `tic` and `toc` measure *elapsed* time. When a computer is doing more than one task at a time, the `cputime` function can be used to measure how much computer time is being used for each task alone. If the `maxNumCompThreads` function is not available or if the computer being used is shared with other users, then `cputime` can be used in Exercise 15.1 and subsequent exercises.

Exercise 15.1. The MATLAB command `inv(A)` computes an approximation for the inverse of a matrix `A`. Do not worry about the methods it uses, but be assured that it uses the most efficient and reliable ones available.

(1) Copy the following code (or download the file `exer1.txt`) to a MATLAB function m-file named `exer1.m` and modify it to produce information for the table below. Be sure to add comments and your name and the date.

```
function elapsedTime=exer1(n)
  % elapsedTime=exer1(n)
  % comments

  % your name and the date

  if mod(n,2)==0
    error('Please use only odd values for n');
  end

  A = magic(n);    % only odd n yield invertible matrices
  b = ones(n,1);   % the right side vector doesn't change the time
  tic;
  Ainv =  ???      % compute the inverse matrix
  xSolution = ???  % compute the solution
  elapsedTime=toc;
end
```

(2) Theory shows that the time required to invert an $n \times n$ matrix should be proportional to n^3. Fill in the following table, where the column entitled "ratio" should contain the ratio of the time for n divided by the time for the preceding value of n.

Remark 15.2. Do not expect timings to be precise. Compute the first line of the table *twice* and use the second value. The first time a function is called involves substantial overhead that later calls do not require. The last row of this table may take several minutes.

Time to compute inverse matrices		
n	time	ratio
161		
321		
641		
1281		
2561		
5121		
10241		—

(3) Are these solution times roughly proportional to n^3?

Exercise 15.2. MATLAB provides a special operator, the backslash (\) operator, that is designed to solve a linear system without computing the inverse. It is used in the following way, for a matrix A and column vector b.

```
x=A\b;
```

It may help you to remember this operator if you think of the A as being "underneath" the slash. The effect of this operator is to find the solution of the system of equations A*x=b.

The backslash looks strange, but there is a method to this madness. You might wonder why you can't just write x=b/A. This would put the column vector b to the left of the matrix A and that is the wrong order of operations. Further, b would have to be a row vector for that expression to make sense.

(1) Copy exer1.m to exer2.m and replace the inverse matrix computation and solution with an expression using the MATLAB "\" command, and fill in the following table

Time to compute solutions		
n	time	ratio
161		
321		
641		
1281		
2561		
5121		
10241		—

(2) Are these solution times roughly proportional to n^3?

(3) Compare the times for the inverse and solution and fill in the following table

Comparison of times	
n	(time for inverse)/(time for solution)
161	
321	
641	
1281	
2561	
5121	
10241	

(4) Theory shows that computation of a matrix inverse should take approximately three times as long as computation of the solution. Are your results consistent with theory?

You should be getting the message: *You should never compute an inverse when all you want to do is solve a system.*

Warning: the "\" symbol in MATLAB will work when the matrix A is *not square*. In fact, sometimes it will work when A actually is a vector. The results are not usually what you expect and no error message is given. Be careful of this potential for error when using "\". A similar warning is true for the / (division) symbol because MATLAB will try to "interpret" it if you use it with matrices or vectors and you will get "answers" that are not what you intend.

Exercise 15.3. Some matrices consist of mostly zero entries. These so-called "sparse" matrices sometimes admit extremely efficient methods for solving them. MATLAB has a special storage scheme for sparse matrices, and the MATLAB function gallery('tridiag',N) yields a sparse matrix that is the same as the result from tridiagonal(N).

(1) For N=10 and b=ones(N,1), use the "\" command to solve the system $Ax = b$ once using the matrix from tridiagonal(N) and once using the matrix from gallery('tridiag',N). Confirm that the solutions are the same as each other. The easiest way to confirm that two vectors are the same is to compute the relative norm of their difference.

(2) Using the same methodology as in Exercise 15.2, fill in the following table using the matrix from gallery('tridiag',N).
   Note: The matrix sizes in this table increase by a factor of *ten*, not two, because sparse matrix storage allows much larger matrices.

Time to compute sparse matrix solutions		
n	time	ratio
10240		
102400		
1024000		
10240000		—

(3) Because you are now solving a sparse system, solution time is not necessarily $O(n^3)$. Estimate p so that sparse solution time is $O(n^p)$. Do not forget that the matrix sizes in the table increase by factors of ten, not two.

(4) Compare the time required to solve a system using the full matrix tridiagonal(10240) with the time for the sparse matrix computation using gallery('tridiag',10240). You should see a big advantage for sparse storage. One method of sparse storage will be discussed in a later chapter.

Exercise 15.4. Some nonsingular matrices result in systems that are very difficult to solve numerically. You should recall that the condition number of a matrix A, defined as $\|A\|_p\|A^{-1}\|_p$ for $p \geq 1$ (and is equal to the size of the largest eigenvalue over the smallest eigenvalue for symmetric matrices and $p = 2$) is a measure of the likelihood of roundoff errors in solving a matrix equation. In this exercise you will construct a system by picking a solution vector x_{known} and then constructing the right side $b = Ax_{\text{known}}$. Then the known solution can be compared with the solution computed using the "\" command to see how bad a particular solution can be. Consider the following MATLAB code: (you may find it convenient to download the file `exer4.txt`)

```
N=10;
A = tridiagonal(N);
xKnown = sqrt( (1:N)' );
b = ???           % compute the right side corresponding to xKnown
xSolution = ???   % compute the solution using "\"
err=norm(xSolution-xKnown)/norm(xKnown)
```

(1) For the matrix defined by `tridiagonal`, fill in the following table. (Recall that the MATLAB command "cond" yields the condition number and the command "det" yields the determinant.)

size	error	determinant	cond. no.
		tridiagonal matrix		
10			
40			
160			

The `tridiagonal` matrix is about as nicely behaved as a matrix can be. It is neither close to singular nor difficult to invert.

(2) For the matrix defined by `hilb`, fill in the following table.

size	error	determinant	cond. no.
		Hilbert matrix		
10			
15			
20			

The Hilbert matrix is so close to being singular that numerical approximations cannot distinguish it from a singular matrix.

(3) For the matrix defined by `frank`, fill in the following table.

Frank matrix			
size	error	determinant	cond. no.
10			
15			
20			

The Frank matrix can be proved to have a determinant equal to 1.0 because its eigenvalues come in $(\lambda, 1/\lambda)$ pairs. Thus, it is *not* close to singular. Nonetheless, roundoff errors conspire to make it appear to be singular on a computer. The condition number can alert you to this possibility.

(4) For the matrix defined by `pascal`, fill in the following table.

Pascal matrix			
size	error	determinant	cond. no.
10			
15			
20			

The Pascal matrix is another matrix that has a determinant of 1, with eigenvalues occurring in $(\lambda, 1/\lambda)$ pairs. Computing the determinant numerically is subject to even more serious roundoff than for the Frank matrix.

In the following sections, you will write your own routine that does the same task as the MATLAB "\" command.

15.5 Gauß factorization

The standard method for computing the inverse of a matrix is Gaußian elimination. You already know this algorithm, perhaps by another name such as row reduction. It is described, for example by [Quarteroni *et al.* (2007)] in Section 3.3, by [Atkinson (1978)] in Sections 8.1 and 8.2, by [Ralston and Rabinowitz (2001)] in Section 9.3 or in Wikipedia [Gaussian Elimination].

First, it is preferable to think of the process as having two separable parts, the "factorization" or "forward substitution" and "back substitution" steps. The important things to recall include:

(i) In the kth step of factorization, one of the remaining equations (rows) is chosen to become the "pivot," and to be associated with x_k. This equation (row) is swapped with the existing k^{th} equation (row).

(ii) The pivot equation is used to eliminate references to x_k in the remaining equations (*i.e.*, put zeros in position k in the rows below the kth row). After this step, the matrix will have zeros in all positions below and to the left of the diagonal (k, k) position without changing similar zeros generated in earlier steps.

(iii) After $n - 1$ steps of factorization, back substitution begins. The last equation involves a single variable, and can be solved for x_n. But then equation $n - 1$ can be solved for x_{n-1}, because x_n is now known, and similarly, the other equations are solved in backwards order.

Different methods for choosing pivots lead to different variants of Gauß factorization. Important ones include:

- The simplest version of Gauß factorization for a matrix A with entries $A_{k,\ell}$ is called *Gauß factorization with no pivoting*, because the pivot for the kth row is simply the diagonal entry $A_{k,k}$. If on step k, the diagonal is zero the method will fail, even though a solution may actually exist. The forward substitution steps begin at the upper left of the matrix and proceed down the diagonal, converting all the entries below the diagonal to zero. The backward substitution steps begin at the lower right and work their way back up the diagonal.

- The method of "Gauß factorization with partial pivoting" chooses as k^{th} pivot the value $A_{\ell,k}$ maximizing $|A_{\ell,k}|$ for $\ell \geq k$. To keep the calculation orderly, this pivot row is actually moved into the kth row of the matrix. In this case, the matrix gradually is transformed into upper triangular shape. For exact arithmetic, if there is a solution, this method is guaranteed to reach it. Moreover, for computer arithmetic, this method has better accuracy properties than no pivoting.

- The method of "Gauß factorization with scaled partial pivoting" chooses as pivot the value $A_{\ell,k}$ maximizing $|A_{\ell,k}|/|A_{\ell,m}|$ for $m \geq \ell$ (*i.e.*, *relative to the rest of row* ℓ) for $\ell \geq k$. This method has better accuracy properties when the matrix is badly scaled, at the expense of more work in choosing the pivot. You will not see this method in this chapter.

- The method of "Gauß factorization with complete pivoting" chooses as pivot for the kth row the coefficient $A_{i,j}$ of largest magnitude for $i \geq k$ and $j \geq k$. This method has more bookkeeping than the partial pivoting method, and doesn't produce much improvement, so it is little used.

You now turn to writing code for the Gauß factorization. The discussion in this chapter assumes that you are familiar with Gaußian elimination (sometimes called "row reduction"), and is entirely self-contained. Because, however, interchanging rows is a source of confusion, you will assume at first that no row interchanges are necessary.

The idea is that you are going to use the row reduction operations to turn a given matrix **A** into an upper triangular matrix (all of whose elements *below* the main diagonal are zero) **U** and at the same time keep track of the multipliers in the lower triangular matrix **L**. These matrices will be built in such a way that **A** could be reconstructed from the product **LU** at any time during the factorization procedure.

Remark 15.3. The fact that $\mathbf{A} = \mathbf{LU}$ even before the computation is completed can be a powerful debugging technique. If you get the wrong answer at the end of a computation, you can test each step of the computation to see where it has gone wrong. Usually you will find an error in the first step, where the error is easiest to understand and fix.

As a first simple example, consider the matrix

$$\mathbf{A} = \begin{pmatrix} 2 & 4 \\ 1 & 9 \end{pmatrix}.$$

There is only one step to perform in the row reduction process: turn the 1 in the lower left into a 0 by subtracting half the first row from the second. Convince yourself that this process can be written as

$$\begin{pmatrix} 1 & 0 \\ -\frac{1}{2} & 1 \end{pmatrix} \begin{pmatrix} 2 & 4 \\ 1 & 9 \end{pmatrix} = \begin{pmatrix} 2 & 4 \\ 0 & 7 \end{pmatrix}. \tag{15.2}$$

If you want to write $\mathbf{A} = \mathbf{LU}$, you need to have **L** be the inverse of the first matrix on the left in (15.2). Thus

$$\mathbf{L} = \begin{pmatrix} 1 & 0 \\ \frac{1}{2} & 1 \end{pmatrix}$$

and

$$\mathbf{U} = \begin{pmatrix} 2 & 4 \\ 0 & 7 \end{pmatrix}$$

are matrices that describe the row reduction step (**L**) and its result (**U**) in such a way that the original matrix **A** can be recovered as $\mathbf{A} = \mathbf{LU}$.

Now, suppose you have a 5 by 5 Hilbert matrix (switching to MATLAB notation) A, and you wish to perform row reduction steps to turn all the entries in the first column below the first row into zero, and keep track of the operations in the matrix L and the result in U. Convince yourself that the following code does the trick.

```
n=5;
Jcol=1;
A=hilb(5);
L=eye(n);      % square n by n identity matrix
U=A;
for Irow=Jcol+1:n
  % compute Irow multiplier and save in L(Irow,Jcol)
  L(Irow,Jcol)=U(Irow,Jcol)/U(Jcol,Jcol);

  % multiply row "Jcol" by L(Irow,Jcol) and subtract from row "Irow"
  % This vector statement could be replaced with a loop
  U(Irow,Jcol:n)=U(Irow,Jcol:n)-L(Irow,Jcol)*U(Jcol,Jcol:n);
end
```

Exercise 15.5.

(1) Use the above code (you may find it convenient to download `exer5.txt`) to compute the first row reduction steps for the matrix `A=hilb(5)`. Print the resulting matrix U. Check that all entries in the first column in the second through last rows are zero or roundoff. (If any are nontrivial, you have an error somewhere. Find the error before proceeding.)

(2) Multiply L*U and confirm for yourself that it equals A. An easy way to do this is to compute `norm(L*U-A,'fro')/norm(A,'fro')`. (As you have seen, the Frobenius norm is faster to compute than the default 2-norm, especially for large matrices.)

(3) Use the code given above as a model for an m-file called `gauss_lu.m` that performs Gaußian reduction without pivoting. The function should have the following outline.

```
function [L,U]=gauss_lu(A)
  % [L,U]=gauss_lu(A) performs an LU factorization of
  % the matrix A using Gaussian reduction.
  % A is the matrix to be factored.
  % L is a lower triangular factor with 1's on the diagonal
  % U is an upper triangular factor.
  % A = L * U

  % your name and the date

  << your code >>

end
```

(4) Use your routine to find L and U for the Hilbert matrix of order 5.
(5) What is `U(5,5)`, to at least four significant figures?
(6) Verify that L is lower triangular (has zeros above the diagonal) and U is upper triangular (has zeros below the diagonal).
(7) Confirm that L*U recovers the original matrix.
(8) The MATLAB expression `R=rand(100,100);` will generate a 100×100 matrix of random entries. *Do not print this matrix — it has 10,000 numbers in it.* Use `gauss_lu` to find its factors LR and UR.

 (a) Use norms to confirm that LR*UR=R, without printing the matrices.
 (b) Use MATLAB functions `tril` and `triu` to confirm that LR is lower triangular and UR is upper triangular, without printing the matrices.

It turns out that omitting pivoting leaves the function you just wrote vulnerable to failure.

Exercise 15.6.

(1) Compute L1 and U1 for the matrix

```
A1 = [ -2    1    0    0    0
        1    0    1   -2    0
        0    0    0    1   -2
        1   -2    1    0    0
        0    1   -2    1    0]
```

The method should fail for this matrix, producing matrices with infinity (inf) and "Not a Number" (NaN) in them.

(2) On which step of the decomposition (values of Irow and Jcol) does the method fail? Why does it fail?

(3) What is the determinant of A1? What is the condition number (cond(A1))? Is the matrix singular or ill-conditioned?

In Gaußian elimination it is entirely possible to end up with a zero on the diagonal, as the previous exercise demonstrates. Since you cannot divide by zero, the procedure fails. It is also possible to end up with small numbers on the diagonal and big numbers off the diagonal. In this case, the algorithm doesn't fail, but roundoff errors can overwhelm the procedure and result in wrong answers. One solution to these difficulties is to switch rows and columns around so that the largest remaining entry in the matrix ends up on the diagonal. Instead of this "full pivoting" strategy, it is almost as effective to switch the rows only, instead of both rows and columns. This is called "partial pivoting." In the following section, permutation matrices are introduced and in the subsequent section they are used to express Gaußian elimination with partial pivoting as a "PLU" factorization.

15.6 Permutation matrices

In general, a matrix that looks like the identity matrix but with two rows interchanged is a matrix that causes the interchange of those two rows when multiplied by another matrix. A permutation matrix can actually be a little more complicated than that, but all permutation matrices are products of matrices, each with a single interchanged pair of rows.

Exercise 15.7. Consider the two permutation matrices

```
P1=[1 0 0 0
    0 0 1 0
    0 1 0 0
    0 0 0 1];

P2=[1 0 0 0
    0 0 0 1
    0 0 1 0
    0 1 0 0];
```

(1) Compute the product of P1 with the magic square matrix of order 4, A=magic(4), A2=P1*A. If the result is not the matrix A with rows two and three interchanged, then you made an error.

(2) Multiplying on the right by a permutation matrix permutes the *columns* instead of the rows. Compute the product of the magic square matrix of order 4, A=magic(4), with P1, A3=A*P1. If the result is not the matrix A with columns two and three interchanged, then you made an error.

(3) What permutation matrix would interchange the first and fourth rows of A?

(4) Compute the product P=P1*P2. The product of two permutation matrices is again a permutation matrix, so P is also a permutation matrix, but it is slightly more complicated than before. You should observe that, despite the fact that P1 and P2 are symmetric permutation matrices, the permutation matrix P is *not* symmetric.

(5) Compute the product A4=P*A and notice that it is the original A with mixed-up rows. In general, a permutation matrix is one in which each row and each column has exactly one 1 in it and all other entries are zero.

15.7　PLU factorization

You already have written gauss_lu.m to do Gauß elimination without pivoting. Now, you are going to add pivoting to the process. The previous section should have convinced you that pivoting is the same thing as multiplying by a permutation matrix. Just as in gauss_lu.m, where you started out with L equal to the identity matrix and saved the row operations in it, you will start off with the identity permutation matrix P=eye(n) and save the row interchanges in it.

Exercise 15.8.

(1) (a) Copy your file gauss_lu.m to a file called gauss_plu.m. Change the signature line to

```
function [P,L,U]=gauss_plu(A)
```

and change the comments to describe the new output matrix P.

(b) Add the initialization statement P = eye(n) near the beginning.

(c) Add the following code to the beginning of the loop on Jcol, (*i.e.,* just before the for Irow= ... statement. This code will be performed before any other processing inside that loop. You may find it convenient to download the file exer8.txt.

```
% First, choose a pivot row by finding the largest entry
% in column=Jcol on or below position k=Jcol.
% The row containing the largest entry will then be switched
% with the current row=Jcol
[ colMax, pivotShifted ] = max ( abs ( U(Jcol:n, Jcol ) ) );

% The value of pivotShifted from max needs to be adjusted.
% See help max for more information.
pivotRow = Jcol+pivotShifted-1;

if pivotRow ~= Jcol      % no pivoting if pivotRow==Jcol
  U([Jcol, pivotRow], :)        = U([pivotRow, Jcol], :);
  L([Jcol, pivotRow], 1:Jcol-1)= L([pivotRow, Jcol], 1:Jcol-1);
  P(:,[Jcol, pivotRow])         = P(:,[pivotRow, Jcol]);
end
```

The reason that pivotShifted must be adjusted to get pivotRow is because, for example, if J(jcol+1,jcol) happens to be the largest entry in the column, then the max function will return pivotShifted=2, not pivotShifted=(jcol+1). This is because (jcol+1) is the second entry in the vector (Jcol:n).

(d) *Temporarily* add the following statement just before the end of the loop on Jcol (just after the loop on Irow). (You can find this code included in exer8.txt.)

```
% TEMPORARY DEBUGGING CHECK
if norm(P*L*U-A,'fro')> 1.e-12 * norm(A,'fro')
  error('If you see this message, there is a bug!')
end
```

(2) Test gauss_plu.m on the matrix A=pascal(5), and check that P*L*U=A. You should observe that P is *not* the identity matrix.

(3) Test gauss_plu.m on the matrix A1 from Exercise 15.6 above, and check that P1*L1*U1=A1. Recall that this was the matrix for which gauss_lu.m failed.

You should not get any error messages and you should not have NaN or inf in the factors.

(4) Remove the "TEMPORARY DEBUGGING CHECK" that you inserted above. This check involves too much computation and will mess up the timings in later exercises.

15.8 PLU solution

You have seen how to factor a matrix $\mathbf{A}$ into the form:

$$\mathbf{A} = \mathbf{PLU}$$

where $\mathbf{P}$ is a permutation matrix, $\mathbf{L}$ is a unit lower triangular matrix, and $\mathbf{U}$ is an upper triangular matrix. To solve a matrix equation involving $\mathbf{A}$, you use this factor information to "peel away" the multiplication a step at a time:

$$\mathbf{A}\mathbf{x} = \mathbf{b}$$

$$\mathbf{PLU}\mathbf{x} = \mathbf{b}$$

$$\mathbf{LU}\mathbf{x} = \mathbf{P}^{-1}\mathbf{b}$$
$$\mathbf{U}\mathbf{x} = \mathbf{L}^{-1}(\mathbf{P}^{-1}\mathbf{b})$$
$$\mathbf{x} = \mathbf{U}^{-1}(\mathbf{L}^{-1}\mathbf{P}^{-1}\mathbf{b}).$$

However, instead of explicitly computing the inverses of the matrices, you use special facts about their form. It's not the inverse matrix itself that you want, but the inverse times the right hand side.

Consider the factored linear system once more. It helps to make up some names for variables and look at the problem in a different way:

$$\mathbf{P}(\mathbf{LU}\mathbf{x}) = \mathbf{b}$$
$$\mathbf{P}(\mathbf{z}) = \mathbf{b}$$

$$\mathbf{L}(\mathbf{U}\mathbf{x}) = \mathbf{z}$$
$$\mathbf{L}(\mathbf{y}) = \mathbf{z}$$

$$\mathbf{U}\mathbf{x} = \mathbf{y}$$

$$\mathbf{x} = \mathbf{U}^{-1}\mathbf{y}$$
$$\mathbf{x} = \mathbf{U}^{-1}\mathbf{L}^{-1}\mathbf{P}^{-1}\mathbf{b}.$$

Notice that all that has been done is to use parentheses to group factors, and name them. But you should now be able to see what an algorithm for solving this problem might look like.

In summary, the **PLU Solution Algorithm** is: To solve $\mathbf{A}\mathbf{x} = \mathbf{b}$ after finding the factors $\mathbf{P}, \mathbf{L}, \mathbf{U}$ and the right hand side $\mathbf{b}$,

(i) Solve $\mathbf{Pz} = \mathbf{b}$;

(ii) Solve $\mathbf{Ly} = \mathbf{z}$; and,

(iii) Solve $\mathbf{Ux} = \mathbf{y}$.

This may look more complicated, but things actually have gotten better, because these are really simple systems to solve:

(i) The solution of $\mathbf{Pz} = \mathbf{b}$ is $\mathbf{z} = \mathbf{P}^T\mathbf{b}$;

(ii) It is easy to solve $\mathbf{Ly} = \mathbf{z}$ because $\mathbf{L}$ is lower triangular and you can start at the top left; and

(iii) It is easy to solve $\mathbf{Ux} = \mathbf{y}$ because $\mathbf{U}$ is upper triangular and you can start at the bottom right.

The next exercise illustrates this process on a simple matrix that can be used as a worked-out example you can use later to check your MATLAB code. The subsequent exercise has you write a MATLAB routine to carry out the process in general.

Exercise 15.9. Consider the simple matrix factorization

$$\mathbf{A} = \mathbf{PLU}$$

$$\begin{pmatrix} 2\ 6\ 12 \\ 1\ 3\ 8 \\ 4\ 4\ 8 \end{pmatrix} = \begin{pmatrix} 0\ 1\ 0 \\ 0\ 0\ 1 \\ 1\ 0\ 0 \end{pmatrix} \begin{pmatrix} 1 & 0 & 0 \\ 0.5 & 1 & 0 \\ 0.25 & 0.5 & 1 \end{pmatrix} \begin{pmatrix} 4\ 4\ 8 \\ 0\ 4\ 8 \\ 0\ 0\ 2 \end{pmatrix}. \tag{15.3}$$

(1) Using the following matrices,

```
P = [0      1      0
     0      0      1
     1      0      0];

L = [1      0      0
     0.5    1      0
     0.25   0.5    1];

U = [4      4      8
     0      4      8
     0      0      2];
```

confirm that $\mathbf{A} = \mathbf{PLU}$.

(2) Using pencil and paper, go through the steps of solving the system, with right hand side b=[28; 18; 16] (note the semicolon to make b a column vector), using the PLU factorization. This will serve as a test problem for the code.

Step 0	Step 1	Step 2	Step 3
b	z	y	x
$\begin{bmatrix} 28 \\ 18 \\ 16 \end{bmatrix}$	$\begin{bmatrix} \ \\ \ \\ \ \end{bmatrix}$	$\begin{bmatrix} \ \\ \ \\ \ \end{bmatrix}$	$\begin{bmatrix} \ \\ \ \\ \ \end{bmatrix}$

In the following exercise, you will be writing a routine `plu_solve.m`, to solve the linear system (15.1) by solving the permutation, lower and upper triangular systems. Because this is a three stage process, you will divide it into four separate m-files: `u_solve.m`, `l_solve.m`, `p_solve.m`, and `plu_solve.m`. The `plu_solve` routine will look like:

```
function x = plu_solve ( P, L, U, b )
  % function x = plu_solve ( P, L, U, b )
  % solve a factored system
  % P=permutation matrix
  % L=lower-triangular
  % U=upper-triangular
  % b=right side
  % x=solution

  % your name and the date

  % Solve P*z=b for z using p_solve
  z = p_solve(P,b);

  % Solve L*y=z for y using l_solve
  y = l_solve(L,z);

  % Solve U*x=y for x using u_solve
  x = u_solve(U,y);
end
```

(You may find it convenient to download the file `plu_solve.txt`.)

Exercise 15.10.

(1) The easiest of these routines by far is `p_solve` because the matrix $\mathbf{P}$ is orthogonal so its solution is $\mathbf{z} = \mathbf{P}^T\mathbf{b}$. This is a one-liner. Replace the question marks with code in the following outline.

```
function z=p_solve(P,b)
  % z=p_solve(P,b)
  % P is an orthogonal matrix
```

```
% b is the right hand side
% z is the solution of P*z=b

% your name and the date

z= ???
end
```

(2) Test your routine using the **P** matrix and **b** vector from the previous exercise. Do not continue until you are confident your file is correct.
(3) The second routine is still not so hard because all you have to do is start at the top and run down the rows. Replace the question marks with code in the following outline. (You may find it convenient to download l_solve.txt.)

```
function y=l_solve(L,z)
  % y=l_solve(L,z)
  % L is a lower-triangular matrix whose diagonal entries are 1
  % z is the right hand side
  % y is the solution of L*y=z

  % your name and the date

  % set n for convenience and simplicity
  n=numel(z);
  % initialize y to zero and make sure it is a column vector
  y=zeros(n,1);

  % first row is really an easy one, especially since the diagonal
  % entries of L are equal to 1
  Irow=1;
  y(Irow)=z(Irow);

  for Irow=2:n
    rhs = z(Irow);
    for Jcol=1:Irow-1
      rhs = rhs - ???
    end
    y(Irow) = ???
  end
end
```

(4) Test this routine using the **L** matrix and **z** vector from the previous exercise. Do not continue until you are confident your file is correct.

(5) The third routine is a little harder because you have to start at the *bottom* and run *up* the rows. Replace the question marks with code in the following outline. (You may find it convenient to download u_solve.txt.)

```
function x = u_solve(U,y)
  % x=u_solve(U,y)
  % U is an upper-triangular matrix
  % y is the right hand side
  % x is the solution of U*x=y

  % your name and the date

  % set n for convenience and simplicity
  n=numel(y);
  % initialize y to zero and make sure it is a column vector
  x=zeros(n,1);

  % last row is the easy one
  Irow=n;
  x(Irow)=y(Irow)/U(Irow,Irow);

  % the -1 in the middle means it is going up from the bottom
  for Irow=n-1:-1:1
    rhs = y(Irow);
    % the following loop could also be written as a single
    % vector statement.
    for Jcol=Irow+1:n
      rhs = rhs - ???
    end
    x(Irow) = ???
  end
end
```

(6) Test this routine using the **U** matrix and **y** vector from previous exercise. Do not continue until you are confident your file is correct.

(7) Now put them all together and verify your solution by calculating

```
x = plu_solve(P,L,U,b)
```

for the matrices and right side vector in the previous exercise. Double-check your work by showing

```
relErr = norm(P*L*U*x-b)/norm(b)
```

is zero or roundoff.

(8) As a final test, solve a large system with randomly generated coefficients and right side. What is `relErr`?

```
A = rand(100,100);
x = rand(100,1);
b = A*x;
```

and the rest of the code as in the previous part of this Exercise. The error `relErr` will usually be a roundoff sized number. There is always some chance that the randomly generated matrix A will end up poorly conditioned or singular, though. If that happens, generate A and x again: they will be different this time.

You might wonder why you don't combine the m-files `gauss_plu.m` and `plu_solve.m` into a single routine. One reason is that there are other things you might want to do with the factorization instead of solving a system. For example, one of the best ways of computing a determinant is to factor the matrix. (The determinant of any permutation matrix is ±1, the determinant of L is 1 because L has all diagonal entries equal to 1, and the determinant of U is the product of its diagonal entries.) Another reason is that the factorization is much more expensive (*i.e.*, time-consuming) than the solve is, and you might be able to solve the system many times using the same factorization. An example of this latter case is presented in the following section.

15.9   A system of ODEs

Chapter 12 discusses numerical solution of certain ordinary differential equations. You saw that you could solve a scalar linear ODE using the backward Euler method but for nonlinear and for systems of equations, you used Newton's method to solve the timestep equations. It turns out that *linear systems* can be solved using matrix solution methods without resort to Newton's method. At each time step in that solution, you need to solve a linear system of equations, and you will use `plu_factor` and `plu_solve` for this task. You will also see the advantage of splitting the solution into a "factor" step and a "solve" step.

Consider a system of ordinary differential equations.

$$\frac{d\mathbf{u}}{dt} = \mathbf{A}\mathbf{u} \tag{15.4}$$

where $\mathbf{A}$ is an $n \times n$ matrix and $\mathbf{x}$ is an n-vector. Since $\mathbf{u}$ is a vector, it has subscripts, but since you are solving an ODE numerically, there will be time steps as well. You will write time levels using a superscript here so that it is a little clearer. Applying the implicit Euler method to (15.4) yields the system of equations

$$(\mathbf{I} - \Delta t\mathbf{A})\mathbf{u}^{k+1} = \mathbf{u}^k$$
$$\mathbf{u}^{k+1} = (\mathbf{I} - \Delta t\mathbf{A})^{-1}\mathbf{u}^k \tag{15.5}$$

where **I** denotes the identity matrix and in MATLAB notation, the vector $\mathbf{u}^k$ is represented as u(:,k).

The following code embodies the procedure described above. Copy it into a *script* m-file called bels.m (for Backward Euler Linear System). This particular system, using the negative of the **tridiagonal** matrix, turns out to be a discretization of the time-dependent heat equation. The variable **u** represents temperature and is a function of time and also of a one-dimensional spatial variable $x_j = j\,dx$ for fixed dx. The plot is a compact way to illustrate the vector **u** with its subscript measured along the horizontal axis. The red line is the initial shape, the green line is what the final limiting shape would be if integration were continued to $t = \infty$, and the blue lines are the shapes at intermediate times.

```
% ntimes = number of temporal steps from t=0 to t=1 .
ntimes=30;
% dt = time increment
dt=1/ntimes;
t(1,1)=0;

% N is the dimension of the space
N=402;

% initial values
u(1:N/3   ,1) = linspace(0,1,N/3)';
u(N/3+1:2*N/3,1) = linspace(1,-1,N/3)';
u(2*N/3+1:N,1) = linspace(-1,0,N/3)';

% discretization matrix is -tridiagonal multiplied
% by a constant to make the solution interesting
A= -N^2/5 * tridiagonal(N);
EulerMatrix=eye(N)-dt*A;

tic;
for k = 1 : ntimes
  t(1,k+1) = t(1,k) + dt;
  [P,L,U] = gauss_plu(EulerMatrix);
  u(:,k+1) = plu_solve(P,L,U,u(:,k));
end
CalculationTime=toc

% plot the solution at sequence of times
plot(u(:,1),'r')
hold on
for k=2:3:ntimes
```

```
  plot(u(:,k))
end
plot(zeros(size(u(:,1)))),'g')
title 'Solution at selected times'
hold off
```

Exercise 15.11.

(1) Double-check to be sure the "debugging check" has been eliminated from gauss_plu.m.

(2) Use cut-and-paste to copy the above code to a script m-file named bels.m or download the file. Record the time this solution takes. Save your solution for comparison by copying it (usave=u).

(3) Next, change bels.m to bels1.m so that it uses the routine gauss_plu.m *outside the loop* and is executed *only once* to factor the matrix EulerMatrix into P, L, and U. Then use plu_solve.m inside the loop to solve the system. Check that, for the case ntimes=1, bels.m and bels1.m yield the same solution in about the same elapsed time.

(4) Use bels1.m with ntimes=30 and record the time. The time spent for this case should be much smaller than for bels.m because factorization takes a lot more arithmetic than solution.

(5) Compare the two solutions by computing the norm of the difference (norm(usave-u,'fro')) to show they agree.

You should be aware that the system matrix in this exercise is sparse, but you have not taken advantage of that fact here. Neither have you taken advantage of the fact that the matrix is symmetric and positive definite (so that pivoting is unnecessary). As you have seen, a lot of time could be saved by using gallery(tridiag,n) and sparse forms of P, L, and U.

If you happened to try Exercise 15.11 using the "\" form, you would find that it is substantially faster than your routines. It is faster because it takes advantage of programming efficiencies that are beyond the scope of the course, not because it uses different methods. One such efficiency would be the replacement of as many loops as possible with vector statements.

15.10 Pivoting

In Exercise 15.8 above, you added pivoting code to gauss_plu.m. That code is reproduced below

```
[ colMax, pivotShifted ] = max ( abs ( U(Jcol:n, Jcol ) ) );

pivotRow = Jcol+pivotShifted-1;
```

```
if pivotRow ~= Jcol      % no pivoting if pivotRow==Jcol
  U([Jcol, pivotRow], :)        = U([pivotRow, Jcol], :);
  L([Jcol, pivotRow], 1:Jcol-1)= L([pivotRow, Jcol], 1:Jcol-1);
  P(:,[Jcol, pivotRow])         = P(:,[pivotRow, Jcol]);
end
```

In this section, you will write your own code to accomplish the same end.

Remark 15.4. Given a projection matrix P that represents a single interchange of two rows, so that PB results in two rows of a matrix B being interchanged, then the product BP^T represents the interchange of two columns.

The key to the proof of correctness of the Gauß factorization strategy is the fact that, at each step, the equality $A = PLU$ holds, where

(i) P is a projection matrix that represents the product of row interchange operations,

(ii) L is a lower triangular matrix with ones on the diagonal that represents the product of "row reduction operations" (one row minus a multiple of a previous row), and

(iii) U is partially upper triangular, with zeros below the diagonal in the first several columns, and each step of the algorithm results in one more column with zeros below the diagonal.

Clearly, when the algorithm is complete, U is (fully) upper triangular.

Suppose you have completed several steps of the Gauß factorization, and on the subsequent step have discovered that pivoting is necessary. Call the pivoting matrix Π. Recall that $\Pi^T\Pi = \Pi\Pi^T = I$ so that you can write

$$A = PLU = P\Pi^T\Pi L\Pi^T\Pi U = \overline{P}\,\overline{L}\,\overline{U}.$$

The matrix $\overline{P} = P\Pi^T$ clearly satisfies Condition (i) above, and the matrix $\overline{U} = \Pi U$ satisfies Condition (iii). For Condition (ii), note that L can be written as

$$L = (I + \Lambda)$$

where Λ is lower triangular, has zeros on the diagonal, and is non-zero only for its first several columns. As a consequence, $\Lambda\Pi^T = \Lambda$ (see the exercise below). Hence

$$\overline{L} = \Pi L\Pi^T = (\Pi(I + \Lambda)\Pi^T = I + \Pi\Lambda,$$

thus confirming Condition (ii).

Exercise 15.12.

(1) Consider the following matrix

```
A = [ 6   1   1   0   1   1
      1   6   1   1   0   1
      1   1   6   1   1   0
      1   1   0   1   1   6
      1   0   1   1   6   1
      0   1   1   6   1   1 ];
```

(You may find it convenient to download `exer12.m`.) Use your `gauss_plu` function on this matrix and confirm that only a single interchange of rows four and six is the result of the pivoting strategy.

(2) Make a copy of `gauss_plu.m` called `new_plu.m`. Change it so that it performs only three reduction steps, to the point just before the pivoting is to take place.

(3) Construct the matrix Π describing the next row interchange (interchange rows 4 and 6) and construct the matrix $\Lambda = L - I$. Show that $\Lambda\Pi^T = \Lambda$. Explain, in no more than one sentence, why this is an example of a generally true relation. This point is central to the proof that the method is valid.

(4) Restore `new_plu.m` so it does its reduction on all rows, not just the first three. Rewrite the pivoting code in it using loops, without using colons, and using the `max` function only for scalar variables. Carefully test your code to be sure results from `new_plu.m` agree with those from `gauss_plu.m`. Explain how you tested that your code is correct.

———————————

Chapter 16

Factorizations

16.1 Introduction

Chapter 15 illustrates that the PLU factorization can be used to solve a linear system provided that the system is square, and that it is nonsingular, and that it is not too badly conditioned. However, handling problems with a bad condition number, or that are singular, or even rectangular, requires a different approach. In this chapter, you will look at two versions of the QR factorization:

$$A = QR$$

where Q is an "orthogonal" matrix, and R is upper triangular. Chapter 18 discusses the "singular value decomposition" (SVD) that factors a matrix into the form $A = USV^T$.

Two variants of the QR factorization will be considered in this chapter: Gram–Schmidt, and Householder. The Householder variant is the more commonly used and is more numerically stable than the Gram–Schmidt variant.

There is a third type of QR factorization based on Givens rotations, but it is more difficult to implement. This factorization is numerically stable and somewhat more costly (requiring more arithmetic operations) than the Householder variant, but its implementation can be made more numerically efficient especially on today's muliti-core computers. It is not discussed here although it is related to the SVD factorization in Chapter 18.

If the matrix A is nonsingular, and if the diagonal elements of R are chosen to be positive, then the QR factorization of A is unique. See, for example, [Atkinson (1978)], page 614. If A is singular or not square, different variants can give different factorizations.

The QR factorization forms the basis of a method for finding eigenvalues of matrices in Chapter 17. It can also be used to solve linear systems, especially poorly-conditioned ones that the PLU factorization can have trouble with. The QR factorization is also a good way to solve least-squares minimization problems. In addition, topics such as the Gram–Schmidt method and Householder matrices have application in other contexts.

Exercise 16.1 addresses the classical Gram–Schmidt algorithm, and Exercise 16.2 addressees the modified (for stability) Gram–Schmidt algorithm. These exercises are independent of one another, except for the comparison between them. Exercise 16.3 discusses the QR factorization arising from the modified Gram–Schmidt algorithm. Exercises 16.4 through 16.6 present the Householder QR algorithm. These exercises are independent of the earlier exercises save for the accuracy comparisons. Exercise 16.7 offers an example of how to use the QR factorization to solve a linear system. Independently, Exercise 16.8 introduces the Cholesky factorization as a final example of factorizations.

16.2   Orthogonal matrices

Definition: An "orthogonal matrix" is a real matrix whose inverse is equal to its transpose.

By convention, an orthogonal matrix is often denoted by the symbol Q. The definition of an orthogonal matrix immediately implies that

$$QQ^T = Q^TQ = I.$$

One way to interpret this equation is that the columns of the matrix Q, when regarded as vectors, form an orthonormal set of vectors, and similarly for the rows. From the definition of an orthogonal matrix, and from the fact that the L^2 vector norm of x can be defined by:

$$\|\mathbf{x}\|_2 = \sqrt{(\mathbf{x}^T\mathbf{x})}$$

and the fact that

$$(A\mathbf{x})^T = \mathbf{x}^T A^T$$

you should be able to deduce that, for orthogonal matrices,

$$\|Q\mathbf{x}\|_2 = \|\mathbf{x}\|_2.$$

Multiplying a *two-dimensional* vector $\mathbf{x}$ by Q doesn't change its L^2 norm, and so $Q\mathbf{x}$ must lie on the circle whose radius is $\|\mathbf{x}\|$. In other words, $Q\mathbf{x}$ is $\mathbf{x}$ rotated around the origin by some angle, or reflected through the origin or about a diameter. That means that a two-dimensional orthogonal matrix represents a *rotation* or *reflection*. Even in N dimensions, orthogonal matrices are often called rotations.

When matrices are complex, the term "unitary" is an analog to "orthogonal." A matrix is unitary if $UU^H = U^HU = I$, where the H superscript refers to the "Hermitian" or "conjugate-transpose" of the matrix. The prime operator in MATLAB implements the Hermitian and the dot-prime operator implements the transpose. A real matrix that is unitary is orthogonal.

16.3   The Gram–Schmidt method

The "Gram–Schmidt method" can be thought of as a process that analyzes a set of vectors $\mathcal{X}$, producing a (possibly smaller) set of vectors $\mathcal{Q}$ that: (a) span the same space; (b) have unit L^2 norm; and, (c) are pairwise orthogonal. Such a set of vectors provides answers to several important questions. For example, the size of the set $\mathcal{Q}$ tells us whether the set $\mathcal{X}$ was linearly independent, and the dimension of the space spanned and the rank of a matrix constructed from the vectors. And of course, the vectors in $\mathcal{Q}$ comprise the desired orthonormal basis.

In this chapter, the Gram–Schmidt process will be used to factor a matrix, but the process itself pops up repeatedly whenever sets of vectors must be made orthogonal. You may see it, for example, in Krylov subspace methods for iteratively solving systems of equations and for eigenvalue problems.

In words, the Gram–Schmidt process goes like this:

(i) Start with no vectors in $\mathcal{Q}$;
(ii) Consider $\mathbf{x}$, the next (possibly the first) vector in $\mathcal{X}$;
(iii) For each vector $\mathbf{q}_i$ already in $\mathcal{Q}$, compute the projection of $\mathbf{q}_i$ on $\mathbf{x}$ (*i.e.*, $\mathbf{q}_i \cdot \mathbf{x}$). Subtract all of these projections from $\mathbf{x}$ to get a vector orthogonal to $\mathbf{q}_i$;
(iv) Compute the norm of (what's left of) $\mathbf{x}$. If the norm is zero (or too small), discard $\mathbf{x}$ from $\mathcal{Q}$; otherwise, divide $\mathbf{x}$ by its norm and move it from $\mathcal{X}$ to $\mathcal{Q}$.
(v) If there are more vectors in $\mathcal{X}$, return to step 2.
(vi) When $\mathcal{X}$ is finally empty, you have an orthonormal set of vectors $\mathcal{Q}$ spanning the same space as the original set $\mathcal{X}$.

Here is a sketch of the Gram–Schmidt process as an algorithm. Assume that n_x is the number of $\mathbf{x}$ vectors:

Classical Gram–Schmidt algorithm

```
nq = 0       % nq will become the number of q vectors
for k = 1 to nx
   y = xk
   for ℓ = 1 to nq        % if nq = 0 the loop is skipped
      rℓk = qℓ · xk
      y = y - rℓk qℓ
   end
   rkk = √(y · y)
   if rkk > 0
      nq = nq + 1
      qnq = y/rkk
   end
end
```

You should be able to match this algorithm to the previous verbal description. Can you see how the L^2 norm of a vector is being computed? Note that the auxiliary vector $\mathbf{y}$ is needed here because without it the vector $\mathbf{x}_k$ would be changed *during* the loop on ℓ instead of only *after* the loop is complete.

The Gram–Schmidt algorithm as described above can be subject to errors due to roundoff when implemented on a computer. Nonetheless, in the first exercise you will implement it and confirm that it works well in some cases. In Exercise 16.2, you will examine an alternative version that has been modified to reduce its sensitivity to roundoff error.

Exercise 16.1.

(1) Implement the Gram–Schmidt process in an m-file called `unstable_gs.m`. Your function should have the outline:

```
function Q = unstable_gs ( X )
  % Q = unstable_gs( X )
  % more comments

  % your name and the date

  << your code >>

end
```

You should regard the **x** vectors as columns of the matrix X. Be sure you understand this data structure because it can be confusing. In the algorithm description above, the **x** vectors are members of a set $\mathcal{X}$, but here these vectors are stored as the columns of a matrix X. Similarly, the vectors in the set $\mathcal{Q}$ are stored as the columns of a matrix Q. The first column of X corresponds to the vector $\mathbf{x}_1$, so that the MATLAB expression X(k,1) refers to the kth component of the vector $\mathbf{x}_1$, and the MATLAB expression X(:,1) refers to the MATLAB column vector corresponding to $\mathbf{x}_1$. Similarly for the other columns.

(2) It is always best to test your code on simple examples for which you know the answers. Test your code using the following input:

```
X = [ 1  1  1
      0  1  1
      0  0  1 ]
```

You should be able to see that the correct result is the identity matrix. If you do not get the identity matrix, find your bug before continuing.

(3) Another simple test you can do in your head is the following:

```
X = [ 1  1  1
      1  1  0
      1  0  0 ]
```

The columns of Q should have L^2 norm 1, and be pairwise orthogonal, and you should be able to confirm these facts "by inspection."

(4) Test your Gram–Schmidt code to compute a matrix Q using the following input:

```
X = [  2  -1   0
      -1   2  -1
       0  -1   2
       0   0  -1 ]
```

If your code is working correctly, you should compute approximately:

```
[  0.8944   0.3586   0.1952
  -0.4472   0.7171   0.3904
   0.0000  -0.5976   0.5855
   0.0000   0.0000  -0.6831 ]
```

You should verify that the columns of Q have L^2 norm 1, and are pairwise orthogonal. If you like programming problems, think of a way to do this check in a single line of MATLAB code.

(5) The matrix Q you just computed is *not* an orthogonal matrix, even though its columns form an orthonormal set. Which one of the defining conditions for an orthogonal matrix fails?

(6) Compute Q1=unstable_gs(X) where X is the Hilbert matrix hilb(10). Is Q1 an orthogonal matrix? Is it close to an orthogonal matrix?

(7) To examine Q1 further, compute B1=Q1'*Q1, where Q1 is the matrix you just computed from a Hilbert matrix. If Q1 were orthogonal, B1 would be the identity. Look at the 3×3 submatrix at the top left of B1 (print the 3×3 submatrix B1(1:3,1:3)). Similarly, look at the 3×3 submatrix at the lower right of B1, and choose a 3×3 submatrix including the diagonal near the middle of B1. You should be able see how roundoff has grown as the algorithm progressed.

It would be foolish to use the unstable form of the Gram–Schmidt factorization when the modified Gram–Schmidt algorithm is only slightly more complicated. The modified Gram–Schmidt algorithm can be described in the following way.

Modified Gram–Schmidt algorithm

$n_q = 0$ `% ` n_q ` will become the number of q vectors`
`for k = 1 to ` n_x
    $y = x_k$
    `for ` ℓ ` = 1 to ` n_q `% if ` $n_q = 0$ ` the loop is skipped`
       $r_{\ell k} = \mathbf{q}_\ell \cdot \mathbf{y}$ `% modification`
       $\mathbf{y} = \mathbf{y} - r_{\ell k}\mathbf{q}_\ell$
    `end`
    $r_{kk} = \sqrt{\mathbf{y} \cdot \mathbf{y}}$
    `if ` $r_{kk} > 0$
       $n_q = n_q + 1$
       $\mathbf{q}_{n_q} = \mathbf{y}/r_{kk}$
    `end`
`end`

The modification seems pretty minor. In exact arithmetic, it makes no difference at all. Suppose, however, that because of roundoff errors a "little bit" of $\mathbf{q}_1$ slips into $\mathbf{q}_3$. In the original algorithm the "little bit" of $\mathbf{q}_1$ will cause $r_{31} = \mathbf{q}_3 \cdot \mathbf{x}_1$ to be slightly less accurate than $r_{31} = \mathbf{q}_3 \cdot \mathbf{y}$ because there is a "more of" $\mathbf{q}_1$ inside $\mathbf{x}_3$ than there is inside $\mathbf{y}$. In the next exercise you will see that the modified algorithm can be less affected by roundoff than the original one.

Exercise 16.2.

(1) Implement the modified Gram–Schmidt process in an m-file named `modified_gs.m`. Your function should have the outline:

```
function Q = modified_gs( X )
  % Q = modified_gs( X )
  % more comments

  % your name and the date

  << your code >>

end
```

Your code should be something like `unstable_gs.m`.

(2) Choose at least three different 3×3 matrices and compare the results from `modified_gs.m` with those from `unstable_gs.m`. The results should agree, up to roundoff. If they do not, fix `modified_gs.m`. Describe your tests completely in a manner that your tests can be replicated.

(3) Consider again the Hilbert matrix `X1=hilb(10)` and compute `Q2=modified_gs(X1)` and `B2=Q2'*Q2`. Show that B2 is "closer" to the identity matrix than B1 (from Exercise 16.1) by computing the Frobenius norm of the differences `norm(B1-eye(10),'fro')` and `norm(B2-eye(10),'fro')`.

16.4 Gram–Schmidt QR factorization

Recall how the process of Gauß elimination could actually be regarded as a process of factorization. This insight enabled the solution of other problems. In the same way, the Gram–Schmidt process is actually carrying out a different factorization that will provide the key to other problems. In this factorization context stop thinking of X as a bunch of vectors, and instead regard it as a matrix. Since it is traditional for matrices to be called A, the following discussion will use A instead of X.

Now, in the Gram–Schmidt algorithm, the quantities $r_{\ell k}$ and r_{kk}, that were computed, used, and discarded, actually record important information. They can be regarded as the non-zero elements of an upper triangular matrix R. The Gram–Schmidt process actually produces a factorization of the matrix A of the form:

$$A = QR.$$

Here, the matrix Q has the same $M \times N$ "shape" as A, so it's only square if A is. The matrix R will be square $(N \times N)$, and upper triangular. In order to produce a factorization of a matrix, the Gram–Schmidt algorithm of the previous section needs some modification. For each vector $\mathbf{x}_k$ at the beginning of a loop, a vector q_k must be produced at the end of the loop, instead of dropping some out. Otherwise, the product QR will be missing some columns. The choice of column is not important because the corresponding row of R is zero, but the additional column of Q must result in Q being orthogonal. It is not hard to make such a choice, but it adds complexity that is beside the point here. For the purpose of this exercise, if r_{kk}, the norm of vector $\mathbf{x}_k$, is zero, then quit with a call to `error()` and an appropriate error message.

Exercise 16.3.

(1) Make a copy of the m-file `modified_gs.m` and call it `gs_factor.m`. Modify it to compute the Q and R factors of a (possibly) rectangular matrix A. It should have the outline

```
function [ Q, R ] = gs_factor ( A )
  % [ Q, R ] = gs_factor ( A )
  % more comments

  % your name and the date
```

```
<< your code >>
```

```
end
```

Remark 16.1. Recall that `[Q,R]=gs_factor` is the syntax that MATLAB uses to return two matrices from the function. When calling the function, you use the same syntax:

```
[Q,R]=gs_factor(A)
```

When you *use* this syntax, it is OK (but not a great idea) to leave out the comma between the Q and the R but leaving out that comma is a syntax error in the signature line of a function m-file. Using the comma all the time is recommended.

(2) Compute the Gram–Schmidt QR factorization of the following matrix.

```
A = [ 1   1   1
      1   1   0
      1   0   0 ]
```

Compare the matrix Q computed using `gs_factor` with the matrix computed using `modified_gs`. They should be the same. Check that $A = QR$ and R is upper triangular. If not, fix your code now.

(3) Compute the Gram–Schmidt QR factorization of a 100×100 matrix of random numbers (`A=rand(100,100)`). Use norms to check the equalities because the matrices are too large to examine "by eye."

- Is it true that $Q^T Q = I$? (Hint: Does `norm(Q'*Q-eye(100),'fro')` equal 0 or roundoff?)
- Is it true that $QQ^T = I$?
- Is Q orthogonal?
- Is the matrix R upper triangular? (Hint: the MATLAB functions `tril` or `triu` can help, so you don't have to look at all those numbers.)
- Is it true that $A = QR$?

(4) Compute the Gram–Schmidt QR factorization of the following matrix.

```
A = [ 0    0    1
      1    2    3
      5    8   13
     21   34   55 ]
```

- Is it true that $Q^T Q = I$?
- Is it true that $QQ^T = I$?
- Is Q orthogonal?
- Is the matrix R upper triangular?
- Is it true that $A = QR$?

The following two sections introduce another QR factorization, one that is very robust and useful.

16.5 Householder matrices

Both the PLU and Gram–Schmidt QR factorizations result in upper-triangular matrices as part of the factorization. The PLU factorization uses row-reduction operations to place zeros below the diagonal, and the Gram–Schmidt QR factorization uses dot products to place zeros below the diagonal. A third approach is presented here, using Householder matrices to turn the lower triangle of a matrix into zeros.

Remark 16.2. A fourth approach uses "Givens rotations" to generate zeros below the diagonal. You will see Givens rotations introduced in Chapter 18.

It turns out that it is possible to take a given vector $\mathbf{d}$ and a given integer k and find a matrix H so that the product $H\mathbf{d}$ is proportional to the vector $\mathbf{e}_k = (0, \ldots, 0, \underbrace{1}_{k}, 0, \ldots, 0)^T$. That is, it is possible to find a matrix H that "zeros out" all but the kth entry of $\mathbf{d}$. This is not a step-by-step process, but zeros out all the elements below the kth in one fell swoop.

This matrix, called a Householder matrix, is given by

$$\mathbf{v} = \frac{\mathbf{d} - \|\mathbf{d}\|\mathbf{e}_k}{\|(\mathbf{d} - \|\mathbf{d}\|\mathbf{e}_k)\|}$$
$$H = I - 2\mathbf{v}\mathbf{v}^T. \tag{16.1}$$

Note that $\mathbf{v}^T\mathbf{v}$ is a scalar (the inner or dot product) but $\mathbf{v}\mathbf{v}^T$ is a matrix (the outer or exterior product). To see why Equation (16.1) is true, denote $\mathbf{d} = (d_1, d_2, \ldots)^T$ and do the following calculation.

$$\begin{aligned}
H\mathbf{d} &= \mathbf{d} - 2\frac{(\mathbf{d} - \|\mathbf{d}\|\mathbf{e}_k)(\mathbf{d} - \|\mathbf{d}\|\mathbf{e}_k)^T}{(\mathbf{d} - \|\mathbf{d}\|\mathbf{e}_k)^T(\mathbf{d} - \|\mathbf{d}\|\mathbf{e}_k)}\mathbf{d} \\
&= \mathbf{d} - \frac{2}{\|\mathbf{d}\|^2 - 2d_k\|\mathbf{d}\|) + \|\mathbf{d}\|^2}(\mathbf{d} - \|\mathbf{d}\|\mathbf{e}_k)(\|\mathbf{d}\|^2 - \|\mathbf{d}\|d_k) \\
&= \mathbf{d} - \frac{2(\|\mathbf{d}\|^2 - \|\mathbf{d}\|d_k)}{(2\|\mathbf{d}\|^2 - 2d_k\|\mathbf{d}\|)}(\mathbf{d} - \|\mathbf{d}\|\mathbf{e}_k) \\
&= \|\mathbf{d}\|\mathbf{e}_k.
\end{aligned}$$

You can find further discussion in [Quarteroni *et al.* (2007)], Section 5.65.1, in [Atkinson (1978)], Section 9.3, and in many other references.

There is a way to use this idea to take any column of a matrix and make those entries below the diagonal entry be zero, as is done in the kth step of Gaußian factorization. This idea will lead to another matrix factorization.

Consider a column vector of length n split into two

$$
\mathbf{b} = \left[\begin{array}{c} b_1 \\ b_2 \\ \vdots \\ b_{k-1} \\ \hline b_k \\ b_{k+1} \\ \vdots \\ b_n \end{array} \right],
$$

and denote part of $\mathbf{b}$ below the line as $\mathbf{d}$, a column vector of length $n - k + 1$.

$$
\mathbf{d} = \left[\begin{array}{c} d_1 \\ d_2 \\ \vdots \\ d_{n-k+1} \end{array} \right] = \left[\begin{array}{c} b_k \\ b_{k+1} \\ \vdots \\ b_n \end{array} \right].
$$

Construct an $(n-k+1) \times (n-k+1)$ matrix H_d that changes $\mathbf{d}$ so it has a non-zero first component and zeros below it. Then set

$$
Hb = \left[\begin{array}{c|c} I & 0 \\ \hline 0 & H_d \end{array} \right] \left[\begin{array}{c} b_1 \\ b_2 \\ \vdots \\ b_{k-1} \\ \hline b_k \\ b_{k+1} \\ \vdots \\ b_n \end{array} \right] = \left[\begin{array}{c} b_1 \\ b_2 \\ \vdots \\ b_{k-1} \\ \hline \|\mathbf{d}\| \\ 0 \\ \vdots \\ 0 \end{array} \right]. \tag{16.2}
$$

The algorithm for computing H_d involves constructing a $(n - k + 1)$-vector $\mathbf{v}$, the same size as $\mathbf{d}$. Once $\mathbf{v}$ has been constructed, it will be expanded to a n-vector

w by adding $k - 1$ leading zeros.

$$\mathbf{w} = \begin{bmatrix} 0 \\ \vdots \\ 0 \\ v_1 \\ v_2 \\ \vdots \\ v_{n-k+1} \end{bmatrix}.$$

It should be clear that $H = I - 2\mathbf{w}\mathbf{w}^T$ satisfies

$$H = \begin{bmatrix} I & 0 \\ 0 & H_d \end{bmatrix}.$$

The following algorithm for constructing **v** and **w** includes a choice of sign that minimizes roundoff errors. This choice results in the sign of $(H\mathbf{b})_k$ in (16.2) being either positive or negative.

Constructing a Householder matrix

(i) Set $\alpha = \pm\|\mathbf{d}\|$ with signum$(\alpha) = -$signum(d_1), and $\alpha > 0$ if $d_1 = 0$. (The sign is chosen to minimize roundoff errors.)

(ii) Set $v_1 = \sqrt{\frac{1}{2}\left[1 - \frac{d_1}{\alpha}\right]}$.

(iii) Set $p = -\alpha v_1$.

(iv) Set $v_j = \frac{d_j}{2p}$ for $j = 2, 3, \ldots, (n - k + 1)$.

(v) Set $\mathbf{w} = \begin{bmatrix} 0 \\ \vdots \\ 0 \\ \mathbf{v} \end{bmatrix}.$

(vi) Set $H = I - 2\mathbf{w}\mathbf{w}^T$.

The vector **v** is specified above in (16.1) and clearly is a unit vector. It is harder to see that the vector **w** as defined in the algorithm above is also a unit vector satisfying (16.1). The calculation confirming that **w** is a unit vector is presented here. You should look carefully at the algorithm to convince yourself that **w** also satisfies (16.1).

Suppose you are given a vector **d**. Define

$$\alpha = \|\mathbf{d}\|$$

$$v_1 = \sqrt{\frac{1}{2}\left(1 - \frac{d_1}{\alpha}\right)}$$

$$p = -\alpha v_1$$

$$v_j = \frac{d_j}{2p} \text{ for } j \geq 2.$$

Then the following calculation shows that $\mathbf{v}$ is a unit vector.

$$\|\mathbf{v}\|^2 = \frac{1}{2}\left(1 - \frac{d_1}{\alpha}\right) + \sum_{j>1}\left(\frac{d_j^2}{4\alpha^2\left(\frac{1}{2}\left(1 - \frac{d_1}{\alpha}\right)\right)}\right)$$

$$= \frac{\alpha^2\left(1 - \frac{d_1}{\alpha}\right)^2 + \sum_{j>1}d_j^2}{2\alpha^2\left(1 - \frac{d_1}{\alpha}\right)}$$

$$= \frac{(\alpha - d_1)^2 + \alpha^2 - d_1^2}{2\alpha(\alpha - d_1)}$$

$$= \frac{(\alpha - d_1) + (\alpha + d_1)}{2\alpha} = 1.$$

In the following exercise you will write a MATLAB function to implement the Householder algorithm.

Exercise 16.4.

(1) (a) Start a function m-file named `householder.m` with outline

```
function H = householder(b, k)
  % H = householder(b, k)
  % more comments

  % your name and the date

  n = size(b,1);
  if size(b,2) ~= 1
    error('householder: b must be a column vector');
  end

  d(:,1) = b(k:n);
  if d(1)>=0
    alpha = -norm(d);
  else
    alpha =  norm(d);
  end

  << more code >>

end
```

You may find it convenient to download `householder.txt`.

(b) Add comments at the beginning, being sure to explain the use of the variable `k`.

(c) In the case that `alpha` is exactly zero, the algorithm will fail because of a division by zero. For the case that `alpha` is zero, return

```
H = eye(n);
```

otherwise, complete the above algorithm.

(2) Test your function on the vector `b=[10;9;8;7;6;5;4;3;2;1];` with k=1 and again with k=4. Check that H is orthogonal and H*b has zeros in positions k+1 and below.

Debugging hints:

- **w** is an $n \times 1$ (column) unit vector.
- Orthogonality of H is equivalent to the vector **w** being a unit vector. If H is not orthogonal, check **w**.
- **v** is a $(n - k + 1) \times 1$ (column) unit vector.
- No individual component of a unit vector can ever be larger than 1.

This `householder` function can be used for the QR factorization of a matrix by proceeding through a series of partial factorizations $A = Q_k R_k$, where Q_0 is the identity matrix, and R_0 is the matrix A. When you begin the kth step of factorization, your factor R_{k-1} is only upper triangular in columns 1 to $k-1$. Your goal on the kth step is to find a better factor R_k that is upper triangular through column k. If you can do this process $n-1$ times, you are done. Suppose, then, that you have partially factored the matrix A, up to column $k-1$. In other words, you have a factorization

$$A = Q_{k-1} R_{k-1}$$

for which the matrix Q_{k-1} is orthogonal, but for which the matrix R_{k-1} is only upper triangular for the first $k-1$ columns. To proceed from your partial factorization, you are going to consider taking a Householder matrix H, and "inserting" it into the factorization as follows:

$$A = Q_{k-1} R_{k-1}$$
$$= Q_{k-1} H^T H R_{k-1}.$$

Then, if you define

$$Q_k = Q_{k-1} H^T$$
$$R_k = H R_{k-1}$$

it will again be the case that:

$$A = Q_k R_k$$

and you are guaranteed that Q_k is still orthogonal. Your construction of H guarantees that R_k is actually upper triangular all the way through column k.

Exercise 16.5. In this exercise you will see how the Householder matrix can be used in the steps of an algorithm that passes through a matrix, one column at a time, and turns the bottom of each column into zeros.

(1) Define the matrix A to be the magic square of order 5, A=magic(5).
(2) Compute H1, the Householder matrix that knocks out the subdiagonal entries in column 1 of A, and then compute A1=H1*A. Are there any non-zero (non-roundoff) values below the diagonal in the first column?
(3) Now compute H2, the Householder matrix that knocks out the subdiagonal entries in column 2 of A1 (not A), and compute A2=H2*A1. This matrix should have subdiagonal zeros (or roundoff) in column 2 as well as in column 1. You should be convinced that you can zero out any subdiagonal column you want. Since zeroing out one subdiagonal column doesn't affect the previous columns, you can proceed sequentially to zero out *all* subdiagonal columns.

16.6   Householder factorization

For a rectangular M by N matrix A, the Householder QR factorization has the form

$$A = QR$$

where the matrix Q is $M \times M$ (hence square and orthogonal) and the matrix R is $M \times N$, and upper triangular (or upper trapezoidal if you want to be more accurate).

If the matrix A is not square, then this definition is different from the Gram–Schmidt factorization you discussed before. The obvious difference is the *shape* of the factors. Here, it's the Q matrix that is square. The other difference, which you'll have to take on faith, is that the Householder factorization is generally more accurate (smaller arithmetic errors), and easier to define compactly.

Householder QR Factorization Algorithm:
```
Q = I;
R = A;
for k = 1:min(m,n)
   Construct the Householder matrix H for column k of the matrix R;
   Q = Q * H';
   R = H * R;
end
```

Exercise 16.6.

(1) Write an m-file h_factor.m. It should have the outline

```
function [ Q, R ] = h_factor ( A )
   % [ Q, R ] = h_factor ( A )
   % more comments

   % your name and the date
```

```
        << your code >>

end
```

Use the routine `householder.m` in order to compute the H matrix that you need at each step.

(2) A simple test is the matrix

```
A = [ 0 1 1
      1 1 1
      0 0 1 ]
```

You should see that simply interchanging the first and second rows of A turns it into an upper triangular matrix. The matrix Q that you get is equivalent to a permutation matrix, except possibly with (-1) in some positions. You can guess what the solution is. Be sure that Q*R is A.

(3) Test your code by computing the QR factorization of the matrix `A=magic(5)`. Visually compare your results with those from your `gs_factor`. **Warning:** Q remains an orthogonal matrix if one of its columns is multiplied by (-1). This is equivalent to multiplying on the right by a matrix that is like the identity matrix except with one (-1) on the diagonal instead of all $(+1)$. Call this matrix J. The product Q*J is clearly orthogonal and can be incorporated into the QR factorization, but R is changed, too, by multiplying on the left by J. This changes the sign of the corresponding row of R. Aside from these signs, the two methods should yield the same results.

(4) There is a MATLAB function for computing the QR factorization called `qr`. Visually compare your results from `h_factor` applied to the matrix `A=magic(5)` with those from `qr`.

(5) Householder factorization is *much* less sensitive to roundoff errors. Test `h_factor` on the same Hilbert matrix, `A=hilb(10)` that you used before in Exercise 16.2. Call the resulting orthogonal matrix Q3. Show that B3=Q3'*Q3 is much closer to the identity matrix than either B2 or B1 (from Exercise 16.2) are.

16.7   The QR method for linear systems

With the Householder QR factorization of a matrix, it is easy to solve linear systems. Remember that the Q factor satisfies $Q^T Q = I$:

$$Ax = b$$
$$QRx = b$$
$$Q^T QRx = Q^T b$$
$$Rx = Q^T b \qquad (16.3)$$

so forming the new right hand side $Q^T b$ and solving the upper triangular system is all that is necessary. Solving the upper triangular system can be done using u_solve.m from Chapter 15.

Exercise 16.7.

(1) Make a copy of your file u_solve.m from Exercise 15.10 (upper-triangular solver).
(2) Write a file called h_solve.m. It should have outline

```
function x = h_solve ( Q, R, b )
  % x = h_solve ( Q, R, b )
  % more comments

  % your name and the date

  << your code >>

end
```

The matrix R is upper triangular, so you can use u_solve to solve (16.3) above. Assume that the QR factors come from the h_factor routine. Set up your code to compute $Q^T b$ and then solve the upper triangular system $Rx = Q^T b$ using u_solve.
(3) When you think your solver is working, test it out on a system as follows. (You may find it convenient to download the file exer7.txt.)

```
n = 5;
A = magic ( n );
x = [ 1 : n ]';  % transpose makes a column vector
b = A * x;
[ Q, R ] = h_factor ( A );
x2 = h_solve ( Q, R, b );
norm(x - x2)/norm(x) % should be close to zero.
```

Now you know another way to solve a square linear system. It turns out that you can also use the QR algorithm to solve non-square systems, but that task is better left to the singular value decomposition in a later chapter.

16.8 Cholesky factorization

In the case that the matrix A is symmetric positive definite, then A can be factored uniquely as LL^T, where L is a lower-triangular matrix with positive values on the diagonal. This factorization is called the "Cholesky factorization" and is both highly efficient (in terms of both storage and time) and also not very sensitive to roundoff errors. Cholesky factorization is often useful in solving least-squares problems because the matrix that arises is usually symmetric and positive definite.

You can find more information about Cholesky factorization in, for example, [Atkinson (1978)], page 524, in [Ralston and Rabinowitz (2001)], page 424, and in Wikipedia [Cholesky decomposition (2020)].

Exercise 16.8. The Cholesky factorization can be described by the following equations.

$$L_{jj} = \sqrt{A_{jj} - \sum_{m=1}^{j-1} L_{jm}^2}$$

$$L_{kj} = \frac{1}{L_{jj}} \left(A_{kj} - \sum_{m=1}^{j-1} L_{km} L_{jm} \right) \qquad \text{for } j < k \leq n.$$

(1) As you recall, the MATLAB command A=pascal(6) generates a symmetric positive definite matrix based on Pascal's triangle, and the command L1=pascal(6,1) generates a lower triangular matrix based on Pascal's triangle. Further, you know that A=L1*L1'. Why is L1 *not* the Cholesky factor of A?

(2) Write a function m-file with outline

```
function L=cholesky(A)
  % L=cholesky(A)
  % more comments

  % your name and the date

  << your code >.

end
```

(3) Test your function on the matrix A=(eye(6)+pascal(6)) by computing L=cholesky(A). Check that L*L'=A.

(4) MATLAB has a built-in version of Cholesky factorization named chol. By default this function returns an *upper* triangular matrix, but chol(A,'lower') returns

a lower triangular matrix. Compare the results from `chol` with those from your `cholesky` function on a matrix of your choice. Describe your testing.

(5) Recover your `l_solve.m` from Exercise 15.10 and modify it so that it does not assume that the diagonal elements are all ones. How did you test your modified function is correct?

(6) Use the following code to generate a 50×50 positive definite symmetric matrix (using Gershgorin's theorem) along with solution and right side vectors. (You may find it convenient to download the file `exer8.txt`.)

```
N = 50;
A = rand(N,N);          % random numbers between 0 and 1
A = .5 * (A + A');      % force A to be symmetric
A = A + diag(sum(A));   % make A positive definite by Gershgorin
x = rand(N,1);          % solution
b = A*x;                % right side
```

(7) Use your `cholesky` function along with your modified `l_solve` and `u_solve` to solve the system $Ax_0 = b$ and show that $\|x_0 - x\|/\|x\|$ is small. It turns out that solving by Cholesky factorization is generally the most efficient way to solve a positive definite symmetric matrix system.

Chapter 17

The eigenvalue problem

17.1 Introduction

This chapter is concerned with several ways to compute eigenvalues and eigenvectors for a real matrix. All methods for computing eigenvalues and eigenvectors are iterative in nature, except for very small matrices. A brief summary discussion of eigenvalues and eigenvectors precedes the exercises.

Exercise 17.1 presents the Rayleigh quotient. Exercise 17.2 presents the power method with an example of why the method should work. Exercise 17.3 includes an elementary stopping criterion and applies the power method to a collection of matrices. The inverse power method is presented in Exercise 17.4, with application in Exercise 17.5. Exercise 17.6 presents the inverse power method for multiple eigenpairs, using orthogonalization to keep the eigenvectors separate. The best method for orthogonalization would be Gram–Schmidt from Chapter 16, but the MATLAB qr function is used here in the interest of keeping the chapters independent. The shifted inverse power method is presented in Exercise 17.7. An alternative method, QR, is introduced in Exercise 17.8 and the special case of a real symmetric matrix is used to illustrate the power of provable convergence testing in Exercise 17.9. Exercise 17.10 presents eigenvalue computation as the preferred method for finding roots of polynomials.

17.2 Eigenvalues and eigenvectors

For any square $N \times N$ matrix A, consider the equation $\det(A - \lambda I) = 0$. This is a polynomial equation of degree N in the variable λ so there are exactly N complex roots (counting multiplicities) that satisfy the equation. If the matrix A is real, then any complex roots occur in conjugate pairs. The roots are known as "eigenvalues" of A. Interesting questions include:

- How can you find one or more of these roots?
- When are the roots distinct?
- When are the roots real?

In textbook examples, the determinant is computed explicitly, the (often cubic) equation is factored exactly, and the roots "fall out." This is not how a real problem will be solved.

If λ is an eigenvalue of A, then $A - \lambda I$ is a singular matrix, and therefore there is at least one non-zero vector $\mathbf{x}$ with the property that $(A - \lambda I)\mathbf{x} = \mathbf{0}$. This equation is usually written

$$A\mathbf{x} = \lambda \mathbf{x}. \tag{17.1}$$

Such a vector is called an "eigenvector" for the given eigenvalue. There may be as many as N linearly independent eigenvectors. In some cases, the eigenvectors are, or can be made into, a set of pairwise orthogonal vectors. If a set of eigenvectors is arranged to form the columns of a matrix E and the corresponding eigenvalues are arranged as a diagonal matrix Λ, then (17.1) is written for the whole set of eigenpairs as

$$AE = E\Lambda. \tag{17.2}$$

(Notice that Λ appears on the right.)

Interesting questions about eigenvectors include:

- How do you compute an eigenvector?
- When will there be a full set of N independent eigenvectors?
- When will the eigenvectors be orthogonal?

In textbook examples, the singular system $(A - \lambda I)\mathbf{x} = \mathbf{0}$ is examined, and by inspection, an eigenvector is determined. This is not how a real problem is solved either.

Some useful facts about the eigenvalues λ of a matrix A:

- A^{-1} has the same eigenvectors as A, and eigenvalues $1/\lambda$;
- For integer n, A^n has the same eigenvectors as A with eigenvalues λ^n;
- $A + \mu I$ has the same eigenvectors as A with eigenvalues $\lambda + \mu$;
- If A is real and symmetric, all its eigenvalues and eigenvectors are real and the eigenvectors can be written as an orthonormal set; and
- If B is invertable, then $B^{-1}AB$ has the same eigenvalues as A, with eigenvectors given as $B^{-1}x$, for each eigenvector x of A.

Methods for finding eigenvalues and eigenvectors can be found, for example, in [Quarteroni *et al.* (2007)], Chapter 5, [Atkinson (1978)] in Chapter 9, [Ralston and Rabinowitz (2001)] in Chapter 10, and in Wikipedia [Eigenvalues and eigenvectors].

17.3   The Rayleigh quotient

If a vector $\mathbf{x}$ is an exact eigenvector of a matrix A, then it is easy to determine the value of the associated eigenvalue: simply take the ratio of a component of $(A\mathbf{x})_k$

and $\mathbf{x}_k$, for any index k that you like. But suppose $\mathbf{x}$ is only an approximate eigenvector, or worse yet, just a wild guess. The ratio of corresponding components for various indices would vary from component to component. It might be appealing to take an average, but the preferred method for estimating the value of an eigenvalue uses the "Rayleigh quotient."

The Rayleigh quotient of a matrix A and vector $\mathbf{x}$ is

$$R(A, \mathbf{x}) = \frac{\mathbf{x} \cdot A\mathbf{x}}{\mathbf{x} \cdot \mathbf{x}}$$
$$= \frac{\mathbf{x}^H A\mathbf{x}}{\mathbf{x}^H \mathbf{x}},$$

where $\mathbf{x}^H$ denotes the Hermitian (complex conjugate transpose) of $\mathbf{x}$. The MATLAB prime (as in $\mathbf{x}$') actually means the complex conjugate transpose, not just the transpose, so you can use the prime here. (In order to take a transpose, without the complex conjugate, you must use the MATLAB function **transpose** or the prime with a dot as in $\mathbf{x}.$'.) Note that the denominator of the Rayleigh quotient is just the square of the 2-norm of the vector. In this chapter you will be mostly computing Rayleigh quotients of unit vectors, in which case the Rayleigh quotient reduces to

$$R(A, \mathbf{x}) = \mathbf{x}^H A\mathbf{x} \qquad \text{for } \|\mathbf{x}\| = 1.$$

If $\mathbf{x}$ happens to be an eigenvector of the matrix A, the Rayleigh quotient must equal its eigenvalue. (Plug $\mathbf{x}$ into the formula and you will see why.) When the real vector $\mathbf{x}$ is an approximate eigenvector of A, the Rayleigh quotient is a very accurate estimate of the corresponding eigenvalue. Complex eigenvalues and eigenvectors require a little care because the dot product involves multiplication by the conjugate transpose. The Rayleigh quotient remains a valuable tool in the complex case, and most of the facts remain true.

If the matrix A is symmetric, then all eigenvalues are real, and the Rayleigh quotient satisfies the following inequality.

$$\lambda_{\min} \le R(A, \mathbf{x}) \le \lambda_{\max}.$$

Exercise 17.1.

(1) Write an m-file called **rayleigh.m** that computes the Rayleigh quotient. Do not assume that x is a unit vector. It should have the following outline.

```
function r = rayleigh ( A, x )
  % r = rayleigh ( A, x )
  % comments

  % your name and the date

  << your code >>

end
```

Be sure to include comments describing the input and output parameters and the purpose of the function.

(2) Download the m-file `eigen_test.m`. This file contains several test problems. Verify that the matrix you get by calling `A=eigen_test(1)` has eigenvalues 1, -1.5, and 2, and eigenvectors $[1; 0; 1]$, $[0; 1; 1]$, and $[1; -2; 0]$, respectively. That is, verify that $A\mathbf{x} = \lambda\mathbf{x}$ for each eigenvalue λ and eigenvector $\mathbf{x}$.

(3) Compute the value of the Rayleigh quotient for the matrix `A=eigen_test(1)` and vectors in the following table.

Some Rayleigh quotients		
x	R(A,x)	
[3; 2; 1]	4.5	
[1; 0; 1]		(is an eigenvector)
[0; 1; 1]		(is an eigenvector)
[1;-2; 0]		(is an eigenvector)
[1; 1; 1]		
[0; 0; 1]		

(4) Starting from the vector `x=[3;2;1]`, you are interested in what happens to the Rayleigh quotient when you look at the sequence of vectors $\mathbf{x}^{(1)\dagger} = \mathbf{x}$, $\mathbf{x}^{(2)\dagger} = \mathbf{A}\mathbf{x}$, $\mathbf{x}^{(3)\dagger} = \mathbf{A}^2\mathbf{x}$, ... Plot the Rayleigh quotients of $\mathbf{x}^{(k)\dagger} = \mathbf{A}^{k-1}\mathbf{x}$ vs. $k = 1, \ldots, 25$. You should observe what appears to be convergence to one of the eigenvalues of A. Which eigenvalue?

The Rayleigh quotient $R(A, \mathbf{x})$ provides a way to approximate eigenvalues from an approximate eigenvector. The following section presents one way to come up with an approximate eigenvector.

17.4   The power method

In many physical and engineering applications, the largest or the smallest eigenvalue associated with a system represents the dominant (and hence most interesting) mode of behavior. For a bridge or support column, the smallest eigenvalue might reveal the maximum load, and the eigenvector represents the shape of the object at the instant of failure under this load. For a concert hall, the smallest eigenvalue of the acoustic equation reveals the lowest resonating frequency. For nuclear reactors, the largest eigenvalue determines whether the reactor is subcritical ($\lambda < 1$ and reaction dies out), critical ($\lambda = 1$ and reaction is sustained), or supercritical ($\lambda > 1$ and reaction grows). Hence, a simple means of approximating just one extreme eigenvalue might be enough for some cases.

---

†Superscripts in parentheses are used to denote iteration number to reduce confusion among iteration number, exponentiation and vector components.

The "power method" tries to determine the largest magnitude eigenvalue, and the corresponding eigenvector, of a matrix, by computing (scaled) vectors in the sequence:

$$\mathbf{x}^{(k)} = A\mathbf{x}^{(k-1)}.$$

Remark 17.1. As elsewhere in this book, the superscript $^{(k)}$ refers to the kth vector in a sequence. When you write MATLAB code, you will typically denote both $\mathbf{x}^{(k)}$ and $\mathbf{x}^{(k-1)}$ with the same MATLAB name x, and overwrite it each iteration. In the code specified below, xold is used to denote $\mathbf{x}^{(k-1)}$ and x is used to denote $\mathbf{x}^{(k)}$ and all subsequent iterates.

Including the scaling and an estimate of the eigenvalue, the **power method** can be described in the following way.

(i) Starting with any non-zero vector **x**, divide by its length to make a unit vector called $\mathbf{x}^{(0)}$;

(ii) $\mathbf{x}^{(k)} = (A\mathbf{x}^{(k-1)})/\|A\mathbf{x}^{(k-1)}\|$;

(iii) Compute the Rayleigh quotient of the iterate (unit vector) $r^{(k)} = (\mathbf{x}^{(k)} \cdot A\mathbf{x}^{(k)})$;

(iv) If not done, increment k and go back to Step (ii).

Exercise 17.2. :

(1) Write an m-file that takes a specified number of steps of the power method. Your file should have following outline. You may find it convenient to download the file power_method.txt

```
function [r, x,rHistory,xHistory]=power_method(A, x ,maxNumSteps)
  % [r, x,rHistory,xHistory]=power_method(A, x ,maxNumSteps)
  % comments

  % your name and the date

  << your code >>

  for k=1:maxNumSteps

    << your code >>

    r= ???   %Rayleigh quotient

    rHistory(k)=r;      % save Rayleigh quotient on each step
    xHistory(:,k)=x;    % save one eigenvector on each step
  end
end
```

On exit, the `xHistory` variable will be a matrix whose columns are a (hopefully convergent) sequence of approximate dominant eigenvectors. The history variables `rHistory` and `xHistory` will be used to illustrate the progress that the power method is making and have no mathematical importance.

Remark 17.2. It is usually the case that when writing code such as this that requires the Rayleigh quotient, you would use `rayleigh.m`, but in this case the Rayleigh quotient is so simple that it is better to just copy the code for it into `power_method.m`.

(2) Using your power method code, try to determine the largest eigenvalue of the matrix `eigen_test(1)`, starting from the vector `[0;0;1]`. (The results for the first few steps are presented to help you check your code out.)

Power method iterations		
Step	Rayleigh q.	x(1)
0	-4	0.0
1	0.20455	0.12309
2	2.19403	
3		
4		
10		
15		
20		
25		

(3) Plot the progress of the method with the following commands

```
plot(rHistory);      title 'Rayleigh quotient vs. k'
plot(xHistory(1,:)); title 'Eigenvector first component vs. k'
```

Exercise 17.3.

(1) (a) Revise your `power_method.m` to a new function named `power_method1.m` with the outline

```
function [r, x, rHistory, xHistory] =
   power_method1 ( A, x , tol )
% [r, x, rHistory, xHistory] = power_method1 ( A, x , tol )
% comments

% your name and the date

maxNumSteps=10000;  % maximum allowable number of steps
```

```
    << your code >>
```

```
end
```

You should have an **if** test with the MATLAB **return** statement in it before the end of the loop. To determine when to stop iterating, add code to effect *all three* of the following criteria:

$$k > 1$$
$$|r^{(k)} - r^{(k-1)}| \leq \epsilon |r^{(k)}| \quad \text{and}$$
$$\|(x^{(k)} - x^{(k-1)})\| \leq \epsilon$$

where the MATLAB variable **tol** represents ϵ. (Recall that $\|x^{(k)}\| = 1$.) The second two expressions say that r and the eigenvector are accurate to a relative error of ϵ (the eigenvector is a unit vector). The denominator of the relative error expression multiplies both sides of the inequality so that numerical errors do not cause trouble when r is small.

Remark 17.3. In real applications, it would not be possible to return the history variables because they would require too much storage. Without the history variables a good way to implement this stopping criterion would be to keep track only of $r^{(k-1)}$ and $x^{(k-1)}$ in separate variables. Just before you update the values of r and x, use MATLAB code such as

```
rold=r;   % save for convergence tests
xold=x;   % save for convergence tests
```

to save values for $r^{(k-1)}$ and $x^{(k-1)}$.

(b) Since the method should take fewer than **maxNumSteps** iterations, the method has failed to find an eigenpair if it takes **maxNumSteps** iterations. Place the statement

```
disp('power_method1: Failed');
```

after the loop. (Normally you would use an **error** function call here, rather than **disp**, but for this chapter you will need to see the intermediate values in **rHistory** and **xHistory** even when the method does not converge at all.)

(2) Using your new power method code, try to determine the largest-in-magnitude eigenvalue, and corresponding eigenvector, of the **eigen_test** matrices by repeatedly taking power method steps. Use a starting vector of all 0's, except for a 1 in the last position, and a tolerance of **1.e-8**. **Hint:** The number of iterations is given by the length of **rHistory**. If the method does not converge, write "failed" in the "no. iterations" column.

Matrix	Rayleigh q.	x(1)	no. iterations
1			
2			
3			
4			
5			
6			
7			

These matrices were chosen to illustrate different iterative behavior. Some converge quickly and some slowly, some oscillate and some don't. One does not converge. Generally speaking, the eigenvalue converges more rapidly than the eigenvector. You can check the eigenvalues of these matrices using the `eig` function.

(3) Plot the histories of Rayleigh quotient and first eigenvector component for cases 3, 5, and 7.

Remark 17.4. The `eig` command will show you the eigenvectors as well as the eigenvalues of a matrix A, if you use the command in the form:

`[ V, D ] = eig ( A )`

The quantities V and D are matrices, storing the n eigenvectors as columns in V and the eigenvalues along the diagonal of D.

17.5   The inverse power method

The inverse power method reverses the iteration step of the power method. Write

$$Ax^{(k)} = x^{(k-1)}$$

or, equivalently,

$$x^{(k)} = A^{-1}x^{(k-1)}.$$

In other words, this looks like doing a power method iteration, but now for the matrix A^{-1}. You can immediately conclude that this process will often converge to the eigenvector associated with the largest-in-magnitude eigenvalue of A^{-1}. The eigenvectors of A and A^{-1} are the same, and if v is an eigenvector of A for the eigenvalue λ, then it is also an eigenvector of A^{-1} for $1/\lambda$ and vice versa. Naturally, you should not compute A^{-1} in numerical work, but solve the system instead.

Remark 17.5. Here is another example where the matrix A can be factored once and the factors reused for many subsequent iterations, although this idea will not be explored here.

This means that you can freely choose to study the eigenvalue problem of either A or A^{-1}, and easily convert information from one problem to the other. The difference is that the inverse power iteration will find us the largest-in-magnitude eigenvalue of A^{-1}, and that's the eigenvalue of A that's smallest in magnitude, while the plain power method finds the eigenvalue of A that is largest in magnitude.

The **inverse power method** iteration is given in the following algorithm.

(1) Starting with any non-zero vector $\mathbf{x}$, divide by its length to make a unit vector called $\mathbf{x}^{(0)}$;
(2) Solve $A\hat{\mathbf{x}} = \mathbf{x}^{(k-1)}$;
(3) Normalize the iterate by setting $\mathbf{x}^{(k)} = \hat{\mathbf{x}}/\|\hat{\mathbf{x}}\|$;
(4) Compute the Rayleigh quotient of the iterate (unit vector) $r^{(k)} = (\mathbf{x}^{(k)} \cdot A\mathbf{x}^{(k)})$; and
(5) If converged, stop; otherwise, go back to step 2.

Exercise 17.4.

(1) Write an m-file that computes the inverse power method, stopping when
$$k > 1$$
$$|r^{(k)} - r^{(k-1)}| \le \epsilon |r^{(k)}|, \text{ and}$$
$$\|x^{(k)} - x^{(k-1)}\| \le \epsilon,$$

where ϵ denotes a specified relative tolerance. (Recall that $\|x^{(k)}\| = 1$.) This is the same stopping criterion you used in power_method1.m.
Your file should have the outline:

```
function  [r, x, rHistory, xHistory] = inverse_power ( A, x, tol)
% [r, x, rHistory, xHistory] = inverse_power ( A, x, tol)
% comments

% your name and the date

<< your code >>

end
```

Be sure to include a check on the maximum number of iterations (10000 is a good choice) so that a non-convergent problem will not go on forever.

(2) Using your inverse power method code, determine the smallest eigenvalue of the matrix A=eigen_test(2),
```
[r x rHistory xHistory]=inverse_power(A,[0;0;1],1.e-8);
```
What are r, x and the number of steps?

(3) As a debugging check, run
```
[rp xp rHp xHp]=power_method1(inv(A),[0;0;1],1.e-8);
```

You should find that r is close to 1/rp, that x and xp are close, and that
xHistory and xHp are close. (You should not expect rHistory and 1./rHp to
agree in the early iterations.) If these relations are not true, you have an error
somewhere. Fix it before continuing. **Warning:** the two methods may take
slightly different numbers of iterations. Compare xHistory and xHp only for
the common iterations.

Exercise 17.5.

(1) Apply the inverse power method to the matrix A=eigen_test(3) starting from
the vector [0;0;1] and with a tolerance of 1.e-8. You will find that it does
not converge. Even if you did not print some sort of non-convergence message,
you can tell it did not converge because the length of the vectors rHistory
and xHistory is precisely the maximum number of iterations you chose. In the
following steps, you will discover why it did not converge and fix it.

(2) Plot the Rayleigh quotient history rHistory and the first component of the
eigenvector history xHistory(1,:). You should find that rHistory seems to
show convergence, but xHistory oscillates in a plus/minus fashion, and the
oscillation is the cause of non-convergence. **Hint:** you may find the plot hard
to interpret. Either zoom in or plot only the final 50 or so values.

(3) If a vector x is an eigenvector of a matrix, then so is $-x$, and with the same eigenvalue. The following code (you may find it convenient to download exer5.txt)
will choose the sign of $x^{(k)}$ so $x^{(k)}$ and $x^{(k-1)}$ point in nearly the same direction. This is done by making the dot product of the two vectors positive. At
the beginning of the loop, define

```
xold=x;
```

and include the following code after updating x in your inverse_power.m. It
will still return an eigenvector, and it will no longer flip signs from iteration to
iteration.

```
% choose the sign of x so that the dot product xold'*x >0
factor=sign(xold'*x);
if factor == 0
  factor=1;
end
x=factor*x;
```

(4) Try inverse_power again. It should converge very rapidly.

(5) Fill in the following table with the eigenvalue of smallest magnitude. Start from
the vector of zeros with a 1 in the last place, and with a tolerance of 1.0e-8.

matrix	Eigenvalue	x(1)	no. iterations
1			
2			
3			
4			
5			
6			
7			

Be careful! One of these still does not converge. You could use a procedure similar to this exercise to discover and fix that one, but it is more complicated and can also be handled by shifting. See Exercise 17.8 also.

17.6 Finding several eigenvectors at once

It is common to need more than one eigenpair, but not all eigenpairs. For example, finite element models of structures often have upwards of a million degrees of freedom (unknowns) and just as many eigenvalues. Only the lowest few are of practical interest. Fortunately, matrices arising from finite element structural models are typically symmetric and negative definite, so the eigenvalues are real and negative and the eigenvectors are orthogonal to one another.

For the remainder of this section, and this section only, you will be assuming that the matrix whose eigenpairs are being sought is symmetric so that the eigenvectors are orthogonal.

If you try to find two eigenvectors by using the power method (or the inverse power method) twice, on two different vectors, you will discover that you get two copies of the same dominant eigenvector. (The eigenvector with eigenvalue of largest magnitude is called "dominant.") Because the matrix is assumed symmetric, you know that the dominant eigenvector is orthogonal to the others, so you could use the power method (or the inverse power method) twice, forcing the two vectors to be orthogonal at each step of the way. This will guarantee they do not both converge to the same vector. If this approach converges to anything, it is probably the dominant and "next-dominant" eigenvectors and their associated eigenvalues.

Orthogonalization is discussed in Chapter 16, and a function called `modified_gs.m` was introduced in Exercise 16.2. This function would be the preferred function to use in the `power_several.m` function in the following exercise. In order to keep these chapters as independent from one another as possible, however, the MATLAB `qr` function can be used instead.

This `qr` function uses methods similar to `h_factor.m` in Exercise 16.6 to factor a possibly rectangular matrix $V = QR$ with Q being an orthogonal matrix and R being upper triangular. The MATLAB command `[Q,R]=qr(V,0)` provides the matrices. The second parameter, 0, is needed because without it the matrix Q will be a square matrix.

Regarding a small set of vectors as the columns of a matrix V, then the columns of an orthogonal matrix Q form an orthonormal set. To orthonormalize a set of vectors arranged as columns of a matrix V, then, the columns of the matrix Q would seem to be an excellent candidate. To be sure that the span of the columns of V is the same as the span of the columns of Q, consider the kth column of the product $QR = V$. Call the kth column of Q the vector $\mathbf{q}_k$, and similarly for the columns of V, then because $QR = V$,

$$\sum_{j=1}^{k} \mathbf{q}_j r_{jk} = \mathbf{v}_k,$$

and the sum only goes up to k because R is upper triangular. This is precisely the statement that the vectors $\mathbf{v}_k$ are in the span of the vectors $\mathbf{q}_k$. Because the number of columns of Q and X are the same, then the columns of Q can be regarded as the orthonormalization of the columns of V.

Remark 17.6. In the applications in this chapter, the columns of the matrix X will always be a linearly independent set. If they were not independent, then the matrix Q from qr *would* be linearly independent, and might introduce some confusion in interpreting the results.

The power method for a symmetric matrix applied to several vectors can be described in the following algorithm:

(i) Start with any linearly independent set of vectors stored as columns of a matrix V, orthonormalize this set to generate a matrix $V^{(0)}$;
(ii) Generate the next iterates by computing $\hat{V} = AV^{(k-1)}$;
(iii) Orthonormalize the iterates $\hat{V}$ and name them $V^{(k)}$;
(iv) Adjust signs of eigenvectors as necessary;
(v) Compute the Rayleigh quotient of the iterates as the vector equal to the *diagonal entries* of the matrix $(V^{(k)})^T AV^{(k)}$;
(vi) If converged, exit; otherwise return to step (ii).

Note: There is no guarantee that the eigenvectors that you find are the ones with largest magnitude! It is possible that your starting vectors were themselves orthogonal to one or more eigenvectors, and you will never recover eigenvectors that are not spanned by the initial ones. In applications, estimates of the inertia of the matrix can be used to mitigate this problem. See, for example, [Sylvester Law of Inertia] for a discussion of inertia of a matrix and how it can be used.

In the following exercise, you will write a MATLAB function to carry out this procedure to find several eigenpairs for a symmetric matrix. The initial set of eigenvectors will be represented as columns of a MATLAB matrix V.

Exercise 17.6.

(1) (a) Begin writing a function mfile named power_several.m with the following outline. You may find it convenient to download the file power_several.txt

```
function  [r, V, numSteps] = power_several ( A, V , tol )
  % [r, V, numSteps] = power_several ( A, V , tol )
  % comments


  % your name and the date


  << your code >>


  end
```

to implement the algorithm described above.

(b) Add comments explaining that V is a matrix whose columns represent the eigenvectors and also that R is a diagonal matrix with the eigenvalues on the diagonal. Since you are assuming A to be a symmetric matrix, add code to check for symmetry and call the MATLAB **error** function if it is not symmetric.

(c) Use either the line

```
V=modified_gs(V);  % orthogonalization with Gram-Schmidt
```

or the line

```
[V,unused]=qr(V,0);  % orthogonalization with qr
                     % The unused matrix is required!)
```

to orthogonalize the new set of vectors V.

(d) Use the following code to make sure the columns of D do not flip sign from iteration to iteration.

```
d=sign(diag(Vold'*V));  % find signs of dot product of columns
D=diag(d);              % diagonal matrix with +-1
V=V*D;                  % Transform V
```

(e) Stop the iterations when *all three* of the following conditions are satisfied.

$$k > 1$$
$$||r^{(k)} - r^{(k-1)}|| \le \epsilon ||r^{(k)}||, \text{ and}$$
$$||V^{(k)} - V^{(k-1)}||_{\text{fro}} \le \epsilon ||V^{(k)}||_{\text{fro}}.$$

(f) Add an error message about too many iterations after end of your loop, similar to the ones in power_method1.m and inverse_power.m. You can use MATLAB 's **error** function this time.

(2) Test `power_several` for `A=eigen_test(4)` starting from the vector of zeros with a 1 in the last place, and with a tolerance of 1.0e-8. It should agree with the results in Exercise 17.3, up to roundoff.

(3) Run `power_several` function for the matrix `A=eigen_test(4)` with a tolerance of 1.0e-8 starting from the vectors

```
V = [ 0  0
      0  0
      0  0
      0  0
      0  1
      1  0];
```

What are the eigenvalues and eigenvectors? How many iterations are needed?

(4) Further test your results by checking that the two eigenvectors you found are actually eigenvectors and are orthogonal. (Test that `AV=VR`, is approximately satisfied, where `V` is your matrix of eigenvectors and `R=diag(r)` is the diagonal matrix of eigenvalues.)

(5) Since this `A` is a small (6×6) matrix, you can find all eigenpairs by using `power_several` starting from the (obviously linearly independent) vectors

```
V=[0  0  0  0  0  1
   0  0  0  0  1  0
   0  0  0  1  0  0
   0  0  1  0  0  0
   0  1  0  0  0  0
   1  0  0  0  0  0];
```

What are the six eigenvalues? How many iterations were needed?

17.7   Using shifts

A major limitation of the power method is that it can only find the eigenvalue of largest magnitude. Similarly, the inverse power iteration seems only able to find the eigenvalue of smallest magnitude. Suppose, however, that you want one (or a few) eigenpairs with eigenvalues that are away from the extremes? Instead of working with the matrix A, you can work with the matrix $A+\sigma I$, where the matrix has been shifted by adding σ to every diagonal entry. This shift makes a world of difference.

The inverse power method finds the eigenvalue of smallest magnitude. Denote the smallest eigenvalue of $A + \sigma I$ by μ. If μ is an eigenvalue of $A + \sigma I$ then $A + \sigma I - \mu I = A - (\mu - \sigma)I$ is singular. In other words, $\lambda = \mu - \sigma$ is an eigenvalue of A, and is the eigenvalue of A that is closest to σ.

In general, the idea of shifting "focuses the attention" of the inverse power method on seeking the eigenvalue closest to any particular value you care to name. This also means that if you have an estimated eigenvalue, you can speed up convergence by using this estimate in the shift. Although various strategies for adjusting the shift during the iteration are known, and can result in very rapid convergence, this strategy will not be investigated here, and only constant shifts will be used.

Do not worry here about the details or the expense of shifting and refactoring the matrix. (A real algorithm for a large problem would be very concerned.) The simple-minded method algorithm for the **shifted inverse power method** looks like this:

(i) Starting with any non-zero vector $\mathbf{x}^{(0)}$, and given a fixed shift value σ:
(ii) Solve $(A - \sigma I)\hat{\mathbf{x}} = \mathbf{x}^{(k-1)}$, and
(iii) Normalize the iterate by setting $\mathbf{x}^{(k)} = \hat{\mathbf{x}}/\|\hat{\mathbf{x}}\|$; and
(iv) Adjust the eigenvector as necessary.
(v) Estimate the eigenvalue as

$$r^{(k)} = \sigma + (\mathbf{x}^{(k)} \cdot (A - \sigma I)\mathbf{x}^{(k)})$$
$$= (\mathbf{x}^{(k)} \cdot A\mathbf{x}^{(k)}).$$

(vi) If converged, exit; otherwise return to step 2.

Exercise 17.7.

(1) Starting from your `inverse_power.m`, write an m-file for the shifted inverse power method. Your file should have the following outline.

```
function  [r, x, numSteps] = shifted_inverse ( A, x, shift, tol)
   % [r, x, numSteps] = shifted_inverse ( A, x, shift, tol)
   % comments

   % your name and the date

   << your code >>

end
```

(2) Compare your results from `inverse_power.m` applied to the matrix `A=eigen_test(2)` starting from `[0;0;1]` and using a relative tolerance of 1.0e-8 with results from `shifted_inverse.m` with a shift of 0. The results should be identical and take the same number of steps.
(3) Apply `shifted_inverse` to the same `A=eigen_test(2)` with a shift of 1.0. You should get the same eigenvalue and eigenvector as with a shift of 0. How many iterations did it take this time?
(4) Apply `shifted_inverse.m` to the matrix `eigen_test(2)` with a shift of 4.73, very close to the largest eigenvalue. Start from the vector `[0;0;1]` and use a

relative tolerance of 1.0e-8. You should get a result that agrees with the one you calculated using power_method but it should take very few steps. (In fact, it gets the right eigenvalue on the first step, but convergence detection is not that fast.)

(5) Using your shifted inverse power method code, you can search for the "middle" eigenvalue of matrix eigen_test(2). The power method gives the largest eigenvalue as about 4.73 and the inverse power method gives the smallest as 1.27. Assume that the middle eigenvalue is near 2.5, start with the vector [0;0;1] and use a relative tolerance of 1.0e-8. What is the eigenvalue and how many steps did it take?

(6) Recapping, what are the three eigenvalues and eigenvectors of A? Do they agree with the results of eig?

The value of the shift is at the user's disposal, and there is no good reason to stick with the initial (guessed) shift. If it is not too expensive to re-factor the matrix $A - \sigma I$, then the shift can be reset after each iteration or after a set of n iterations, for n fixed.

One possible choice for σ would be to set it equal to the current eigenvalue estimate on each iteration. The result of this choice is an algorithm that converges at an extremely fast rate, and is the predominant method in practice.

17.8   The QR method

So far, the methods you have seen seem suitable for finding one or a few eigenvalues and eigenvectors at a time. In order to find more eigenvalues, we'd have to do a great deal of special programming. Moreover, these methods are surely going to have trouble if the matrix has repeated eigenvalues, distinct eigenvalues of the same magnitude, or complex eigenvalues. The "QR method" is a method that handles these sorts of problems in a uniform way, and computes all the eigenvalues, but not the eigenvectors, at one time.

It turns out that the QR method is equivalent to the power method starting with a basis of vectors and with Gram–Schmidt orthogonalization applied at each step, as you did in Exercise 17.6. A nice explanation can be found in Prof. Greg Fasshauer's class notes, [Fasshauer (2006)].

The **QR method** can be described in the following way.

(1) Given a matrix $A^{(k-1)}$, construct its QR factorization; and
(2) Set $A^{(k)} = RQ$;
(3) If A^k has converged, stop, otherwise go back to step 1.

The construction of the QR factorization means that $A^{(k-1)} = QR$, and the matrix factors are reversed to construct $A^{(k)}$.

The sequence of matrices $A^{(k)}$ is orthogonally similar to the original matrix A, and hence has the same eigenvalues. This is because $A^{(k-1)} = QR$, and $A^{(k)} = RQ = Q^T QRQ = Q^T A^{(k-1)} Q$.

- If the matrix A is symmetric, the sequence of matrices $A^{(k)}$ converges to a diagonal matrix.
- If the eigenvalue are all real, the lower triangular portions of A converge to zero and diagonals converge to eigenvalues.

In addition, the method can be modified in a way you will not consider here so that it converges to an "almost" upper triangular matrix, where the main subdiagonal will have non-zero entries only when there is a complex conjugate pair of eigenvalues.

Exercise 17.8.

(1) Write an m-file for the QR method for a matrix A. You should use the MATLAB qr function ([Q,R]=qr(A)) to perform the QR factorization, but you can also use gs_factor or h_factor from Chapter 16. Your file should have the following outline.

```
function  [e,numSteps] = qr_method ( A, tol)
  % [e,numSteps] = qr_method ( A, tol)
  % comments

  % your name and the date

  << your code >>

end
```

e is a vector of eigenvalues of A. Since the lower triangle of the matrix $A^{(k)}$ is supposed to converge, terminate the iteration when

$$\|A^{(k)}_{\text{lower triangle}} - A^{(k-1)}_{\text{lower triangle}}\|_{\text{fro}} \leq \epsilon \|A^{(k)}_{\text{lower triangle}}\|_{\text{fro}}$$

where ϵ is the relative tolerance. You can use tril(A) to recover just the lower triangle of a matrix. Include a check so the number of iterations never exceeds some maximum (10000 is a good choice), and use the MATLAB "error" function to stop iterating if it is exceeded.

(2) The eigen_test(2) matrix is 3×3 and you have found its eigenvalues in Exercises 17.3, 17.4, and 17.7. Check that you get the same values using qr_method. Use a relative tolerance of 1.0e-8.

(3) Try out the QR iteration for the eigen_test matrices except number 5. Use a relative tolerance of 1.0e-8. Report the results in the following table.

Matrix	largest eigenvalue	smallest eigenvalue	number of steps
1			
2			
3			
4			
(5)	skip	skip	skip
6			
7			

(4) For matrix 7, what is the difference between your computed largest eigenvalue and the one from the MATLAB `eig` function? For the smallest eigenvalue? How does the size of these errors compare with the tolerance of 1.0e-8?

(5) Matrix 5 has 4 distinct eigenvalues of equal norm, and the QR method does not converge. Temporarily reduce the maximum number of iterations to 50 and change `qr_method.m` to print e at each iteration to see why it fails. What happens to the iterates? As with the case of changing signs, it is possible to "handle" this case in the code so that it converges, but you will not pursue it here. Return the maximum number of iterations to its large value and make sure nothing is printed each iteration before continuing.

(6) Just as with the inverse power method, shifting can help with the QR method. For the matrix `A=eigen_test(5)`, compute the four eigenvalues of `A+(1+i)*eye(4)` and subtract `(1+i)` to get the eigenvalues of A. How many steps does it take, and what are the four eigenvalues of A?

Serious implementations of the QR method do *not* work the way our version works. The code resembles the code for the singular value decomposition you will see in Chapter 18 and the matrices Q and R are never explicitly constructed. The algorithm also uses shifts to try to speed convergence and ceases iteration for already-converged eigenvalues.

In the following section you consider the question of convergence of the basic QR algorithm.

17.9   Convergence of the QR algorithm

Throughout this chapter you have used a very simple convergence criterion, and you saw in the previous exercise that this convergence criterion can result in errors that are too large. In general, convergence is a difficult subject and fraught with special cases and exceptions to rules. To give you a flavor of how convergence might be reliably addressed, consider the case of the QR algorithm applied to a real, symmetric matrix with distinct eigenvalues, no two of which are the same in magnitude. For such matrices, the following theorem can be shown.

Theorem 17.1. *Let A be a real symmetric matrix of order n and let its eigenvalues satisfy*

$$0 < |\lambda_1| < |\lambda_2| < \cdots < |\lambda_n|.$$

Then the iterates $\{A^{(k)}\}$ of the QR method will converge to a diagonal matrix D whose entries are the eigenvalues of A. Furthermore,

$$\|D - A^{(k)}\| \leq C \max_j \left| \frac{\lambda_j}{\lambda_{j+1}} \right| \|D - A^{(k-1)}\|.$$

The theorem points out a method for detecting convergence. Supposing you start out with a matrix A satisfying the above conditions, and you call the matrix at the kth iteration $A^{(k)}$. If $D^{(k)}$ denotes the matrix consisting only of the diagonal entries of $A^{(k)}$, then the Frobenius norm of $A^{(k)} - D^{(k)}$ must go to zero as a power of $\rho = \max |\lambda_j|/|\lambda_{j+1}|$, and if D is the diagonal matrix with the true eigenvalues on its diagonal, then $D^{(k)}$ must converge to D as a geometric series with ratio ρ. In the following algorithm, the diagonal entries of $A^{(k)}$ (these converge to the eigenvalues) will be the only entries considered. The off-diagonal entries converge to zero, but there is no need to make them small so long as the eigenvalues are converged.

The following algorithm includes a convergence estimate as part of the QR algorithm.

(i) Define $A^{(0)} = A$ and $e^{(0)} = \text{diag}(A)$.
(ii) Compute Q and R uniquely so that $A^{(k-1)} = QR$, set $A^{(k)} = RQ$ and $e^{(k)} = \text{diag}(A^{(k)})$. (This is one step of the QR algorithm.)
(iii) Define a vector of absolute values of eigenvalues $\lambda^{(k)} = |e^{(k)}|$, sorted in ascending order. (MATLAB has a function, sort, that sorts a vector in ascending order.) Compute $\rho^{(k)} = \max_j (\lambda_j^{(k)}/\lambda_{j+1}^{(k)})$. Since the $\lambda^{(k)}$ are sorted, $\rho^{(k)} < 1$.
(iv) Compute the eigenvalue convergence estimate

$$\frac{\|e^{(k)} - e^{(k-1)}\|}{(1 - \rho^{(k)})\|e^{(k)}\|}.$$

Return to step (ii) if it exceeds the required tolerance.
Programming hint: Your code should not do a division for this step! If the tolerance value is ϵ, your code should check that

$$\|e^{(k)} - e^{(k-1)}\| < \epsilon(1 - \rho^{(k)})\|e^{(k)}\|.$$

The reason is that if $\rho^{(k)}$ happens to equal 1, there is no division by zero and there is no need to check for that special case.

You may wonder where the denominator $(1 - \rho)$ came from. Consider the geometric series

$$S = 1 + \rho + \rho^2 + \rho^3 + \cdots + \rho^k + \rho^{k+1} + \cdots = \frac{1}{(1 - \rho)}.$$

Truncating this series at the $(k+1)^{\text{th}}$ term and calling the truncated sum S_j, you have

$$S - S_j = \rho^{j+1}(1 + \rho + \rho^2 + \dots) = \frac{\rho^{j+1}}{1 - \rho}.$$

But, $\rho^{j+1} = S_{j+1} - S_j$, so

$$S - S_j = \frac{S_{j+1} - S_j}{1 - \rho}.$$

Thus, the error made in truncating a geometric series is the difference of two partial sums divided by $(1 - \rho)$.

Exercise 17.9.

(1) Copy your `qr_method.m` file to `qr_convergence.m` and change the outline to

```
function [e, numSteps]=qr_convergence(A,tol)
% [e, numSteps]=qr_convergence(A,tol)
% comments

% your name and the date

<< your code >>

end
```

Modify the code to implement the above algorithm.

(2) Test your code on `eigen_test(1)` with a tolerance of 1.e-8. Do the three eigenvalues approximately agree with results from Exercise 17.8? The number of iterations required might be slightly larger than in Exercise 17.8.

(3) Apply your algorithm to `eigen_test(8)`, with a tolerance of 1.0e-8. Include the eigenvalues and the number of iterations in your summary.

(4) Compare your computed eigenvalues with ones computed by `eig`. How close are your eigenvalues? Is the relative error near the tolerance?

(5) If you decrease the tolerance by a factor of ten to 1.0e-9, does the relative error decrease by a factor of ten as well? This test indicates the quality of the error estimate.

(6) Apply `qr_convergence` to `A=eigen_test(7)` with a tolerance of 1.0e-8. How many iterations does it take?

(7) Use `eMat=eig(A)` to compute "exact" eigenvalues for `eigen_test(7)`. The entries of eMat might not be in the same order, so both should be sorted in order to be compared. What is the relative error

```
norm(sort(e)-sort(eMat))/norm(e)
```

where e is the vector of eigenvalues from `qr_convergence`?

(8) How many iterations does `qr_method` take to reach the relative tolerance of 1.0e-8 for the matrix `A=eigen_test(7)`? Which convergence method yields more accurate results? If your paycheck depended on achieving a particular accuracy, which method would you use?

(9) Convergence based on provable results is very reliable if the assumptions are satisfied. Try to use your `qr_convergence` function on the nonsymmetric matrix

```
A = [ .01  1
      -1  .01];
```

It should fail. Explain how you know it failed and what went wrong.

17.10   Roots of polynomials

Have you ever wondered how all the roots of a polynomial can be found? It is fairly easy to find the largest and smallest real roots using, for example, Newton's method, but then what? If you know a root r exactly, of course, you can eliminate it by dividing by $(x - r)$ (called "deflation"), but in the numerical world you never know the root exactly, and deflation with an inexact root can introduce large errors.

Suppose you are given a monic $(a_1 = 1)$ polynomial of degree N

$$p(x) = x^N + a_2 x^{N-1} + a_3 x^{N-2} + \ldots + a_N x + \ldots a_{N+1}. \qquad (17.3)$$

Consider the $N \times N$ matrix with ones on the lower subdiagonal and the negatives of the coefficients of the polynomial in the last column:

$$A = \begin{pmatrix}
0 & 0 & 0 & \cdots & 0 & -a_{N+1} \\
1 & 0 & 0 & \cdots & 0 & -a_N \\
0 & 1 & 0 & \cdots & 0 & -a_{N-1} \\
0 & 0 & 1 & \cdots & 0 & -a_{N-2} \\
 & & & \ddots & \vdots & \vdots \\
0 & 0 & 0 & \cdots & 1 & -a_2
\end{pmatrix}.$$

It is easy to see that $det(A - \lambda I) = (-1)^N p(\lambda)$ so that the eigenvalues of A are the roots of p.

Exercise 17.10.

(1) Write a MATLAB function m-file with outline

```
function r=myroots(a)
  % r=myroots(a)
  % comments

  % your name and the date
```

```
    << your code >>

end
```

that will take a vector of coefficients and find all the roots of the polynomial $p(x)$ defined in (17.3) by constructing the companion matrix and using the MATLAB `eig` function to compute its eigenvalues.

(2) To test `myroots`, choose at least three different vectors $r^{(j)}$ of length 5 or more. Have at least one of the vectors with complex values. Use the MATLAB `poly` function to find the coefficients of the three polynomials whose roots are the values $r^{(j)}$ and then use `myroots` to find the roots of those polynomials. You should find that your computed roots are close to the chosen values in each $r^{(j)}$.

————————————————

Chapter 18

Singular value decomposition

18.1   Introduction

Suppose A is an $m \times n$ matrix. Then there exist two unitary (or orthogonal in the case A is real) matrices: an $m \times m$ matrix U and a $n \times n$ matrix V, and a $m \times n$ "diagonal" matrix S with non-negative numbers on its diagonal sorted by size and zeros elsewhere, so that

$$A = USV^H, \tag{18.1}$$

where the superscript ^{H} denotes the Hermitian (or conjugate transpose) of a matrix, and the diagonal entries of S are $S_{kk} = \sigma_k \geq 0$, for $k = 1 \ldots \min(m, n)$ and all the rest zero. The triple of matrices (U, S, V) is called the "singular value decomposition" (SVD) and the diagonal entries of S are called the "singular values" of A. The columns of U and V are called the left and right "singular vectors" of A respectively. You can get more information from a very nice Wikipedia article [Singular Value Decomposition]. There is an illustrative proof of the existence of the SVD in a simple case beginning in Section 2 (page 5) of [Stewart (1992–04)].

In this chapter, you will see some applications of the SVD, including

- The right singular vectors corresponding to vanishing singular values of A span the nullspace of A.
- The right singular vectors corresponding to positive singular values of A span the domain of A.
- The left singular vectors corresponding to positive singular values of A span the range of A.
- The rank of A is the number of positive singular values of A.
- The singular values characterize the "relative importance" of some basis vectors in the domain and range spaces over others. This fact can be used as a compression algorithm for images.
- The Moore–Penrose "pseudo-inverse" of A is computed from the SVD, making it possible to solve the system $Ax = b$ in the least-squares sense.

Numerical methods for finding the singular value decomposition will also be addressed in this chapter. One "obvious" algorithm involves finding the eigenvalues of $A^H A$, but this is not really practical because of roundoff difficulties caused by squaring the condition number of A. MATLAB includes a function called svd with signature [U,S,V]=svd(A) to compute the singular value decomposition and it will be used here, too. This function uses the Lapack subroutine dgesvd, so if you were to need it in a Fortran or C program, it would be available by linking against the Lapack library, [Anderson *et al.* (1999)].

You may be interested in a six-minute video made by Cleve Moler (the MATLAB founder) in 1976 at what was then known as the Los Alamos Scientific Laboratory. Most of the film is computer animated graphics. Moler recently retrieved the film from storage, digitized it, and uploaded it to YouTube [Moler (1976)]. You may also be interested in a description of the making of the film, and its use as background graphics in the first Star Trek movie, in Moler's blog [Moler (2012)].

18.2   The singular value decomposition

The matrix S in (18.1) is $m \times n$ and is of the form

$$S_{k\ell} = \begin{cases} \sigma_k & \text{for } k = \ell \\ 0 & \text{for } k \neq \ell \end{cases}$$

and $\sigma_1 \geq \sigma_2 \geq \sigma_3 \geq \cdots \geq \sigma_r > \sigma_{r+1} = 0 = \cdots = 0$. Since U and V are unitary (and hence nonsingular), it is easy to see that the number r is the rank of the matrix A and is necessarily no larger than $\min(m,n)$. The "singular values," $\sigma_k, k = 1 \ldots r$ are real and positive and are the eigenvalues of the Hermitian matrix $A^H A$.

If the matrix A is $n \times n$, then the rank of the nullspace of A is $n - r$. The first r columns of U are an orthonormal basis of the range of A, and the last $n - r$ columns of V are an orthonormal basis for the nullspace of A. Of course, in the presence of roundoff some judgment must be made as to what constitutes a zero singular value, but the SVD is the most reliable (if expensive) way to discover the rank and nullity of a matrix. This is especially true when there is a large gap between the "smallest of the significant singular values" and the "largest of the insignificant singular values."

The SVD can also be used to solve a matrix system. Assuming that the matrix is non-singular, all singular values are strictly positive, and the SVD can be used to solve a system.

$$b = Ax$$
$$b = USV^H x$$
$$U^H b = SV^H x \qquad (18.2)$$
$$S^+ U^H b = V^H x$$
$$VS^+ U^H b = x.$$

Where S^+ is the diagonal matrix whose diagonal entries are $1/\sigma_k$ for $\sigma_k > 0$ and zero otherwise. It turns out that Equation (18.2) is too expensive to be a good way to solve nonsingular systems, but is an excellent way to "solve" systems that are singular or almost singular! To this end, the matrix S^+ is the matrix with the same shape as S^T and whose elements are given as

$$S^+_{k\ell} = \begin{cases} 1/\sigma_k & \text{for } k = \ell \text{ and } \sigma_k > 0 \\ 0 & \text{otherwise} \end{cases}.$$

Further, if A is close to singular, a similar definition but with diagonal entries $1/\sigma_k$ for $\sigma_k > \epsilon$ for some ϵ can work very nicely. In these latter cases, the "solution" is the least-squares best fit solution and the matrix $A^+ = VS^+U^H$ is called the Moore–Penrose pseudo-inverse of A.

Exercise 18.1. In this exercise you will use the MATLAB svd function to solve for the best fit linear function of several variables through a set of points. This is an example of "solving" a rectangular system.

Imagine that you have been given many "samples" of related data involving several variables and you wish to find a linear relationship among the variables that approximates the given data in the best-fit sense. This can be written as a rectangular system

$$\begin{aligned} a_1 d_{1,1} + a_2 d_{1,2} + a_3 d_{1,3} + \ldots + d_{1,n+1} &= 0 \\ a_1 d_{2,1} + a_2 d_{2,2} + a_3 d_{2,3} + \ldots + d_{2,n+1} &= 0 \\ a_1 d_{3,1} + a_2 d_{3,2} + a_3 d_{3,3} + \ldots + d_{3,n+1} &= 0 \\ \vdots \quad\quad \vdots \quad\quad \vdots \quad\quad \ldots \quad\quad \vdots \end{aligned} \tag{18.3}$$

where d_{ij} represents the data and a_i the desired coefficients.

The system (18.3) can be written in matrix form. Recall that the values a_j are the unknowns. For a moment, denote the vector $(\mathbf{a})_j = a_j$, the matrix $(D)_{ij} = d_{ij}$ and the vector $(\mathbf{b})_i = -d_{i,n+1}$, all for $i = 1, \ldots, M$ and $j = 1, \ldots, N$, where $M \geq N$. With this notation, (18.3) can be written as the matrix equation $D\mathbf{a} = \mathbf{b}$, where D is (usually) a rectangular matrix. A "least-squares" solution of $D\mathbf{a} = \mathbf{b}$ is a vector $\mathbf{a}$ satisfying

$$\|D\mathbf{a} - \mathbf{b}\|^2 = \min_{\mathbf{x}} \|D\mathbf{x} - \mathbf{b}\|^2,$$

and, since there could, in general, be many such vectors, you require that $\mathbf{a}$ have the smallest norm of all such vectors. It turns out that the vector $\mathbf{a}$ is determined as $D^+\mathbf{b}$.

To see why this is so, suppose that the rank of D is r, so that there are precisely r positive singular values of D. Then $D = USV^T$ and

$$\begin{aligned} \|D - \mathbf{b}\|^2 &= \|USV^T\mathbf{x} - \mathbf{b}\|^2 \\ &= \|SV^T\mathbf{x} - U^T\mathbf{b}\|^2 \\ &= \|S\mathbf{z} - U^T\mathbf{b}\|^2, \quad \text{where } \mathbf{z} = V^T\mathbf{x} \\ &= \sum_{j=1}^{r} (\sigma_j z_j - (U^T\mathbf{b})_j)^2 + \sum_{j=r+1}^{N} ((U^T\mathbf{b})_j)^2 \end{aligned}$$

because S is a matrix with diagonal entries $S_{ii} = \sigma_i$, $i = 1, \ldots, r$, with all other entries being zero. Clearly, the sum is minimized when the first term is zero, $z_j = (U^T\mathbf{b})_j/\sigma)j$. The other components, z_j for $j = r+1, \ldots, N$ are arbitrary, but $\|\mathbf{z}\|$ will be minimized when they are taken to be zero. Thus, the minimum norm solution for the coefficients $\mathbf{a}$ is given by $\mathbf{a} = V\mathbf{z} = VS^+U^T\mathbf{b} = D^+\mathbf{b}$.

As you have seen several times before, you will first generate a data set with known solution and then use **svd** to recover the known solution. Place MATLAB code for the following steps into a script m-file called **exer1.m**. You may find it convenient to download the file **exer1.txt** and build on it.

(1) Generate a data set consisting of twenty "samples" of each of four variables using the following MATLAB code.

```
N=20;
d1=rand(N,1);
d2=rand(N,1);
d3=rand(N,1);
d4=4*d1-3*d2+2*d3-1;
```

It should be obvious that these vectors satisfy the equation

$$4d_1 - 3d_2 + 2d_3 - d_4 = 1,$$

but, if you were solving a real problem, you would not be aware of this equation, and it would be your objective to discover it, *i.e.*, to find the coefficients in the equation.

(2) Introduce small "errors" into the data. The **rand** function produces numbers between 0 and 1.

```
EPSILON=1.e-5;
d1=d1.*(1+EPSILON*rand(N,1));
d2=d2.*(1+EPSILON*rand(N,1));
d3=d3.*(1+EPSILON*rand(N,1));
d4=d4.*(1+EPSILON*rand(N,1));
```

You have now constructed the "data" for the following least-squares calculation.

(3) Imagine the four vectors **d1,d2,d3,d4** as data given to you, and construct the matrix consisting of the four column vectors, **A=[d1,d2,d3,d4]**.

(4) Use the MATLAB **svd** function

```
[U,S,V]=svd(A);
```

What are the four non-zero values of S?

(5) To confirm that you have everything right, compute **norm(A-U*S*V','fro')**. This number should be roundoff-sized.

(6) Construct the matrix S^+ (call it `Splus`) so that

$$S_{k\ell}^+ = \begin{cases} 1/S(k,k) & \text{for } k = \ell \text{ and } S(k,k) > 0 \\ 0 & \text{for } k \neq \ell \text{ or } S(k,k) = 0 \end{cases}.$$

(7) You are seeking the coefficient vector x that

$$x_1 d_1 + x_2 d_2 + x_3 d_3 + x_4 d_4 = 1.$$

Find this vector by setting `b=ones(N,1)` (the coefficients in Equation (18.3) have been moved to the right and are $-d_{k5} = 1$) using Equation (18.2) above. Your solution should be close to the known solution `x=[4;-3;2;-1]` but not exact because of the random perturbations.

Sometimes the data you are given turns out to be deficient because the variables supposed to be independent are actually related. This dependency will cause the coefficient matrix to be singular or nearly singular. When the matrix is singular, the system of equations is actually redundant, and one equation can be eliminated. This yields fewer equations than unknowns and any member of an affine subspace can be legitimately regarded as "the" solution.

Dealing with dependencies of this sort is one of the greatest difficulties in using least-squares fitting methods because it is hard to know which of the equations is the right one to eliminate. The singular value decomposition is the best way to deal with dependencies. In the following exercise you will construct a deficient set of data and see how to use the singular value decomposition to find the solution.

Exercise 18.2.

(1) Copy your m-file `exer1.m` to `exer2.m` and modify it in the following way.

    (a) Replace the line

        `d3=rand(N,1);`

    with the line

        `d3=d1+d2;`

    This makes the data deficient and introduces nonuniqueness into the solution for the coefficients.

    (b) After using `svd` to find `U`, `S`, and `V`, print `S(1,1)`, `S(2,2)`, `S(3,3)`, `S(4,4)`. You should see that `S(4,4)` is substantially smaller than the others. Set `S(4,4)` to zero and construct S^+ according to

$$S_{k\ell}^+ = \begin{cases} 1/S(k,k) & \text{for } k = \ell \text{ and } S(k,k) > 0 \\ 0 & \text{for } k \neq \ell \text{ or } S(k,k) = 0. \end{cases}$$

    Use your S^+ to find the coefficient vector `x` as in (18.2). The solution you found is probably not `x=[4;-3;2;-1]`.

(2) Look at V(:,4). This is the vector associated with the singular value that you set to zero. Since the original system is rank deficient, there must be a non-trivial nullspace. Any solution can be represented as the sum of some particular solution and a vector from the nullspace and the vector V(:,4) spans the nullspace. V(:,4) is associated with the extra relationship d1+d2−d3=0 that was introduced in the first step. What multiple of V(:,4) can be added to your solution to approximately yield [4;-3;2;-1]?

Remark 18.1. The singular value decomposition allows you to discover deficiencies in the data without knowing they are there (or even if there are any) in the first place. This idea can be the basis of a fail-safe method for computing least-squares coefficients.

In the following exercise you will see how the singular value decomposition can be used to "compress" a graphical figure by representing the figure as a matrix and then using the singular value decomposition to find the closest matrix of lower rank to the original. This approach can form the basis of efficient compression methods.

Exercise 18.3.

(1) The file TarantulaNebula.jpg provided with this workbook is a copy of one from a Hubble telescope web site.[1] Credit for this photograph goes to NASA, ESA, E. Sabbi (STScI). It presents a dramatic picture of this extra-galactic formation. It is included among the files for this chapter. Download the provided file, not the one from spacetelescope.org.
(2) MATLAB provides various image processing utilities. In this case, read the image in using the following command.

```
nasacolor=imread('TarantulaNebula.jpg');
```

The variable **nasacolor** will be a $768 \times 1024 \times 3$ matrix of integer values between 0 and 255.
(3) Display the color plot using the command

```
image(nasacolor)
```

(4) The third subscript of the array **nasa** refers to the red, green, and blue color components. To simplify this exercise, turn it into a grayscale with ordinary double precision values 0-255 using the following commands:

```
nasa=sum(nasacolor,3,'double');   %sum up red+green+blue
m=max(max(nasa));                 %find the max value
nasa=nasa*255/m;                  %make this be bright white
```

----
[1]https://cdn.spacetelescope.org/archives/images/wallpaper1/heic1402a.jpg

The result from these commands is that **nasa** is an ordinary 768 × 1024 array of double precision numbers.

Remark 18.2. This is easy but inaccurate way to create a grayscale image from an rgb image. It is good enough for your purpose here.

(5) MATLAB has the notion of a "colormap" that determines how the values in the matrix will be colored. Use the command

```
colormap(gray(256));
```

to set a grayscale colormap. Now you can show the picture with the command

```
image(nasa)
title('Grayscale NASA photo');
```

(6) Use the command [U,S,V]=svd(nasa); to perform the singular value decomposition. Plot the singular values on a semilog scale: semilogy(diag(S)). You should observe that the values drop off very rapidly to less than 2% of the maximum in fewer than 50 values.

(7) Construct the three new matrices

```
nasa100=U(:,1:100)*S(1:100,1:100)*V(:,1:100)';
nasa50=U(:,1:50)*S(1:50,1:50)*V(:,1:50)';
nasa25=U(:,1:25)*S(1:25,1:25)*V(:,1:25)';
```

These matrices are of lower rank than the **nasa** matrix, and can be stored in a more efficient manner. (The **nasa** matrix is 768 × 1024 and requires 786,432 numbers to store it. In contrast, nasa50 requires subsets of U and V that are 768 × 50 and the diagonal of S, for a total of 77,568. This is better than ten to one savings in space.)

(8) Plot the three matrices **nasa100**, **nasa50** and **nasa25** as images, including descriptive titles. You should only very slight differences between the original and the one with 100 singular values, some noticeable differences with 50 singular values while you should see serious degradation of the image in the case of 25 singular values. Please include the 25 singular value case.

In the next sections you will see two different methods for computing the SVD.

18.3 Two numerical algorithms

The easiest algorithm for SVD is to note the relationship between it and the eigenvalue decomposition: singular values are the square roots of the eigenvalues of $A^H A$

or AA^H. Indeed, if $A = USV^H$, then

$$A^H A = (VSU^H)(USV^H) = V(S^2)V^H, \text{ and}$$
$$AA^H = (USV^H)(VSU^H) = U(S^2)U^H.$$

So, if you can solve for eigenvalues and eigenvectors, you can find the SVD.

Unfortunately, this is not a good algorithm because forming the product $A^H A$ roughly squares the condition number, so that the eigenvalue solution is not likely to be accurate. Of course, it will work fine for small matrices with small condition numbers and you can find this algorithm presented in many web pages. Don't believe everything you see on the internet.

A more practical algorithm is a Jacobi algorithm that is given in a 1989 report by James Demmel and Krešimir Veselić, that can be found at [Demmel and Veselić (1989)]. The algorithm is a one-sided Jacobi iterative algorithm that appears as Algorithm 4.1, page 32, of that report. This algorithm amounts to the Jacobi algorithm for finding eigenvalues of a symmetric matrix. (See, for example, [Wilkinson (1965)].)

The algorithm *implicitly* computes the product AA^T and then uses a sequence of "Jacobi rotations" to diagonalize it. A Jacobi rotation is a 2×2 matrix rotation that annihilates the off-diagonal term of a symmetric 2×2 matrix. You can find more in a nice Wikipedia article [Jacobi Rotation].

Given a 2×2 matrix M with

$$M^T M = \begin{bmatrix} \alpha & \gamma \\ \gamma & \beta \end{bmatrix}.$$

It is possible to find a "rotation by angle θ matrix"

$$\Theta = \begin{bmatrix} \cos\theta & -\sin\theta \\ \sin\theta & \cos\theta \end{bmatrix}$$

with the property that $\Theta^T M^T M \Theta = D$ is a diagonal matrix. Θ can be found by multiplying out the matrix product, expressing the off-diagonal matrix component in terms of $t = \tan\theta$, and setting it to zero to arrive at an equation

$$t^2 + 2\zeta t - 1 = 0$$

where $\zeta = (\beta - \alpha)/(2\gamma)$. The quadratic formula gives $t = \tan\theta$, and then $\sin\theta$ and $\cos\theta$ can be recovered. There is no need to recover θ itself.

The following algorithm repeatedly passes down the diagonal of the implicitly-constructed matrix AA^T, choosing pairs of indices $k < j$, constructing the 2×2 submatrix from the intersection of rows and columns, and using Jacobi rotations to annihilate the off-diagonal term. Unfortunately, the Jacobi rotation for the pair $k = 2$, $j = 10$ messes up the rotation from $k = 1$, $j = 10$, so the process must be repeated until convergence. At convergence, the matrix S of the SVD has been implicitly generated, and the right and left singular vectors are recovered by multiplying all the Jacobi rotations together. The following algorithm carries out this process.

Note: For the purpose of this chapter, that matrix A will be assumed to be square. Rectangular matrices introduce too much algorithmic complexity.

Algorithm: Given a convergence criterion ϵ, a *square* matrix A, a matrix U, that starts out as $U = A$ and ends up as a matrix whose columns are left singular vectors, and another matrix V that starts out as the identity matrix and ends up as a matrix whose columns are right singular vectors.

repeat

    for all pairs $k < j$

        (*compute* $\begin{bmatrix} \alpha & \gamma \\ \gamma & \beta \end{bmatrix} \equiv$ *the* (k, j) *submatrix of* $U^T U$)

        $\alpha = \sum_{\ell=1}^{n} U_{\ell k}^2$

        $\beta = \sum_{\ell=1}^{n} U_{\ell j}^2$

        $\gamma = \sum_{\ell=1}^{n} U_{\ell k} U_{\ell j}$

        (*compute the Jacobi rotation that diagonalizes* $\begin{bmatrix} \alpha & \gamma \\ \gamma & \beta \end{bmatrix}$)

        $\zeta = (\beta - \alpha)/(2\gamma)$

        $t = \mathrm{signum}(\zeta)/(|\zeta| + \sqrt{1 + \zeta^2})$

        $c = 1/\sqrt{1 + t^2}$

        $s = ct$

        (*update columns k and j of U*)

        for $\ell = 1$ **to** n

          $T = U_{\ell k}$

          $U_{\ell k} = cT - sU_{\ell j}$

          $U_{\ell j} = sT + cU_{\ell j}$

        endfor

        (*update the matrix V of right singular vectors*)

        for $\ell = 1$ **to** n

          $T = V_{\ell k}$

          $V_{\ell k} = cT - sV_{\ell j}$

          $V_{\ell j} = sT + cV_{\ell j}$

        endfor

    endfor

  until all $|\gamma|/\sqrt{\alpha\beta} \leq \epsilon$

The computed singular values are the norms of the columns of the final U and the computed left singular vectors are the normalized columns of the final U. As mentioned above, the columns of V are the computed right singular vectors.

Exercise 18.4.

(1) Copy the following code skeleton to a function m-file named `jacobi_svd.m`. (You may find it convenient to download the file `jacobi_svd.txt`.) Complete the code so that it implements the above algorithm.

```
function [U,S,V]=jacobi_svd(A)
  % [U,S,V]=jacobi_svd(A)
```

```
% A is original square matrix
% Singular values come back in S (diag matrix)
% orig matrix = U*S*V'
%
% One-sided Jacobi algorithm for SVD
% lawn15: Demmel, Veselic, 1989,
% Algorithm 4.1, page 32

% your name and the date

TOL=1.e-8;
MAX_STEPS=40;

n=size(A,1);
U=A;
V=eye(n);
for steps=1:MAX_STEPS
  converge=0;
  for j=2:n
    for k=1:j-1
      % compute [alpha gamma;gamma beta]=(k,j) submatrix of U'*U
      alpha=???  %might be more than 1 line
      beta=???   %might be more than 1 line
      gamma=???  %might be more than 1 line
      converge=max(converge,abs(gamma)/sqrt(alpha*beta));

      % compute Jacobi rotation that diagonalizes
      %    [alpha gamma;gamma beta]
      if gamma ~= 0
        zeta=(beta-alpha)/(2*gamma);
        t=sign(zeta)/(abs(zeta)+sqrt(1+zeta^2));
      else
        % if gamma=0, then zeta=infinity and t=0
        t=0;
      end
      c=???
      s=???

      % update columns k and j of U
      T=U(:,k);
      U(:,k)=c*T-s*U(:,j);
      U(:,j)=s*T+c*U(:,j);
```

```
        % update matrix V of right singular vectors

        ???

      end
    end
    if converge < TOL
      break;
    end
  end
  if steps >= MAX_STEPS
    error('jacobi_svd failed to converge!');
  end

  % the singular values are the norms of the columns of U
  % the left singular vectors are the normalized columns of U
  for j=1:n
    singvals(j)=norm(U(:,j));
    U(:,j)=U(:,j)/singvals(j);
  end
  S=diag(singvals);
end
```

(2) Apply your version of `jacobi_svd` to the 2×2 matrix `A=U*S*V'` generated from the following three matrices.

```
U=[0.6  0.8
   0.8 -0.6];
V=sqrt(2)/2*[1   1
             1  -1];
S=diag([5 4]);
```

It is easy to see that U and V are orthogonal matrices, so that the matrices U, S and V comprise the SVD of A. You may get a "division by zero" error, but this is harmless because it comes from **gamma** being zero, which will cause **zeta** to be infinite and **t** to be zero anyhow.

Your algorithm should essentially reproduce the matrices U, S and V. You might find that the diagonal entries of S are not in order, so that the U and V are similarly permuted, or you might observe that certain columns of U or V have been multiplied by (-1). Be sure, however, that an *even* number of factors of (-1) have been introduced.

If your code is not correct, you can debug your code in the following way. First, note that there is only a single term, k=1 and j=2 in the double for loop.

(a) Bring up your code in the editor and click on the dash to the left of the last line of code, the one that starts off "V(:,j)=". This will cause a red dot to appear, indicating a "breakpoint." (If you are not using the MATLAB desktop, you can accomplish the same sort of thing using the dbstop command, placed just after the statement V(:,j)=.)

(b) Now, try running the code once again. You will find the code has stopped at the breakpoint.

(c) Take out a piece of paper and calculate the value of alpha. There are only two terms in this sum. Did your code value agree with your hand calculation? If not, double-check *both* calculations and fix the one that is in error.

(d) Do the same for beta and gamma.

(e) Similarly, check the values of s and c.

(f) Press the "Step" button to complete the calculation of V. Now multiply (you can do this at the MATLAB command prompt) U*V'. Do you get A back? It can be shown that the algorithm *always* maintains the condition A=U*V' when it is correctly programmed. If not, you probably have something wrong in the two statements defining V, or, perhaps, your code updating U is in error. Find the error. Do not forget that at this point in the iteration, U and V have not yet converged to their final values!

(3) Use your jacobi_svd to find the SVD, U1, S1 and V1 of the matrix

```
A1 = [ 1  3  2
       5  6  4
       7  8  9];
```

Check that U1 and V1 are orthogonal matrices and that A1=U1*S1*V1' up to roundoff. In addition, print U1, S1 and V1 using format long and visually compare them with the singular values you get from the MATLAB svd function. You should find they agree almost to roundoff, despite the coarse tolerance inside jacobi_svd. You also will note that the values do not come out sorted, as they do from svd. It is a simple matter to sort them, but then you would have to permute the columns of U1 and V1 to match. Again, some columns of U1 and V1 might be multiplied by (-1) and, if so, there must be an *even* number of those sign changes.

You now have an implementation of one SVD algorithm. This algorithm is a good one, with some realistic applications, and is one of the two algorithms available for the SVD in the GNU scientific library, [Galassi *et al.* (2009)]. In the following sections, you will see a different algorithm.

18.4   The "standard" algorithm

The most commonly used SVD algorithm is found in MATLAB and in the Lapack linear algebra library, [Anderson *et al.* (1999)], and is presented in [Demmel and Kahan (1990)].

The standard algorithm is a two-step algorithm. It first reduces the matrix to bidiagonal form and then finds the SVD of the bidiagonal matrix. Reduction to bidiagonal form is accomplished using Householder transformations, a topic you have already seen. Finding the SVD of a bidiagonal matrix is an iterative process that must be carefully performed in order to minimize both numerical errors and the number of iterations required. Many optimization choices are made to speed up each iteration. There is not time to discuss all these details, so only a simplified version of Demmel and Kahan's zero-shift algorithm will be covered here.

In the Jacobi algorithm in the previous section, you saw how the two matrices U and V can be constructed by multiplying the various rotation matrices as the iterations progress. This procedure is the same for the standard algorithm, so, in the interest of simplicity, most of the rest of this chapter will be concerned only with the singular values themselves.

The following algorithm is taken from [Golub and Van Loan (1996)], Algorithm 5.4.2. The first step in the algorithm is to reduce the matrix to bidiagonal form. You have seen in Chapter 16 how to use Householder matrices to reduce a matrix to upper-triangular form. The idea, you recall, is to start with matrices U_0 and V_0 equal to the identity matrix and $B_0 = A$, so that $A = U_0 B_0 V_0^H$. Then,

(i) For each column k of A, you choose a Householder matrix $\overline{H}_k$ in such a way that:

    (a) The matrix $\overline{B}_k = \overline{H}_k B_{k-1}$ has zeros *below* the diagonal in column k; and,

    (b) $U_k = U_{k-1}\overline{H}_k^H$, so that $A = U_k \overline{B}_k V_{k-1}^H$ for each k.

(ii) Apply a similar procedure for *row* k of A, choosing H_k in such a way that:

    (a) The matrix $B_k = \overline{B}_k H_k$ has zeros to the *right* of the *superdiagonal*; and,

    (b) V_k maintains the identity $A = U_k B_k V_k^H$.

Remark 18.3. You cannot reduce the matrix to diagonal form this way because the Householder matrices for the rows would change the diagonal entries originally found for the columns and ruin the original factorization.

Exercise 18.5. Consider the following incomplete MATLAB code, which is very similar to the h_factor function from Exercise 16.6. (The matrices called Q and R there are called U and B here.) (You may find it convenient to download bidiag_reduction.txt.)

```
function [U,B,V]=bidiag_reduction(A)
  % [U B V]=bidiag_reduction(A)
  % Algorithm 5.4.2 in Golub & Van Loan, Matrix Computations
```

```
% Third Edition, Johns Hopkins University Press 1996.
% Finds an upper bidiagonal matrix B so that A=U*B*V'
% with U,V orthogonal.  A is an m x n matrix

[m,n]=size(A);
B=A;
U=eye(m);
V=eye(n);
for k=1:n-1
   % eliminate non-zeros below the diagonal
   % Keep the product U*B unchanged
   H=householder(B(:,k),k);
   B=H*B;
   U=U*H';

   % eliminate non-zeros to the right of the
   % superdiagonal by working with the transpose
   % Keep the product B*V' unchanged

   << more code >>

   end
end
```

(1) Copy the above code to a file `bidiag_reduction.m`. Complete the statements with the question marks in them. Remember that `A=U*B*V'` remains true at all steps in the algorithm.

(2) Recover your version of `householder.m` from Exercise 16.4.

(3) Test your code on the matrix `A=pascal(5)`. Be sure that the condition that `A=U*B*V'` is satisfied, that the matrix B is bidiagonal, and that the matrices U and V are orthogonal. Describe how you checked these facts and include the results.

(4) Test your code on a randomly-generated matrix

   `A=rand(100,100);`

   Be sure that the condition that `A=U*B*V'` is satisfied, that the matrix B is bidiagonal, and that U and V are orthogonal matrices. Describe how you checked these facts and include the results of your tests. **Hint:** You can use the MATLAB `diag` function to extract any diagonal from a matrix or to reconstruct a matrix from a diagonal. Use "`help diag`" to find out how. Alternatively, you can use the `tril` and `triu` functions.

An important piece of Demmel and Kahan's algorithm is a very efficient way of generating a 2×2 "Givens" rotation matrix that annihilates the second component of a vector. The algorithm is presented on page 13 of their paper and is called "rot." It is important to have an efficient algorithm because this step is the central step in computing the SVD. A 10% (for example) reduction in time for **rot** translates into a 10% reduction in time for the SVD.

Algorithm (Demmel, Kahan) [c,s,r]=rot(f,g)

This algorithm computes the cosine, c, and sine, s, of a rotation angle that satisfies the following condition.

$$\begin{bmatrix} c & s \\ -s & c \end{bmatrix} \begin{bmatrix} f \\ g \end{bmatrix} = \begin{bmatrix} r \\ 0 \end{bmatrix}$$

if $f = 0$ **then**
 $c = 0$, $s = 1$, and $r = g$
else if $|f| > |g|$ **then**
 $t = g/f$, $t_1 = \sqrt{1 + t^2}$
 $c = 1/t_1$, $s = tc$, and $r = ft_1$
else
 $t = f/g$, $t_1 = \sqrt{1 + t^2}$
 $s = 1/t_1$, $c = ts$, and $r = gt_1$
endif

Remark 18.4. The two different expressions for c, s and r are mathematically equivalent. The choice of which to do is based on minimizing roundoff errors.

Exercise 18.6.

(1) Based on the above algorithm, write a MATLAB function m-file named givens_rot.m with the outline

```
function [c,s,r]=givens_rot(f,g)
% [c s r]=givens_rot(f,g)
% more comments

% your name and the date

<< your code >>

end
```

(2) Test it on the vector [f;g] with f=1 and g=0. This vector already has 0 in its second component so you should get that c=1, s=0, and r=1.
(3) Test it on the vector [0;2]. What are the values of c, s and r? Perform the matrix product

```
[c s;-s c]*[f;g]
```

and check that the resulting vector is `[r;0]`.

(4) Test it on the vector `[1;2]`. What are the values of c, s and r? Perform the matrix product

```
[c s;-s c]*[f;g]
```

and check that the resulting vector is `[r;0]`.

(5) Test it on the vector `[-3;2]`. What are the values of c, s and r? Perform the matrix product

```
[c s;-s c]*[f;g]
```

and check that the resulting vector is `[r;0]`.

The standard algorithm is based on repeated "sweeps" through the bidiagonal matrix. Suppose B is a bidiagonal matrix, so that it looks like the following.

$$\begin{matrix} d_1 & e_1 & 0 & 0 & \ldots & 0 \\ 0 & d_2 & e_2 & 0 & \ldots & 0 \\ 0 & 0 & d_3 & e_3 & \ldots & 0 \\ \vdots & \vdots & \ddots & \ddots & \ddots & \vdots \\ 0 & 0 & \ldots & 0 & d_{n-1} & e_{n-1} \\ 0 & 0 & \ldots & 0 & 0 & d_n \end{matrix}$$

In its simplest form, a sweep begins at the top left of the matrix and runs down the rows to the bottom. For row i, the sweep first annihilates the e_i entry by multiplying on the right by a rotation matrix. This action introduces a non-zero value $A_{i+1,i}$ immediately *below* the diagonal. This new non-zero is then annihilated by multiplying on the left by a rotation matrix but this action introduces a non-zero value $B_{i,i+3}$, outside the upper diagonal, but the next row down. Conveniently, this new non-zero is annihilated by the same rotation matrix that annihilates e_{i+1}. (The proof uses the fact that e_i was zero as a consequence of the first rotation, see the paper.) The algorithm continues until the final step on the bottom row that introduces no non-zeros. This process has been termed "chasing the bulge" and will be illustrated in the following exercise.

The following MATLAB code implements the above algorithm. This code will be used in the following exercise to illustrate the algorithm.

```
function B=msweep(B)
  % B=msweep(B)
  % Demmel & Kahan zero-shift QR downward sweep
  % B starts as a bidiagonal matrix and is returned as
  % a bidiagonal matrix
```

```
n=size(B,1);
for k=1:n-1
  [c s r]=givens_rot(B(k,k),B(k,k+1));

  % Construct matrix Q and multiply on the right by Q'.
  % Q annihilates both B(k-1,k+1) and B(k,k+1)
  % but makes B(k+1,k) non-zero.
  Q=eye(n);
  Q(k:k+1,k:k+1)=[c s;-s c];
  B=B*Q';

  [c s r]=givens_rot(B(k,k),B(k+1,k));

  % Construct matrix Q and multiply on the left by Q.
  % Q annihilates B(k+1,k) but makes B(k,k+1) and
  % B(k,k+2) non-zero.
  Q=eye(n);
  Q(k:k+1,k:k+1)=[c s;-s c];
  B=Q*B;

  end
end
```

In this algorithm, there are two orthogonal (rotation) matrices, Q, employed. To see what their action is, consider the piece of B consisting of rows and columns $i - 1$, i, $i + 1$, and $i + 2$. Multiplying on the right by the transpose of the first rotation matrix has the following consequence.

$$\begin{bmatrix} \alpha & * & \beta & 0 \\ 0 & * & * & 0 \\ 0 & 0 & * & * \\ 0 & 0 & 0 & \gamma \end{bmatrix} \begin{bmatrix} 1 & 0 & 0 & 0 \\ 0 & c & s & 0 \\ 0 & -s & c & 0 \\ 0 & 0 & 0 & 1 \end{bmatrix} = \begin{bmatrix} \alpha & \beta & 0 & 0 \\ 0 & * & 0 & 0 \\ 0 & * & * & * \\ 0 & 0 & 0 & \gamma \end{bmatrix}$$

where asterisks indicate (possible) non-zeros and Greek letters indicate values that do not change. The fact that two entries are annihilated in the third column is not obvious and is proved in the paper.

The second matrix Q multiplies on the left and has the following consequence.

$$\begin{bmatrix} 1 & 0 & 0 & 0 \\ 0 & c & s & 0 \\ 0 & -s & c & 0 \\ 0 & 0 & 0 & 1 \end{bmatrix} \begin{bmatrix} \alpha & \beta & 0 & 0 \\ 0 & * & 0 & 0 \\ 0 & * & * & * \\ 0 & 0 & 0 & \gamma \end{bmatrix} = \begin{bmatrix} \alpha & \beta & 0 & 0 \\ 0 & * & * & * \\ 0 & 0 & * & * \\ 0 & 0 & 0 & \gamma \end{bmatrix}.$$

The important consequence of these two rotations is that the row with three non-zeros in it (the "bulge") has moved from row $i - 1$ to row i, (in this example, from row 1 to row 2) while all other rows still have two non-zeros. This action is illustrated graphically in the following exercise.

Exercise 18.7.

(1) Download the file named msweep.txt and change its name to msweep.m. (The "m" stands for "matrix.")
(2) The following lines of code will set roundoff-sized matrix entries in B to zero and use MATLAB 's spy routine to display all non-zeros. The **pause** statement makes the function stop and wait until a key is pressed. This will give you time to look at the plot. (You may find it convenient to download the file exer7.txt.)

```
% set almost-zero entries to true zero
% display matrix and wait for a keypress
B(find(abs(B)<1.e-13))=0;
spy(B)
disp('Plot completed. Strike a key to continue.')
pause
```

Insert *two* copies of this code into msweep.m, one after each of the statements that start off "B=."
(3) Apply your msweep function to the matrix

```
B=[1 11  0  0  0  0  0  0  0  0
   0  2 12  0  0  0  0  0  0  0
   0  0  3 13  0  0  0  0  0  0
   0  0  0  4 14  0  0  0  0  0
   0  0  0  0  5 15  0  0  0  0
   0  0  0  0  0  6 16  0  0  0
   0  0  0  0  0  0  7 17  0  0
   0  0  0  0  0  0  0  8 18  0
   0  0  0  0  0  0  0  0  9 19
   0  0  0  0  0  0  0  0  0 10];
```

The sequence of plots shows the "bulge" (extra non-zero entry) migrates from the top of the matrix down the rows until it disappears off the end. Please include any one of the plots with your work. If B=msweep(B); what is B(10,10)?

———————————————

Demmel and Kahan present a streamlined form of this algorithm in their paper. Their algorithm is presented below. It is not only faster but is more accurate because there are no subtractions to introduce cancellation and roundoff inaccuracy.

The largest difference in speed comes from the representation of the bidiagonal matrix B as a pair of *vectors*, d and e, the diagonal and superdiagonal entries, respectively. In addition, the intermediate product, BQ', is not explicitly formed.

Algorithm (Demmel, Kahan)

This algorithm begins and ends with the two vectors d and e, representing the diagonal and superdiagonal of a bidiagonal matrix. The vector d has length n.

$c_{\text{old}} = 1$

$c = 1$

for $k = 1$ to $n - 1$

    $[c, s, r] = \text{givens\_rot}(c\, d_k, e_k)$

    if $(k \neq 1)$ **then** $e_{k-1} = r\, s_{\text{old}}$ **endif**

    $[c_{\text{old}}, s_{\text{old}}, d_k] = \text{givens\_rot}(c_{\text{old}}r, d_{k+1}s)$

end for

$h = c\, d_n$

$e_{n-1} = h\, s_{\text{old}}$

$d_n = h\, c_{\text{old}}$

Exercise 18.8.

(1) Based on the above algorithm, write a function m-file named `vsweep.m` ("v" for "vector") with outline

```
function [d,e]=vsweep(d,e)
  % [d e]=vsweep(d,e)
  % comments

  % your name and the date

  << your code >>

end
```

(2) Use the same matrix as in the previous exercise to test `vsweep`. Compare the results of `vsweep` and `msweep` to be sure they are the same. You will probably want to remove the extra plotting statements from `msweep`. (You may find it convenient to download the file `exer8.txt`.)

```
B=[1 11  0  0  0  0  0  0  0  0
   0  2 12  0  0  0  0  0  0  0
   0  0  3 13  0  0  0  0  0  0
   0  0  0  4 14  0  0  0  0  0
   0  0  0  0  5 15  0  0  0  0
   0  0  0  0  0  6 16  0  0  0
   0  0  0  0  0  0  7 17  0  0
   0  0  0  0  0  0  0  8 18  0
   0  0  0  0  0  0  0  0  9 19
   0  0  0  0  0  0  0  0  0 10];
d=diag(B);
e=diag(B,1);
```

(3) Use the following code to compare results of `msweep` and `vsweep` and time them as well. Are the results the same? What are the times? You should find that `vsweep` is substantially faster.

```
n=400;
d=2*rand(n,1)-1; %random numbers between -1 and 1
e=2*rand(n-1,1)-1;
B=diag(d)+diag(e,1);
tic;B=msweep(B);mtime=toc
tic;[d e]=vsweep(d,e);vtime=toc
norm(d-diag(B))
norm(e-diag(B,1))
```

It turns out that repeated sweeps tend to reduce the size of the superdiagonal entries. You can find a discussion of convergence in Demmel and Kahan, and the references therein, but the basic idea is to watch the values of e and when they become small, set them to zero. Demmel and Kahan prove that this does not perturb the singular values much. In the following exercise, you will see how the superdiagonal size decreases.

Exercise 18.9.

(1) Consider the same matrix you have been using:

```
d=[1;2;3;4;5;6;7;8;9;10];
e=[11;12;13;14;15;16;17;18;19];
```

Use the following code to plot the absolute values $|e_j| : j = 1, \ldots, 9$ for each of 15 repetitions of the algorithm.

```
for k=1:15
  [d,e]=vsweep(d,e);
  plot(abs(e),'*-')
  hold on
  disp('Strike a key to continue.')
  pause
end
hold off
```

starting from the values of `e` and `d` above. (You may find it convenient to download the file `exer9.txt`.) You should observe that some entries converge quite rapidly to zero, and they all eventually decrease.

(2) Plot $\|e\|$ *versus* iteration number for the first 40 repetitions, starting from the same values as previously:

```
d=[1;2;3;4;5;6;7;8;9;10];
e=[11;12;13;14;15;16;17;18;19];
```

(3) Plot $\|e\|$ *versus* iteration number *on a semilog plot* (semilogy) for the first 100 repetitions, starting from the above values. You should see that the convergence eventually becomes linear. You can estimate this convergence rate graphically by plotting the function $y = cr^n$ (where c can be taken to be 1.0 in this case), r can be found by trial and error, and n denotes iteration number. When this line and the line of $\|e\|$ appear parallel, you have found an estimate for r. To no more than two significant digits, what is the value of r you found?

You should have observed that one entry converges very rapidly to zero, and that overall convergence to zero is asymptotically linear. Frequently, one of the two ends converges more rapidly. Further, a judiciously chosen shift can make convergence even more rapid. These issues will not be investigated here, but you can be sure that the software appearing, for example, in Lapack (MATLAB uses Lapack) for the SVD takes advantage of these and other acceleration methods.

In the following exercise, the superdiagonal is examined during the iteration. As the values become small, they are set to zero, *and the iteration continues with a smaller matrix*. When all the superdiagonals are zero, the iteration is complete.

Exercise 18.10.

(1) Copy (or download) the following code to a function m-file named bd_svd.m ("bidiagonal svd").

```
function [d,iterations]=bd_svd(d,e)
  % [d,iterations]=bd_svd(d,e)
  % more comments

  % your name and the date

  TOL=100*eps;
  n=length(d);
  maxit=500*n^2;

  % The following convergence criterion is discussed by
  % Demmel and Kahan.
  lambda(n)=abs(d(n));
  for j=n-1:-1:1
    lambda(j)=abs(d(j))*lambda(j+1)/(lambda(j+1)+abs(e(j)));
  end
  mu(1)=abs(d(1));
```

```
for j=1:n-1
  mu(j+1)=abs(d(j+1))*mu(j)/(mu(j)+abs(e(j)));
end
sigmaLower=min(min(lambda),min(mu));
thresh=max(TOL*sigmaLower,maxit*realmin);


iUpper=n-1;
iLower=1;
for iterations=1:maxit
  % reduce problem size when some zeros are
  % on the superdiagonal

  % Don't iterate further if value e values are small
  % how many small values are near the bottom right?
  j=iUpper;
  for i=iUpper:-1:1
    if abs(e(i))>thresh
      j=i;
      break;
    end
  end
  iUpper=j;
  % how many small values are near the top left?
  j=iUpper;
  for i=iLower:iUpper
    if abs(e(i))>thresh
      j=i;
      break;
    end
  end
  iLower=j;

  if (iUpper==iLower & abs(e(iUpper))<=thresh) | ...
        (iUpper<iLower)
    % all done, sort singular values
    d=sort(abs(d),1,'descend');  %change to descending sort
    return
  end

  % do a sweep
  [d(iLower:iUpper+1),e(iLower:iUpper)]= ...
```

```
            vsweep(d(iLower:iUpper+1),e(iLower:iUpper));
    end
    error('bd_svd: too many iterations!')
end
```

(2) Consider the same matrix you have used before

```
d=[1;2;3;4;5;6;7;8;9;10];
e=[11;12;13;14;15;16;17;18;19];
B=diag(d)+diag(e,1);
```

What are the singular values that `bd_svd` produces? How many iterations did `bd_svd` need? What is the largest of the differences between the singular values `bd_svd` found and those that the MATLAB function `svd` found for the matrix B?

(3) Generate a random matrix

```
d=rand(30,1);
e=rand(29,1);
B=diag(d)+diag(e,1);
```

Use `bd_svd` to compute its singular values. How many iterations does it take? What is the largest of the differences between the singular values `bd_svd` found and those that the MATLAB function `svd` found for the matrix B?

Remark 18.5. In the above algorithm, you can see that the singular values are forced to be positive by taking absolute values. This is legal because if a negative singular value arises then multiplying both it and the corresponding column of U by negative one does not change the unitary nature of U and leaves the singular value positive.

You saw in the previous exercise that the number of iterations can get large. It turns out that the number of iterations is dramatically reduced by the proper choice of shift, and strategies such as sometimes choosing to run sweeps up from the bottom right to the top left instead of always down as you did, depending on the matrix elements. Nonetheless, the presentation here should give you the flavor of the algorithm used.

In the following exercise, you will put your two functions, `bidiag_reduction` and `bd_svd` together into a function `mysvd` to find the singular values of a full matrix.

Exercise 18.11.

(1) Write a function m-file named `mysvd.m` with the outline

```
function [d,iterations]=mysvd(A)
  % [d,iterations]=mysvd(A)
  % more comments

  % your name and the date

  << your coe >>

end
```

This function should:

(a) Use `bidiag_reduction` to reduce the matrix A to a bidiagonal matrix, B;
(b) Extract d and e, the diagonal and superdiagonal from B; and,
(c) Use `bd_svd` to compute the singular values of A.

(2) Given the matrix `A=magic(10)`, use `mysvd` to find its singular values. How many iterations are required? How do the singular values compare with the singular values of A from the MATLAB function `svd`?

(3) The singular values can be used to indicate the rank of a matrix. What is the rank of A?

––––––––––––––––––––

Chapter 19

Iterative methods

19.1 Introduction

The methods for solving matrix equations that you have seen already are called "direct" methods because they involve an algorithm that is guaranteed to terminate in a predictable number of steps. There are alternative methods, called "iterative" because they involve a repetitive algorithm that terminates when some specified tolerance is met, as with the eigenpair and singular value methods in earlier chapters. In this chapter, you will focus on the conjugate gradient method applied to matrices arising from elliptic partial differential equations. The conjugate gradient method is a typical example of so-called "Krylov subspace" methods that form the backbone of modern computational solution methods for large-scale computing. More theoretical discussion can be found in, for example, [Quarteroni *et al.* (2007)] in Chapter 4, [Trefethen and Bau (1997)] in Part VI, [Barrett *et al.* (1994)] in Chapters 2 and 3, and [Layton and Sussman (2020)].

The conjugate gradient method can also be regarded as a direct method because *in exact arithmetic* it terminates after N steps for an $N \times N$ matrix. In computer arithmetic, however, the termination property is lost and is irrelevant in many cases because the iterates often converge in far fewer than N steps.

Iterative methods are most often applied to matrices that are "sparse" in the sense that their entries are mostly zeros. If most of the entries are zero, the matrices can be stored more efficiently than in the usual form by omitting the zeros, and it is possible to take advantage of this efficient storage using iterative methods but not so easily with direct methods. Further, since partial differential equations often give rise to large sparse matrices, iterative methods are often used to solve systems arising from PDEs.

This chapter focusses on the conjugate gradient method (regarded as an iterative method), and you will deploy it using matrices stored in their usual form and also in a compact form. You will see examples of the rapid convergence of the method and you will solve a matrix system (stored in compact form) that is so large that you could not even construct and store the matrix in its usual form in central memory, let alone invert it.

In this chapter, Exercises 19.1 through 19.5 are a sequence devoted to the conjugate gradient method using usual matrix storage. Exercises 19.6 through 19.9 consider the more compact storage by diagonals, Exercise 19.10 introduces sparse storage in MATLAB with an example, and Exercise 19.11 discusses preconditioning with incomplete Cholesky factors.

19.2   Poisson equation matrices

In two dimensions, the Poisson equation can be written as

$$\frac{\partial^2 u}{\partial x^2} + \frac{\partial^2 u}{\partial y^2} = \rho \tag{19.1}$$

where u is the unknown function and ρ is a given function. Boundary conditions are necessary, and you will be considering Dirichlet boundary conditions, *i.e.*, $u = 0$ at all boundary points.

Suppose that the unit square $[0,1] \times [0,1]$ is broken into $(N+1)^2$ smaller, nonoverlapping squares, each of side $h = 1/(N+1)$. Each of these small squares will be called "elements" and their corner points will be called "nodes." The combination of all elements making up the unit square will be called a "mesh." An example of a mesh is shown in Figure 19.1.

The nodes in Figure 19.1 have coordinates (x, y) given by

$$x = jh \quad \text{for } j = 0, 1, \dots, 11$$
$$y = kh \quad \text{for } k = 0, 1, \dots, 11. \tag{19.2}$$

You will be generating the matrix associated with the Poisson equation in this chapter, using one of the simplest discretizations of this equation, based on the finite difference expression for a second derivative

$$\frac{d^2\phi}{d\xi^2} = \frac{\phi(\xi + \Delta\xi) - 2\phi(\xi) + \phi(\xi - \Delta\xi)}{\Delta\xi^2} + O(\Delta\xi^2). \tag{19.3}$$

Using (19.3) twice, once for $\xi = x$ and once for $\xi = y$, in (19.1) at a mesh node (x, y) yields the expression

$$\frac{u(x+h,y) + u(x,y+h) + u(x-h,y) + u(x,y-h) - 4u(x,y)}{h^2} = \rho(x,y).$$

Denoting the point (x, y) as the "center" point, the point $(x+h, y)$ as the "right" point, $(x - h, y)$ as the "left" point, $(x, y + h)$ as the "above" point, and $(x, y - h)$ as the "below" point yields the expression

$$\frac{1}{h^2} u_{\text{right}} + \frac{1}{h^2} u_{\text{left}} + \frac{1}{h^2} u_{\text{above}} + \frac{1}{h^2} u_{\text{below}} - 4\frac{1}{h^2} u_{\text{center}} = \rho_{\text{center}}. \tag{19.4}$$

Some authors denote these five points as "center", "north", "east", "south", and "west," referring to the compass directions.

The matrix equation can be generated from (19.4) by numbering the nodes in some way and forming a vector from all the nodal values of u. Then, (19.4)

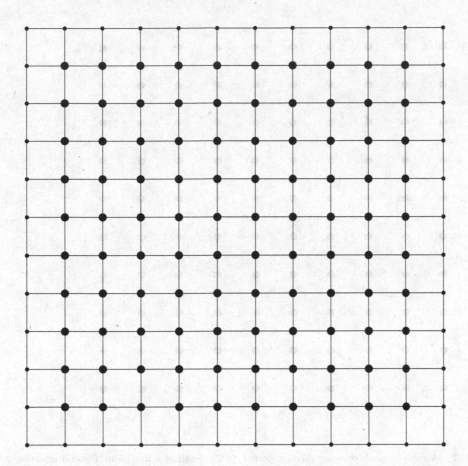

Fig. 19.1 A sample mesh, with interior nodes indicated by larger dots and boundary nodes by smaller ones.

represents one row of a matrix equation $P\mathbf{u} = \rho$. It is immediate that there are at most five non-zero terms in each row of the matrix P, no matter what the size of N. The diagonal term is $-4/h^2$, the other four, if present, are each equal to $1/h^2$, and all remaining terms are zero. You will be constructing such a matrix using MATLAB code.

The equation in the form (19.4) yields the "stencil" of the differential matrix, and is sometimes illustrated graphically as in Figure 19.2 below. Other stencils are possible and are associated with other difference schemes.

In order to construct matrices arising from the Poisson differential equation, it is necessary to give each node at which a solution is desired a number. You will be considering only the case of Dirichlet boundary conditions ($u = 0$ at all boundary points), so you are only interested in interior nodes illustrated in Figure 19.1. These

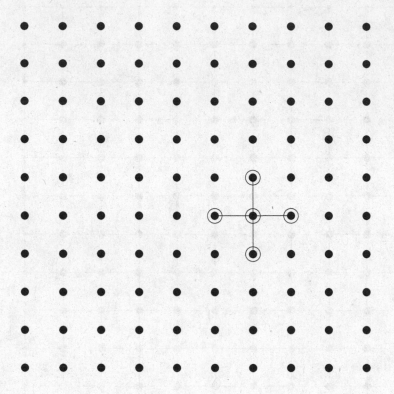

Fig. 19.2   A sample mesh showing interior points and indicating a five-point Poisson equation stencil.

nodes will be counted consecutively, starting with 1 on the lower left and increasing up and to the right to 100 at the upper right, as illustrated in Figure 19.3. This is just one possible way of numbering nodes, and many other ways are often used. Changing the numbering scrambles the rows and columns of the resulting matrix.

Exercise 19.1.

(1) Write a script m-file, named **meshplot.m** that, for N=10, generates two vectors, x(1:N^2) and y(1:N^2) so that the points with coordinates (x(n),y(n)) for n=1,...,N^2 are the interior node points illustrated in Figures 19.1, 19.2 and 19.3 and defined in (19.2). The following outline employs two loops on j and k rather than a single loop on n. The double loop approach leads to somewhat simpler code. Your plot should look similar to Figure 19.2, but without the stencil. You may find it convenient to download the file **meshplot.txt**.

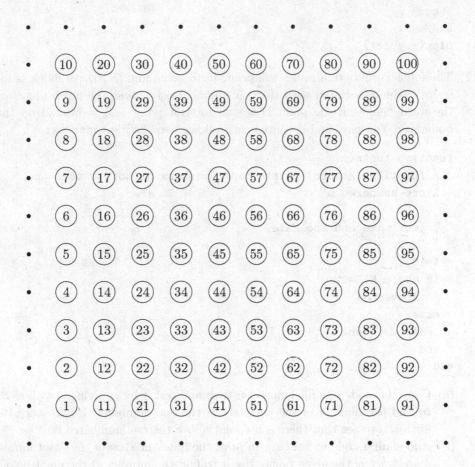

Fig. 19.3 Node numbering.

```
N=10;
h=1/(N+1);
n=0;
% initialize x and y as column vectors
x=zeros(N^2,1);
y=zeros(N^2,1);
for j=1:N
  for k=1:N
    % xNode and yNode should be between 0 and 1
    xNode=???
    yNode=???
    n=n+1;
    x(n)=xNode;
```

```
    y(n)=yNode;
  end
end
plot(x,y,'o');
```

(2) The following function m-file will print, for n given and N=10, the index (subscript) value associated with the point immediately *above* the given point, or
the word "none" if the point is not an interior point and is instead on the
boundary. You may find it convenient to download the file tests.txt.

```
function tests(n)
  % function tests(n) prints index numbers of points near the
  % one numbered n

  % your name and the date.

  N=10;
  if mod(n,N)~=0
    nAbove=n+1
  else
    nAbove='none'
  end
end
```

(a) Copy this code to a file named tests.m and test it by applying it with n=75
to find the number of the point above the one numbered 75. Test it also
for n=80 to see that there is no point above the one numbered 80.

(b) Add similar code to tests.m to print the index nBelow of the point immediately *below* the given point. Use it to find the number of the point below
the one numbered 75, and also apply it to a point that has no interior point
below it.

(c) Add similar code to tests.m to print the index nLeft of the point immediately *to the left of* the given point. Use it to find the number of the point
to the left of the one numbered 75, and also apply it to a point that has no
interior point to its left.

(d) Add similar code to tests.m to print the index nRight of the point immediately *to the right of* the given point. Use it to find the number of the
point to the right of the one numbered 75, and also apply it to a point that
has no interior point to its right.

You are now ready to construct a matrix A derived from the Poisson equation
with Dirichlet boundary conditions on a uniform mesh. Here, you are considering

only one of the infinitely many possibilities for such a matrix. The matrix P defined above turns out to be *negative* definite, so the matrix A you will use in the following exercises is its negative, $A = -P$.

Exercise 19.2.

(1) Recall that there are no more than five non-zero entries per row in the matrix A and that they are the values $4/h^2$, and $-1/h^2$ repeated no more than four times. (Recall that A is the negative of the matrix P above.) Use the following skeleton, and your work from the previous exercise, to write a function m-file to generate the matrix A. You may find it convenient to download the file poissonmatrix.txt.

```
function A=poissonmatrix(N)
  % A=poissonmatrix(N)
  % more comments

  % your name and the date

  h=1/(N+1);
  A=zeros(N^2,N^2);
  for n=1:N^2
    % center point value
    A(n,n)    =4/h^2;

    % "above" point value
    if mod(n,N)~=0
      nAbove=n+1;
      A(n,nAbove)=-1/h^2;
    end

    % "below" point value
    if ???
      nBelow= ???;
      A(n,nBelow)=-1/h^2;
    end

    % "left" point value
    if ???
      nLeft= ???;
      A(n,nLeft)=-1/h^2;
    end
```

```
      % "right" point value
      if ???
        nRight= ???;
        A(n,nRight)=-1/h^2;
      end
  end
```

In the following parts of the exercise, take N=10 and A=poissonmatrix(N).

(2) It may not be obvious from your construction, but the matrix A is symmetric. Consider A(ell,m) for ell=n and m=nBelow, and also for ell=nBelow and m=n (n is "above" nBelow). Think about it. Check that norm(A-A','fro') is essentially zero.

Debugging hint: If it is not zero, follow these instructions to fix your poissonmatrix function.

  (a) Set nonSymm=A-A', and choose one set of indices ell,m for which nonSymm is not zero. You can use the variable browser (double-click on nonSymm in the "variables" pane) or the max function in the following way:

```
      [rowvals]=max(nonSymm);
      [colval,m]=max(rowvals);
      [maxval,ell]=max(nonSymm(:,m));
```

  (b) Look at A(ell,m) and A(m,ell). The most likely case is that one of these is zero and the other is not. Use your tests.m code to check which of them *should* be non-zero, and then look to see why the other is not.

  (c) It is not likely that A(ell,m) and A(m,ell) are both non-zero but disagree because all non-zero, non-diagonal entries are equal to -1/h^2. Check these values.

(3) The Gershgorin circle theorem (see, for example, [Quarteroni *et al.* (2007)] or the interesting Wikipedia article [Gershgorin Circle Theorem]) says that the eigenvalues of a matrix A are contained in the union of the disks in the complex plane centered at the points $z_n = A_{n,n}$, and having radii $r_n = \sum_{k \neq n} |A_{n,k}|$. By this theorem, your matrix A is non-negative definite. A theorem based on irreducible diagonal dominance [Varga (1962)], Theorem 1.8, shows that A is positive definite. For a test vector v=rand(100,1), compute v'*A*v. This quantity should be positive. If it is not, fix your poissonmatrix function. A more satisfying verification would include several repetitions of this test with different random vectors.

(4) Positive definite matrices must have positive determinants. Check that det(A) is positive. If it is not, fix your poissonmatrix function.

(5) As above in Exercise 14.9 a simple calculation shows that the functions are eigenfunctions of the Dirichlet problem for Poisson's equation on the unit square for each choice of k and ℓ, with eigenvalues $\lambda_{k,\ell} = -(k^2 + \ell^2)\pi^2$. It turns out that, for each value of k and ell, the vector E=sin(k*pi*x).*sin(ell*pi*y)

is an eigenvector of the matrix A, where x and y are the vectors from Exercise 19.1. The eigenvalue is L=2*(2-cos(k*pi*h)-cos(ell*pi*h))/h^2, where h=y(2)-y(1). Choose k=1 and ell=2. What is L and, for aid in grading your work, E(10)? (E(10) should be negative and less than 1 in absolute value.) Show by computation that L*E=A*E, up to roundoff. If this is not true for your matrix A, go back and fix poissonmatrix.

(6) It can be shown that A is an "M-matrix" (see, [Quarteroni *et al.* (2007)], [Varga (1962)], [M-matrix]) because the off-diagonal entries are all non-positive and the diagonal entry meets or exceeds the negative of the sum of the off-diagonal entries. It turns out that M-matrices have inverses *all of whose entries are strictly positive.* Use the inv function in MATLAB to find the inverse of A and show that its smallest entry (and hence each entry) is positive. (It might be small, but it must be positive.)

(7) Finally, you can check your construction against MATLAB itself. The MATLAB function gallery('poisson',N) returns a matrix that is h^2*A. Compute norm(gallery('poisson',N)/h^2-A,'fro') to see if it is essentially zero. (**Note:** The gallery function returns the Poisson matrix in a special compressed form that is different from the usual form and also not the same as the one discussed later.) If this test fails, fix poissonmatrix.

You now have a first good matrix to use for the study of the conjugate gradient iteration in the following section. In the following exercise you will generate another.

Exercise 19.3.

(1) Copy your poissonmatrix.m to another one called anothermatrix.m (or use "Save As" on the "File" menu). Modify the comments to describe this new function, and change some components in the following way.

```
A(n,n)      = 8*n+2*N+2;
A(n,nAbove)=-2*n-1;
A(n,nBelow)=-2*n+1;
A(n,nLeft) =-2*n+N;
A(n,nRight)=-2*n-N;
```

This matrix has the same structure as that from poissonmatrix but the values are different.

(2) This matrix is also symmetric. Check that, for N=12, norm(A-A','fro') is essentially zero. **Debug hint:** If this norm is *not* essentially zero for N=12, but *is* essentially zero for N=10, then you have the number 10 coded in your function somewhere. Try searching for "10". (If you have that error in anothermatrix.m you probably have the same error in poissonmatrix.m.)

(3) This matrix is positive definite, just as the Poisson matrix is. For N=12 and one test vector v=rand(N^2,1); check that v'*A*v is positive.

(4) Plot the structure of the matrices and see they are the same. First, examine the structure of poissonmatrix with the commands

```
spy(poissonmatrix(12),'b*');  % blue spots
hold on
```

Next, see that anothermatrix has non-zeros everywhere poissonmatrix does by overlaying its plot on top of the previous.

```
spy(anothermatrix(12),'y*');  % yellow spots cover all blue
```

Finally, confirm that anothermatrix has no new non-zeros by overlaying the plot of poissonmatrix.

```
spy(poissonmatrix(12),'r*');  % red spots cover all yellow
```

(5) To help the grader confirm that your code is correct, what is the result of the following calculation?

```
N=15;
A=anothermatrix(N);
x=((1:N^2).^2)';
sum(A*x)
```

19.3   The conjugate gradient algorithm

The conjugate gradient method is an important iterative method and one of the earliest examples of methods based on Krylov subspaces. Given a matrix A and a vector x, the space spanned by the set $\{x, Ax, A^2x, \ldots\}$ is called a Krylov (sub)space.

One very important point to remember about the conjugate gradient method (and other Krylov subspace methods) is that *only the matrix-vector product* is required. One can write and debug the method using ordinary matrix arithmetic and then apply it using other storage methods, such as the compact matrix storage used below.

A second very important point to remember is that the conjugate gradient algorithm *requires the matrix to be positive definite*. You will note that the algorithm below contains a step with $p^{(k)} \cdot Ap^{(k)}$ in the denominator. This quantity is guaranteed to be non-zero only if A is positive (or negative) definite. If A is negative definite, the equation can be multiplied by (-1) as you have done with the Poisson equation, so nothing is lost by assuming A is positive definite.

The conjugate gradient algorithm can be described in the following way. Starting with an initial guess vector $\mathbf{x}^{(0)}$,

$$\mathbf{r}^{(0)} = \mathbf{b} - A\mathbf{x}^{(0)}$$

For $k = 1, 2, \ldots, m$

$\quad \rho_{k-1} = \mathbf{r}^{(k-1)} \cdot \mathbf{r}^{(k-1)}$

$\quad$ if ρ_{k-1} is zero, stop: the solution has been found.

$\quad$ if k is 1

$\quad\quad \mathbf{p}^{(1)} = \mathbf{r}^{(0)}$

$\quad$ else

$\quad\quad \beta_{k-1} = \rho_{k-1}/\rho_{k-2}$

$\quad\quad \mathbf{p}^{(k)} = \mathbf{r}^{(k-1)} + \beta_{k-1}\mathbf{p}^{(k-1)}$

$\quad$ end

$\quad \mathbf{q}^{(k)} = A\mathbf{p}^{(k)}$

$\quad \gamma_k = \mathbf{p}^{(k)} \cdot \mathbf{q}^{(k)}$

$\quad$ if $\gamma_k \leq 0$, stop: A is not positive definite

$\quad \alpha_k = \rho_{k-1}/\gamma_k$

$\quad \mathbf{x}^{(k)} = \mathbf{x}^{(k-1)} + \alpha_k\mathbf{p}^{(k)}$

$\quad \mathbf{r}^{(k)} = \mathbf{r}^{(k-1)} - \alpha_k\mathbf{q}^{(k)}$

end

In this description, the vectors $\mathbf{r}^{(k)}$ are the "residuals." There are many alternative ways of expressing the conjugate gradient algorithm. This expression is equivalent to most of the others. (Two of the "if" tests guard against dividing by zero.)

In the following exercise you will first write a function to perform the conjugate gradient algorithm, You will test it without using hand calculations.

Exercise 19.4.

(1) Write a function m-file named `cgm.m` with signature

```
function x=cgm(A,b,x,m)
  % x=cgm(A,b,x,m)
  % more comments

  % your name and the date
```

that performs m iterations of the conjugate gradient algorithm for the problem A*y=b, with starting vector x. Do not create vectors for the variables α_k, β_k, and ρ_k, instead use simple MATLAB scalars `alpha` for α_k, `beta` for β_k, `rhoKm1` for ρ_{k-1}, and `rhoKm2` for ρ_{k-2}. Similarly, denote the vectors $\mathbf{r}^{(k)}$, $\mathbf{p}^{(k)}$, $\mathbf{q}^{(k)}$, and $\mathbf{x}^{(k)}$ by the MATLAB variables r, p, q, and x, respectively. For example, part of your loop might look like

```
rhoKm1=0;
for k=1:m
  rhoKm2=rhoKm1;
  rhoKm1=dot(r,r);
```

```
    ???
    r=r-alpha*q;
end
```

(2) Use N=5, b=(1:N^2)', A=poissonmatrix(N), xExact=A\b. Use a perturbed value near the exact solution x=xExact+sqrt(eps)*rand(N^2,1) as initial guess, and take ten steps. Call the result y. You should get that norm(y-xExact) is much smaller than norm(x-xExact). (If you cannot converge when you start near the exact solution, you will *never* converge!)

> **Note 1:** If your result is not small, look at your code. ρ_{k-1} should be very small, and γ_k should be larger than ρ_{k-1}, and α_k should also be very small. The resulting $\|\mathbf{r}^{(k)}\|$ should be very small.
>
> **Note 2:** The choice sqrt(eps) is a good choice for a small but nontrivial number.

(3) If $M \geq N^2$ the set of vectors $\{x, Ax, A^2x, \ldots, A^{M-1}x, A^M x\}$ *must* be linearly dependent because there are more vectors than the dimension of the space. Consequently, the conjugate gradient algorithm terminates after M=N^2 steps on an $N \times N$ matrix. (Finite termination requires that roundoff errors not be significant, so it will only hold for small values of N.) As a second test, choose N=5, xExact=rand(N^2,1) and A= poissonmatrix(N). Set b=A*xExact and then use cgm, starting from the zero vector, to solve the system in N^2 steps. If you call your solution x, show that norm(x-xExact)/norm(xExact) is essentially zero.

(4) You know that your routine solves a system of size M=N^2 in M steps. Choose N=31;xExact=(1:N^2)';, the matrix A=poissonmatrix(N), and b=A*xExact. How close would the solution x calculated using cgm on A be if you started from the zero vector and took only one hundred steps? (Use the relative 2-norm.)

The point of the last part of the previous exercise is that there is no need to iterate to the theoretical limit. Conjugate gradients serves quite well as an *iterative* method, especially for large systems. Regarding it as an iterative method, however, requires some way of deciding when to stop before doing M iterations on an $M \times M$ matrix. Assuming that $\|\mathbf{b}\| \neq 0$, a reasonable stopping criterion is when the relative residual error $\|\mathbf{b} - A\mathbf{x}^{(k)}\|/\|\mathbf{b}\|$ is small. (If $\|\mathbf{b}\| = 0$, then $\mathbf{b}$ is trivial, and so is the solution.) Without an estimate of the condition number of A, this criterion does not bound the solution error, but it will do for your purpose. A simple induction argument will show that $\mathbf{r}^{(k)} = \mathbf{b} - A\mathbf{x}^{(k)}$, so it is not necessary to perform a matrix-vector multiplication to determine whether or not to stop the iteration. Conveniently, the quantity ρ_{k-1} that is computed at the beginning of the loop is the square of the norm of $\mathbf{r}^{(k-1)}$, so the iteration can be terminated *without any extra calculation*.

In the following exercise, you will modify your `cgm.m` file by eliminating the parameter `m` in favor of a specified tolerance on the relative residual.

Exercise 19.5.

(1) Make a copy of your `cgm.m` function and call it `cg.m`. Change its name on the signature line, but leave the parameter `m` for the moment. Modify it by computing $\|b\|^2$ and calling the value `normBsquare`, before the loop starts. Then, *temporarily* add a statement following the computation of `rhoKm1` to compute the quantity `relativeResidual=sqrt(rhoKm1/normBsquare)`. Follow it with the commands `semilogy(k,relativeResidual,'*');hold on;` so that you can watch the progression of the relative residual.

(2) As in the last part of the previous exercise, choose `N=31;xExact=(1:N^2)'`, the matrix `A=poissonmatrix(N)`, and `b=A*xExact`. Run it for 200 iterations starting from `zeros(N^2,1)`. You will have to use the command `hold off` after the function is done. You should observe uneven but rapid convergence to a small value long before 200 (many, many fewer than N^2 with $N = 31^2$) iterations. Estimate from the plot how many iterations it would take to meet a convergence criterion of 10^{-12}.

(3) Change the signature of your `cg` function to

```
[x,k]=cg(A,b,x,tolerance)
```

so that `m` is no longer input. Replace the value `m` in the `for` statement with the number of rows in `A` because you know convergence should be reached at that point. Compute `targetValue=tolerance^2*normBsquare`. Replace the `semilogy` plot with the exit commands

```
if rhoKm1 < targetValue
   return
end
```

There is no longer any need for `relativeResidual` or the plot.

(4) As in the last part of the previous exercise, choose `N=31;xExact=(1:N^2)'`, the matrix `A=poissonmatrix(N)`, and `b=A*xExact`. Start from `zeros(N^2,1)`.

    (a) Run your changed `cg` function with a tolerance of 1.0e-10. How many iterations does it take?

    (b) What is the true error (`norm(x-xExact)/norm(xExact)`)?

    (c) Run it with a tolerance of 1.0e-12. How many iterations does it take this time?

    (d) What is the true relative error?

(5) To see how another matrix might behave, choose `N=31, xExact=(1:N^2)'`, the matrix `A=anothermatrix(N)`, and `b=A*xExact`. How many iterations does it take to reach a tolerance of 1.e-10? What is the true relative error?

As you can see, the stopping criterion you have programmed is OK for some matrices, but not all that reliable. Better stopping criteria are available, but there is no generally accepted "best" one.

Leaving the discussion of stopping criteria to consider alternative storage strategies, in the following section you will turn to an example of an efficient storage schemes for some sparse matrices.

19.4   Compact storage

A matrix A generated by either `poissonmatrix` or `anothermatrix` has only five non-zero entries in each row. When numbered as above, it has exactly five non-zero diagonals: the main diagonal, the super- and sub-diagonals, and the two diagonals that are N columns away from the main diagonal. All other entries are zero. It is prudent to store A in a much more compact form by storing only the diagonals and none of the other zeros. Further, A is symmetric, so it is only necessary to store *three* of the five diagonals. While not all matrices have such a nice and regular structure as these, it is very common to encounter matrices that are sparse (mostly zero entries). In the exercises below, you will be using a very simple compact storage, based on diagonals, but you will get a flavor of how more general compact storage methods can be used.

Remark 19.1.

- Since an $M \times 3$ matrix is rectangular, there will still be some extraneous zeros in this "by diagonals" storage strategy.
- Since `poissonmatrix` contains only two different numbers, it can be stored in an even more efficient manner, but you will not be taking advantage of this strategy here.

Since there are only three non-zero diagonals, you will see how to use a $M \times 3$ rectangular matrix to store a matrix that would be $M \times M$ if ordinary storage were used. Further, since the matrices from `poissonmatrix` and `anothermatrix` are always $N^2 \times N^2$, you will only consider the case that $M = N^2$. The "main diagonal" is the one consisting of entries $A_{k,k}$, the "superdiagonal" is the one consisting of entries $A_{k,k+1}$ and non-zero diagonal farthest from the main diagonal consists of entries $A_{k,k+N}$, where $N^2 = M$. For the purpose of this discussion, call the latter diagonal the "far diagonal."

Conveniently, MATLAB provides a function to extract these diagonals from an ordinary square matrix and to construct an ordinary square matrix from diagonals. It is called `diag` and it works in the following way. Assume that A is a square matrix with entries $A_{k,\ell}$. Then

$$\mathbf{v} = \mathtt{diag(A)} \quad \text{then } v_k = A_{k,k} \text{ for } k = 1, 2, \ldots$$
$$\mathbf{w} = \mathtt{diag(A,n)} \quad \text{then } w_k = A_{k,k+n} \text{ for } k = 1, 2, \ldots$$

where n can be negative, so `diag` returns a column vector whose elements are diagonal entries of a matrix. The same `diag` function also performs the reverse operation. Given a *vector* **v** with entries v_k, then `A=diag(v,n)` returns a *matrix* **A** consisting of all zeros except that $A_{k,k+n} = v_k$, for $k = 1, 2, \ldots,$ `length(v)`. You must be careful to have the lengths of the diagonals and vectors correct.

In the following exercise, you will rewrite `poissonmatrix` and `anothermatrix` to use storage by diagonals.

Exercise 19.6.

(1) Copy your `anothermatrix.m` file to one called `anotherdiags.m` (or use "Save as"). Change the name on the signature line and change the comments to indicate the resulting matrix will be stored by diagonals.

The superdiagonal corresponds to `nAbove` and the far diagonal corresponds to `nRight`. Modify the code so that the matrix **A** has three columns, the first column is the main diagonal, the second column is the superdiagonal, and the third column is the far diagonal. The superdiagonal will have one fewer entry than the main diagonal, so leave a zero as its last entry. The far diagonal will have N fewer entries than the main diagonals, so leave zeros in the last N positions.

(2) Your function should pass the following test. You may find it convenient to download the file `exer6.txt`.

```
N=10;
A=anothermatrix(N);
Adiags=anotherdiags(N);
size(Adiags,2)-3                        % should be zero
norm(Adiags(:,1)-diag(A))               % should be zero
norm(Adiags((1:N^2-1),2)-diag(A,1))     % should be zero
norm(Adiags((1:N^2-N),3)-diag(A,N))     % should be zero
```

If not all four numbers are zero, fix your code. If you are debugging, it is easier to use N=3.

(3) Repeat the steps to write `poissondiags.m` by starting from `poissonmatrix.m`. Be sure to test your work.

If you look at the conjugate gradient algorithm, you will see that the algorithm only made use of the matrix is to multiply the matrix by a vector. If you plan to store a matrix by diagonals, you need to write a special "multiplication by diagonals" function.

Exercise 19.7.

(1) Start off a function m-file named `multdiags.m` in the following way: (You may find it convenient to download the file `multdiags.txt`.)

```
function y=multdiags(A,x)
% y=multdiags(A,x)
% more comments

% your name and the date

M=size(A,1);
N=round(sqrt(M));
if M~=N^2
  error('multdiags: matrix size is not a squared integer.')
end
if size(A,2) ~=3
  error('multdiags: matrix does not have 3 columns.')
end

% the diagonal product
y=A(:,1).*x;
% the superdiagonal product
for k=1:M-1
  y(k)=y(k)+A(k,2)*x(k+1);
end
% the subdiagonal product
???
% the far diagonal product
???
% the far subdiagonal product
???
```

Fill in the three groups of lines indicated by `???`.
Hint: To see the logic, think of the product `B*x`, where B is all zeros except for one of the diagonals.

(2) Choose `N=3`, `x=ones(9,1)`, and compare the results using `multdiags` to multiply `anotherdiags` by x with the results of an ordinary multiplication by the matrix from `anothermatrix`. The results should be the same. If they are not, follow these debugging steps.
Debugging hints:

- All the components of x are `=1` in this case, so the only places your code might be wrong in this test are in the subscripts of A.

- Look at the rows one at a time. Row k of the full matrix product for B=anothermatrix(3) can be written as

```
x=ones(9,1);
B(k,:)*x
```

- For k=1 the product is

```
B(1,1)*x(1)+B(1,2)*x(2)+B(1,4)*x(4)
```

Check that you have this correct.
- For k=2 the product is

```
B(2,1)*x(1)+B(2,2)*x(2)+B(2,3)*x(3)+B(2,5)*x(5)
```

(This product has only the subdiagonal term in it.) Check that you have this correct and continue.

(3) Choose N=3, x=(10:18)', and compare the results using `multdiags` to multiply `anotherdiags` by x with the results of an ordinary multiplication by the matrix from `anothermatrix`. The results should be the same.

(4) Choose N=12, x=rand(N^2,1), and compare the results using `multdiags` to multiply `anotherdiags` by x with the results of an ordinary multiplication by the matrix from `anothermatrix`. The results should be the same.

Remark 19.2. It is generally true that componentwise operations are faster than loops. The `multdiags` function can be made to execute faster by replacing the loops with componentwise operations. This sounds complicated, but there is an easy trick. Consider the superdiagonal code:

```
% the superdiagonal product
for k=1:M-1
  y(k)=y(k)+A(k,2)*x(k+1);
end
```

Look carefully at the following code. It is equivalent, very similar in appearance and will execute more quickly.

```
k=1:M-1;  % k is a vector!
y(k)=y(k)+A(k,2).*x(k+1);
```

Then manually replacing the vector k and eliminating it generates the fastest code:

```
y(1:M-1)=y(1:M-1)+A(1:M-1,2).*x(2:M);
```

You are not required to make these changes here, but you might be interested in trying. Students who are using octave rather than MATLAB are encouraged to speed up the code following this suggestion.

19.5   Conjugate gradient by diagonals

Now is the time to put conjugate gradients together with storage by diagonals. This will allow you to handle much larger matrices than using ordinary square storage.

Exercise 19.8.

(1) Copy your `cg.m` file to `cgdiags.m`. Modify each product of A times a vector to use `multdiags`. There are only two products: one before the `for` loop and one inside it. You do not need to change any other lines.

(2) Solve the same problem twice, once using **anothermatrix** and **cg** and again using **anotherdiags** and **cgdiags**. In each case, use N=3, b=ones(N^2,1), tolerance=1e-10, and start from zeros(N^2,1). You should get exactly the same number of iterations and exactly the same answer in the two cases.

(3) Solve a larger problem twice, with N=10 and b=rand(N^2,1). Again, you should get the same answer in the same number of iterations.

(4) Construct a much larger problem to which you know the solution using the following sample code

```
N=500;
xExact=rand(N^2,1);
Adiags=poissondiags(N);
b=multdiags(Adiags,xExact);
tic;[y,n]=cgdiags(Adiags,b,zeros(N^2,1),1.e-6);toc
```

How many iterations did it take? How much time did it take?
Debugging hint: If you get a memory error message, "Out of memory. Type HELP MEMORY for your options," you most likely are attempting to construct the full Poisson matrix using normal storage, an impossible task (see the next exercise). Examine your **poissondiags.m** code carefully with this in mind.

(5) You can compute the relative norm of the error because you have both the computed and exact solutions. What is it? How does it compare with the tolerance? Recall that the tolerance is the relative *residual* norm and may result in the error being either larger or smaller than the tolerance.

In the previous exercise, you saw that the conjugate gradient works as well using storage by diagonals as using normal storage. You also saw that you could solve a matrix of size $2.5 \cdot 10^5$, stored by diagonals. This matrix is actually quite large, and you would not be able to even construct it, let alone solve it by a direct method, using ordinary storage.

Exercise 19.9.

(1) Roughly, how many entries are in the `Adiags` from the previous exercise? If one number takes eight bytes, how many bytes of central memory does `Adiags` take up?

(2) Roughly, what would the total size of `poissonmatrix(N)` (for N=500) be? If one number takes eight bytes of central memory, how many bytes of central memory would it take to store that matrix? While the cost of RAM memory for PCs varies greatly, use a price of about US$2 for 1GB of RAM memory (2^{30} or roughly 10^9 bytes). How much would it cost to purchase enough memory to store that matrix? A home computer might cost $500, does your estimate for memory cost alone exceed this amount?

(3) In Exercise 15.2, you found that the time it takes to solve a matrix equation using ordinary Gaußian elimination is proportional to M^3, where M is the size of the matrix. If it takes about $T = 1$ minute to solve a system of size $M = 10^4$. Find C so that $T = CM^3$ holds.

(4) How long would it take to use Gaußian elimination to solve a matrix equation using `poissonmatrix(N)` (N=500)? Please convert your result from seconds to days. How does this compare with the time it actually took to solve such a matrix equation in the previous exercise?

Exercise 19.10. MATLAB has a built-in compact storage method for general sparse matrices. It is called "sparse storage," and it is a version of "compressed column storage." It is not difficult to see how it works by looking at the following example. As you saw above, the `gallery('poisson',N)` constructs a matrix that is `h^2` times the matrix from `poissonmatrix`.

(1) Use the following command to print out a sparse matrix

```
A=gallery('toeppen',8,1,2,3,4,5)
```

and you should be able to see how it is stored. From what you just printed what do you think `A(2,1)`, `A(3,2)` and `A(5,1)` are?

(2) What are all the non-zero entries in column 4? What are all the non-zero entries in row 4? Please report both indices and values in your answer.

(3) Consider the full matrix

```
B=[ 1 0 0 2
    0 0 3 0
    0 4 0 0
    5 0 6 0];
```

If B were stored using the MATLAB sparse storage method, what would be the result of printing B (as you printed A).

19.6 ICCG

The conjugate gradient method has many advantages, but it can be slow to converge, a disadvantage it shares with other Krylov-based iterative methods. Since Krylov subspace methods are perhaps the most widely-used iterative methods, it is not surprising that there is a large effort to devise ways to make them converge faster. The largest class of such convergence-enhancing methods is called "preconditioning" and is sometimes explained as solving the system $M^{-1}Ax = M^{-1}b$ instead of $Ax = b$, where M is a suitably chosen, easily inverted matrix. More precisely, it solves $\left(M^{-1/2}AM^{-1/2}\right)\left(M^{1/2}x\right) = M^{-1/2}b$. This more complicated formula is important because even if M and A are symmetric and positive definite, $M^{-1}A$ might not be, but conjugate gradients requires symmetry and positive definiteness.

The preconditioned conjugate gradient algorithm can be described in the following way. Starting with an initial guess vector $\mathbf{x}^{(0)}$,

$\mathbf{r}^{(0)} = \mathbf{b} - A\mathbf{x}^{(0)}$
For $k = 1, 2, \ldots, m$
    Solve the system $M\mathbf{z}^{k-1} = r^{k-1}$.
    $\rho_{k-1} = \mathbf{r}^{(k-1)} \cdot \mathbf{z}^{(k-1)}$
    if ρ_{k-1} is zero, stop: the solution has been found.
    if k is 1
        $\mathbf{p}^{(1)} = \mathbf{z}^{(0)}$
    else
        $\beta_{k-1} = \rho_{k-1}/\rho_{k-2}$
        $\mathbf{p}^{(k)} = \mathbf{z}^{(k-1)} + \beta_{k-1}\mathbf{p}^{(k-1)}$
    end
    $\mathbf{q}^{(k)} = A\mathbf{p}^{(k)}$
    $\gamma_k = \mathbf{p}^{(k)} \cdot \mathbf{q}^{(k)}$
    if $\gamma_k \leq 0$, stop: A is not positive definite
    $\alpha_k = \rho_{k-1}/\gamma_k$
    $\mathbf{x}^{(k)} = \mathbf{x}^{(k-1)} + \alpha_k\mathbf{p}^{(k)}$
    $\mathbf{r}^{(k)} = \mathbf{r}^{(k-1)} - \alpha_k\mathbf{q}^{(k)}$
end

In this form, the residuals $\mathbf{r}^k$ can be shown to be M^{-1}-orthogonal, in contrast with the residuals being orthogonal in the usual sense in the earlier, un-preconditioned, conjugate gradient algorithm.

One of the earliest effective preconditioners discovered was the so-called "incomplete Cholesky" preconditioner. Recall that a positive definite symmetric matrix A can always be factored as $A = U^T U$, where U is an upper-triangular matrix with positive diagonal entries. This is the Cholesky factorization. Although there are several different preconditioners with the same name, the one that is easiest to describe and use here is simply to take

$$\tilde{U}_{mn} = \begin{cases} U_{mn} & \text{if } A_{mn} \neq 0 \\ 0 & \text{otherwise} \end{cases}.$$

For sparse matrices A, this results in a upper triangular factor $\tilde{U}$ with the *same sparsity pattern* as A. If you were using matrix storage by diagonals as in the exercises above, you could store $\tilde{U}$ using the same sort of three column matrix that is used for the matrix A itself. Of course, $A \neq \tilde{U}^T\tilde{U}$. Choosing $M = \tilde{U}^T\tilde{U}$ in the preconditioned conjugate gradient results in the "incomplete Cholesky conjugate gradient" method (ICCG).

For large, sparse matrices A it is generally impossible to construct the true Cholesky factor U, and, if you can construct it, it is usually better to simply solve the system using it rather than some iterative method. On the other hand, there are ways of producing matrices $\tilde{U}$ that have the same sparsity pattern as A and perform almost as well as $\tilde{U}$. For your purpose, you will be using $\tilde{U}$ and seeing how it improves convergence of the conjugate gradient method.

Exercise 19.11.

(1) Modify your `cg.m` m-file to `precg.m` with signature

```
function [x,k]=precg(U,A,b,x,tolerance)
% [x,k]=precg(U,A,b,x,tolerance)
% more comments

% your name and the date
```

and where U is assumed to be an upper-triangular matrix, not necessarily the Cholesky factor of A, and M=U'*U. (Use the MATLAB backslash operator twice, once for U' and once for U. Be sure to use parentheses so that MATLAB does not misinterpret your formulæ.) Change the convergence criterion so that `dot(r,r) < targetValue` is consistent with `cg`. Test it with U being the identity matrix, with A=poissonmatrix(10), the exact solution xExact=ones(100,1), b=A*xExact, and starting from x=zeros(100,1) with tolerance=1.e-10. You should get exactly the same values for x and k as produced by `cg`.

(2) Look carefully at the algorithm above for the preconditioned conjugate gradient. If $\mathbf{x}^0 = \mathbf{0}$ and if $M = A$, in symbols (not numbers), what are the values of $\mathbf{r}^0$, ρ_0, γ_1, and α_1? Your calculation should prove that $k = 2$ at convergence.

(3) The MATLAB `chol(A)` function returns the upper triangular Cholesky factor. Using the following values

```
N=10;
xExact=ones(N^2,1); % exact solution
A=poissonmatrix(N);
b=A*xExact;          % right side vector
```

```
tolerance=1.e-10;
U=chol(A);              % Cholesky factor of A
x=zeros(N^2,1);
```

Use `precg` to solve this problem. Check that the value of k is 2 at convergence, and that the solution is correct. If not, this is precisely the case $M = A$ that you did symbolically above, so you can check where your code goes wrong.

(4) The incomplete Cholesky factor, $\widetilde{U}$, can be easily constructed in MATLAB with the following code.

```
U=chol(A);
U(find(A==0))=0;
```

(Or, equivalently, `U(A==0)=0`.) Consider the system $Ax = b$ starting from $x^0 = 0$ for A given by `anothermatrix(50)` and b given as all ones and with a tolerance of 1.e-10. Find the exact solution as `xExact=A\b`. Compare the numbers of iterations and true errors between the conjugate gradient and ICCG methods. You should observe a significant reduction in the number of iterations without loss of accuracy.

Note: If you measured the times for `precg` and `cg` in the exercise, you might observe that there is little or no improvement in running time for `precg`, despite the reduction in number of iterations. If you try it with larger values of N (for example, N=75), you will see more of an advantage, and it will increase with increasing N. Alternatively, if you implemented it using storage by diagonals as above, you would find the advantage shows up for smaller values of N.

Instructions to Access the Supplementary Templates
To access the supplementary templates for this book, please follow the instructions below:

1. Register an account/login at https://www.worldscientific.com.

2. Go to: https://www.worldscientific.com/r/12340-supp to activate the access.

3. Access the templates from:

https://www.worldscientific.com/worldscibooks/10.1142/12340#t=suppl.

For subsequent access, simply log in with the same login details in order to access. For enquiries, please email: sales@wspc.com.sg.

Bibliography

Anderson, E., Bai, Z., Bischof, C., Blackford, S., Demmel, J., Dongarra, J., Du Croz, J., Greenbaum, A., Hammarling, S., McKenney, A., and Sorensen, D. (1999). *LAPACK Users' Guide*, 3rd edn. (Society for Industrial and Applied Mathematics, Philadelphia, PA), ISBN 0-89871-447-8 (paperback), http://www.netlib.org/lapack/.

ANSYS (2020). https://www.ansys.com/products/platform/ansys-meshing.

Atkinson, K. (1978). *An Introduction to Numerical Analysis* (Wiley, New York).

Barrett, R. *et al.* (1994). *Templates for the Solution of Linear Systems: Building Blocks for Iterative Methods* (Society for Industrial and Applied Mathematics, Philadelphia, PA), ISBN 0-89871-328-5.

Brent, R. P. (1970). *An Algorithm with Guaranteed Convergence for Finding a Zero of a Function*, chap. 4 (Prentice-Hall, Englewood Cliffs, NJ), ISBN ISBN 0-13-022335-2, pp. 61–80.

Brent's Method (2019). http://en.wikipedia.org/wiki/Brent%27s_method.

Chandra, P. and Weisstein, E. W. (2020). Fibonacci number, https://mathworld. wolfram.com/FibonacciNumber.html.

Chebyshev (2020). http://en.wikipedia.org/wiki/Chebyshev_polynomials.

Cholesky decomposition (2020). http://en.wikipedia.org/wiki/Cholesky_decomposition.

Coons, S. A. (1967). Surfaces for computer-aided design of space forms, Report MAC-TR-41, Project MAC, MIT.

Davidenko (originator), D. F. (2011a). Continuation method (to a parametrized family, for non-linear operators), http://www.encyclopediaofmath.org/index.php?title=Continuation_method_(to_a_parametrized_family,_for_non-linear_operators)&oldid=13519.

Davidenko (originator), D. F. (2011b). Method of variation of the parameter, http://www.encyclopediaofmath.org/index.php?title=Parameter,_method_of_variation_of_the&oldid=16671.

Demmel, J. and Kahan, K. (1990). Accurate singular values of bidiagonal matrices, *SIAM J. Sci. Stat. Comput.* **11**, pp. 873–912, http://www.netlib.org/lapack/lawnspdf/lawn03.pdf.

Demmel, J. and Veselić, K. (1989). Jacobi's method is more accurate than qr, Report, Netlib, http://www.netlib.org/lapack/lawnspdf/lawn15.pdf.

Eaton, J. W. *et al.* (2019). Gnu octave, https://www.gnu.org/software/octave/.

Eigenvalues and eigenvectors (2020). https://en.wikipedia.org/wiki/Eigenvalues_and_eigenvectors.

Fasshauer, G. (2006). Chapter 11, the qr algorithm, Class notes for 477 and 577, IIT, `http://www.math.iit.edu/~fass/477577_Chapter_11.pdf`.

Fitzpatrick, R. (2006). Monte-carlo methods, `http://farside.ph.utexas.edu/teaching/329/lectures/node109.html`.

Frank, W. L. (1958). Computing eigenvalues of complex matrices by determinant evaluation and by methods of danilewski and wielandt, *SIAM Journal* **6**, pp. 378–392.

Fritsch, F. N. and Carlson, R. E. (1980). Monotone piecewise cubic interpolation, *SIAM J. Numer. Anal.* **17**, pp. 238–246.

Galassi, M., Davies, J., Theiler, J., Gough, B., Jungman, G., Alken, P., Booth, M., Rossi, F., and Ulerich, R. (2009). *Gnu Scientific Library Reference Manual* (Network Theory, Ltd.), ISBN 978-0954612078, `https://www.gnu.org/software/gsl/`.

Gaussian Elimination (2020). `http://en.wikipedia.org/wiki/Gaussian_elimination`.

Gershgorin Circle Theorem (2019). `https://en.wikipedia.org/wiki/Gershgorin_circle_theorem`.

Golub, G. H. and Van Loan, C. F. (1996). *Matrix Computations, 3rd edn* (Johns Hopkins University Press), ISBN 8-8018-5413-X, `https://www.academia.edu/35209124/Golub_G_H_Van_Loan_C_F_Matrix_Computations`.

Golub, G. H. and Wilkinson, J. H. (1975). Ill-conditioned eigensystems and the computation of the jordan canonical form, Report STAN-CS-75-478, Stanford, `http://i.stanford.edu/pub/cstr/reports/cs/tr/75/478/CS-TR-75-478.pdf`.

Golub, G. H. and Wilkinson, J. H. (1976). Ill-conditioned eigensystems and the computation of the jordan canonical form, *SIAM Review* **18**, pp. 578–619.

Hindmarsh, A. C. (2006). Serial fortran solvers for ode initial value problems, Report UCRL-ID-113855, Lawrence Livermore National Laboratory, `https://computing.llnl.gov/casc/odepack/`.

Hindmarsh, A. C., Brown, P. N., Grant, K. E., Lee, S. L., Serban, R., Shumaker, D. E., and Woodward, C. S. (2005). Suite of nonlinear and differential/algebraic equation solvers, *ACM Trans. Math. Softw.* **31**, pp. 363–396, doi:http://dx.doi.org/10.1145/1089014.1089020, `https://github.com/LLNL/sundials`.

Hubbard, J. H., Schleicher, D., and Sutherland, S. (2001). How to find all roots of complex polynomials by newton's method, *Inventiones Mathematicæ* **146**, pp. 1–33.

Jacobi Rotation (2020). `https://en.wikipedia.org/wiki/Jacobi_rotation`.

Kanamaru, T. (2007). Van der Pol oscillator, *Scholarpedia* **2**, 1, p. 2202, doi:10.4249/scholarpedia.2202, `http://www.scholarpedia.org/article/Van_der_Pol_oscillator`, revision #138698.

Layton, W. and Sussman, M. (2020). *Numerical Linear Algebra* (World Scientific, Hakensack, NJ), ISBN 978-981-122-389-1.

LeVeque, R. J. (2007). *Finite Difference Methods for Ordinary and Partial Differential Equations: Steady State and Time Dependent Problems* (SIAM, Philadelphia), ISBN 978-0-898716-29-0, `http://www.siam.org/books/ot98/sample/OT98Chapter7.pdf`.

M-matrix (2019). `https://en.wikipedia.org/wiki/M-matrix`.

MathWorks (2019). MATLAB, `https://www.mathworks.com/products/matlab.html`.

Moler, C. (1976). `http://youtu.be/R9UoFyqJca8`.

Moler, C. (2012). `http://blogs.mathworks.com/cleve/2012/12/10/1976-matrix-singular-value-decomposition-film/`.

MSC (2020). Patran, `https://www.mscsoftware.com/product/patran`.

Newton Fractal (2017). `http://en.wikipedia.org/wiki/Newton_fractal`.

Newton's Method (2020). `http://en.wikipedia.org/wiki/Newton's_method`.

Octave (2019). Octave online, `https://octave-online.net/`.

Ortega, J. M. and Rheinboldt, W. C. (1970). *Iierative Solution of Nonlinear Equations in Several Variables* (Academic Press, New York).

Pascal matrix (2020). `http://en.wikipedia.org/wiki/Pascal_matrix`.

Platte, R. B., Trefethen, L. N., and Kuijlaars, A. B. (2011). Impossibility of fast stable approximation of analytic functions from equispaced samples, *SIAM Review* **53**, pp. 308–318, doi:http://www.siam.org/journals/sirev/53-2/77470.html, `https://epubs.siam.org/doi/abs/10.1137/090774707`.

Quarteroni, A., Sacco, R., and Saleri, F. (2007). *Numerical Mathematics* (Springer), ISBN 978-3-540-34658-6.

Ralston, A. and Rabinowitz, P. (2001). *A First Course in Numerical Analysis* (Dover Publications, Mineola, New York), ISBN 0-486-41454-X.

Runge, C. (1901). über empirische funktionen und die interpolation zwischen äquidistanten ordinaten, *Z. Math. Phys.* **46**, pp. 224–243.

Rutishauser, H. (1962). Algorithm 125 weightcoeff, *Comm. ACM* **5**, pp. 510–511, `https://www.deepdyve.com/lp/association-for-computing-machinery/algorithm-125-weightcoeff-fyo8u9e0dB`.

Singular Value Decomposition (2020). `http://en.wikipedia.org/wiki/Singular_value_decomposition`.

Stewart, G. W. (1992–04). On the early history of the singular value decomposition, Report, Minnesota Digital Conse3rvancy, `http://hdl.handle.net/11299/1868`.

Sylvester Law of Inertia (2020). `https://en.wikipedia.org/wiki/Sylvester%27s_law_of_inertia`.

Tathan, S. (2017). `http://www.chiark.greenend.org.uk/%7esgtatham/newton`.

Trefethen, L. and Bau, D. (1997). *Numerical Linear Algebra* (Society for Industrial and Applied Mathematics, Philadelphia, PA), ISBN 0-89871-361-7.

van der Pol, B. and van der Mark, J. (1928). The heartbeat considered as a relaxation oscillation, and an electrical model of the heart, *Phil. Mag. Suppl.* **6**, pp. 763–775.

Varga, R. S. (1962). *Matrix Iterative Analysis* (Prentice-Hall, Englewood Cliffs, NJ).

Verschelde, J. (1999). `http://www.math.uic.edu/%7ejan/srvart/node4.html`.

Wang, L.-K. and Schulte, M. J. (2005). Decimal floating-point square root using newton-raphson iteration, *Proceedings of the 16th International Conference on Application-Specific Systems, Architecture and Processors (ASAP 05)* `http://citeseerx.ist.psu.edu/viewdoc/download?rep=rep1&type=pdf&doi=10.1.1.65.7220`.

Weisstein, E. W. (2020a). Cardioid, `https://mathworld.wolfram.com/Cardioid.html`.

Weisstein, E. W. (2020b). Monte carlo integration, `https://mathworld.wolfram.com/MonteCarloIntegration.html`.

Weisstein, E. W. (2020c). Steinmetz solid, `https://mathworld.wolfram.com/SteinmetzSolid.html`.

Wilkinson, J. H. (1965). *The Algebraic Eigenvalue Problem* (Clarendon Press, Oxford), ISBN 978-0-8018-5414-9.

Index

Printed in the United States
by Baker & Taylor Publisher Services

Printed in the United States
by Baker & Taylor Publisher Services